An Introduction to Ore Geology

GEOSCIENCE TEXTS

SERIES EDITOR

A. HALLAM

Lapworth Professor of Geology
University of Birmingham

GEOSCIENCE TEXTS VOLUME 2

An Introduction to Ore Geology

ANTHONY M. EVANS

BSc, PhD, CEng, MIMM, MIGeol, FGS
Senior Lecturer in Mining Geology
University of Leicester

BLACKWELL SCIENTIFIC PUBLICATIONS
OXFORD LONDON EDINBURGH
BOSTON MELBOURNE

To Jo, Nick and Caroline

Editorial offices:
Osney Mead, Oxford, OX2 0EL
8 John Street, London, WC1N 2ES
9 Forrest Road, Edinburgh, EH1 2QH
52 Beacon Street, Boston, Mass. 02108, USA
214 Berkeley Street, Carlton
Victoria 3053, Australia

First published 1980

Distribution in the United States of North America, its dependencies and Canada by
Elsevier North Holland Inc

Set by DMB Services (Typesetting)
Oxford
Printed by Billing and Sons Ltd
Guildford, and bound by
Kemp Hall Bindery
Oxford

British Library
Cataloguing in Publication Data

Evans, Anthony M
An introduction to ore geology. - (Geoscience texts; Vol.2).
1. Ore - deposits
I. Title II. Series
553 TN263

ISBN 0-632-00574-2 Pbk
ISBN 0-632-00665-X

Contents

Preface

This book is an attempt to provide a textbook in ore geology for second and third year undergraduates which, in these days of inflation, could be retailed at a reasonable price. The outline of the book follows fairly closely the undergraduate course in this subject at Leicester University which has evolved over the last twenty years. It assumes that the student will have adequate practical periods in which to handle and examine hand specimens, and thin and polished sections of the common ore types and their typical host rocks. Without such practical work students often develop erroneous ideas of what an orebody looks like, ideas often based on a study of mineralogical and museum specimens. In my opinion, it is essential that the student handles as much run-of-the-mill ore as possible during his course and makes a start on developing such skills as visual assaying, the ability to recognize wall rock alteration, using textural evidence to decide on the mode of genesis, and so on.

In an attempt to keep the reader aware of financial realities I have introduced some mineral economics into Chapter 1 and sprinkled grade and tonnage figures here and there throughout the book. It is hoped that this will go some way towards meeting that perennial complaint of industrial employers that the new graduate has little or no commercial awareness, such as a realization that companies in the West operate on the profit motive. This little essay into mineral economics only scratches the surface of the subject, and the intending practitioner of mining geology would be well advised to accompany his study of ore geology by dipping into such journals as *World Mining,* the *Mining Journal,* the *Engineering and Mining Journal* and the *Mining Magazine,* to watch the latest trends in metal and mineral prices and to gain knowledge of mining methods and recent orebody discoveries.

In order to produce a reasonably priced book, a strict word limit had to be imposed. As a result, the contents are necessarily selective and no doubt some teachers of this subject will feel that important topics have either received rather scanty treatment or have been omitted altogether. To these folk I offer my apologies, and hope that they will send me their ideas for improving the text, always remembering that if the price is to be kept down additions must be balanced by subtractions!

I would like to thank Mr Robert Campbell of Blackwell Scientific Publications for his help and encouragement, and not least for his tact in leaving me to get on with the job. My colleagues Dr J. G. Angus and Dr J. O'Leary read some of the chapters and made helpful suggestions for their improvement, and I thank them for their kindness. To my wife I owe an inestimable debt for the care with which she checked my manuscript and then produced the typescript.

Part I
Principles

'Here is such a vast variety of phenomena and these many of them so delusive, that 'tis very hard to escape imposition and mistake'.

These words, written about ore deposits by John Woodward in 1695, are every bit as true today as when he wrote them.

1

Introductory Definitions and Discussions

Ore, gangue and protore

'What is ore geology?' Unfortunately, it is not possible to give an unequivocal answer to this question if one wishes to go beyond saying that it is a branch of economic geology. The difficulty is that there are a number of distinctly different definitions of ore. A definition which has been current in capitalist economies for nearly a century runs as follows: 'Ore is a metalliferous mineral, or an aggregate of metalliferous minerals, more or less mixed with gangue, which from the standpoint of the miner can be won at a profit, or from the standpoint of the metallurgist can be treated at a profit. The test of yielding a metal or metals *at a profit* seems to be the only feasible one to employ.' Thus wrote J. F. Kemp in 1909. There are many similar definitions of ore which all emphasize (a) that it is material from which we extract a metal, and (b) that this operation must be a profit-making one. Thus there are geologists who divide minerals of economic importance into two main groups: the ore minerals from which a metal (or metals) is extracted, and the industrial minerals in which the mineral itself is used for one or more industrial purposes. Examples of ore minerals are chalcopyrite and galena from which we extract copper and lead respectively, and important among industrial minerals are baryte and asbestos. Over the last two decades there has, however, been a tendency among those who search for and mine industrial minerals to call them 'ore' and to measure 'ore reserves'. Some professional institutions have widened their definitions of ore to take note of this trend. This book will, however, follow the more common usage and consequently will be almost entirely confined to a study of metallic deposits.

A further complication, which we may note in passing, is that in socialist economies ore is often defined as mineral material that can be mined for the benefit of mankind. Such an altruistic definition is necessary to cover those examples in both capitalist and socialist countries where minerals are being worked at a loss. Such operations are carried on for various good or bad reasons depending on one's viewpoint! These include a government's reluctance to allow large isolated mining communities to be plunged into unemployment because a mine or mines have become unprofitable, a need to earn foreign currency and other reasons.

A definition about which there is little argument is that of gangue. This is simply the unwanted material, minerals or rock, with which ore minerals are usually intergrown. Mines commonly possess mineral dressing plants in which the raw ore is milled before the separation of the ore minerals from the gangue minerals by various processes, which provide ore concentrates, and tailings which are made up of the gangue.

Another word that must be introduced at this stage is protore. This is mineral material in which an initial but uneconomic concentration of metals has occurred that may by further natural processes be upgraded to the level of ore. Economically mineable aggregates of ore minerals are termed orebodies, ore shoots or ore deposits.

Economic considerations

The concentration of a metal in an orebody is called its grade, usually expressed as a percentage or in parts per million (ppm). The process of determining these concentrations is called assaying. Various economic and sometimes political considerations will determine the lowest grade of ore which can be produced from an orebody, this is termed the cut-off grade. In order to delineate the boundaries of an orebody in which the level of mineralization gradually decreases to a background value many samples will have to be collected and assayed. The boundaries thus established are called assay limits. Being entirely economically determined, they may not be marked by any particular geological feature.

Grades vary from orebody to orebody and, clearly, the lower the grade, the greater the tonnage of ore required to provide an economic deposit. The tendency with many metals during this century has been to mine lower and lower grade ores. This has led to the development of more large-scale operations with outputs of 40 000 tonnes of ore per day being not unusual. In some ores several metals are present, the sale of one may help finance the mining of another. For example, silver and cadmium can be by-products of the mining of lead-zinc ores and uranium is an important by-product of many South African gold ores.

The properties of a mineral govern the ease with which existing technology can extract and refine certain metals and this may affect the cut-off grade. Thus nickel is far more readily recovered from sulphide than from silicate ores and sulphide ores can be worked down to about 0.5%, whereas silicate ores must assay about 1.5% to be economic.

The price of metals is a vital factor. Prices vary from metal to metal and for most of them daily fluctuations may occur. The prices of many metals are governed by supply and demand and the prices on the London Metal Exchange are quoted daily by many newspapers, whilst more comprehensive guides to current prices can be found in the *Mining Journal, World Mining* and other technical journals. The prices of most of the common metals have not kept pace with inflation as can be seen from the examples in Fig. 1.1. This has had drastic effects on the level of recent mineral exploration activity, the profitability of many mines and the economy of whole nations, such as Zambia and Chile which are heavily dependent on their mineral industries.

The shape and nature of ore deposits also affects the workable grade. Large, low grade deposits which occur at the surface can be worked by cheap open pit methods, whilst thin tabular vein deposits will necessitate more expensive underground methods of extraction, although they can generally be worked in much smaller volumes so that a relatively small initial capital outlay is required. Big mining operations have now reached the stage, thanks to inflation, of requiring enormous initial capital investments. For example, to develop the large RTZ copper-gold mine at Bougainville, Papua New Guinea, would cost around £500 million at 1979 prices.

There are of course many other economic factors such as transport costs, availability of labour and power supplies, equipment costs, geographical location,

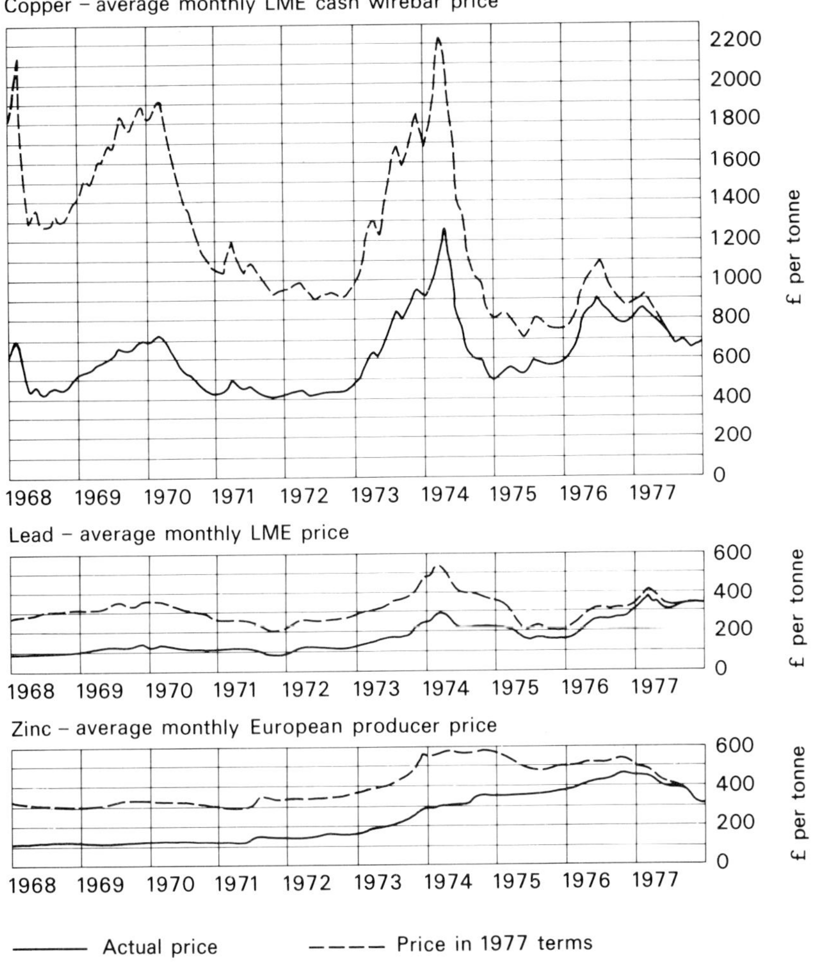

Fig. 1.1. Average prices of copper, lead and zinc during a recent decade showing actual prices, and prices in 1977 terms. (Modified from an RTZ annual report.)

taxation systems, etc., which mean that what constitutes an orebody in one part of the world may be quite uneconomic in another part.

In delineating and working an orebody the mining geologist often has to classify his ore reserves into three classes: proved, probable and possible. Proved ore has been so thoroughly sampled that we can be certain of its outline, tonnage and average grade, within certain limits. Elsewhere in the orebody, sampling from drilling and development workings may not have been so thorough, but there may be enough information to be reasonably sure of its tonnage and grade, this is probable ore. On the fringes of our exploratory workings we may have enough information to infer that ore extends for some way into only partially explored ground and that it may amount to a certain volume and grade of possible ore. In most countries, these, or equivalent, words have nationally recognized definitions and legal connotations. The practising geologist must therefore know the local definitions thoroughly and make sure that he uses them correctly.

Geochemical considerations

It is traditional in the mining industry to divide metals into groups with special names. These are as follows:

(a) *precious metals:* gold, silver, platinum group;

(b) *non-ferrous metals:* copper, lead, zinc, tin, aluminium (the first four being commonly known as the *base metals*);

(c) *iron and ferroalloy metals:* iron, manganese, nickel, chromium, molybdenum, tungsten, vanadium, cobalt;

(d) *minor metals and related non-metals:* antimony, arsenic, beryllium, bismuth, cadmium, magnesium, mercury, selenium, tantalum, tellurium, titanium, zirconium, etc.;

(e) *fissionable metals:* uranium, thorium, (radium).

For the formation of an orebody the element or elements concerned must be enriched to a considerably higher level than their normal crustal abundance. The degree of enrichment is termed the concentration factor and typical values are shown in Table 1.1.

Table 1.1. Concentration factors.

	Average crustal abundance (%)	Average minimum exploitable grade (%)	Concentration factor
Aluminium	8	30	3.75
Iron	5	25	5
Copper	0.005	0.4	80
Nickel	0.007	0.5	71
Zinc	0.007	4	571
Manganese	0.09	35	389
Tin	0.0002	0.5	2500
Chromium	0.01	30	3000
Lead	0.001	4	4000
Gold	0.000 000 4	0.000 01	25

2

The Nature and Morphology of the Principal Types of Ore Deposits

A good way to start an argument among mining geologists is to suggest that a deposit held by common consensus to be syngenetic is in fact epigenetic! These words are clearly concerned with the manner in which deposits have come into being and like all matters of genesis they are fraught with meaning and are

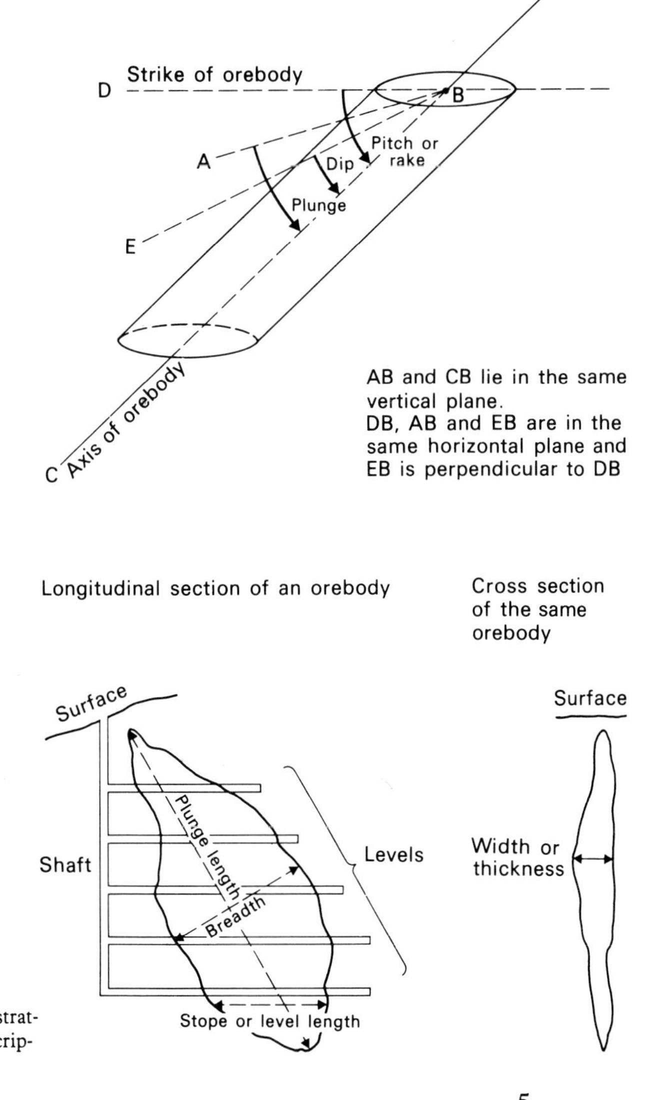

Fig. 2.1. Diagrams illustrating terms used in the description of orebodies.

frequently heard on the lips of mining geologists wherever they gather. What do these magic words mean? A syngenetic deposit is one which has formed at the same time as the rocks in which it occurs. It is sometimes part of a stratigraphic succession like an iron-rich sedimentary horizon. An epigenetic deposit, on the other hand, is one believed to have come into being after the host rocks in which it occurs. A good igneous analogy is a dyke; an example among ore deposits is a vein. Before discussing their nature we must learn some of the terms used in describing orebodies.

If an orebody viewed in plan is longer in one direction than the other we can designate this long dimension as its strike (Fig. 2.1). The inclination of the orebody perpendicular to the strike will be its dip and the longest dimension of the orebody its axis. The plunge of the axis is measured in the vertical plane ABC but its pitch or rake can be measured in any other plane; a usual choice is the plane containing the strike but if the orebody is fault controlled then the pitch may be measured in the fault plane. The meanings of other terms are self-evident from the figure.

It is possible to classify orebodies in the same way as we divide up igneous intrusions according to whether they are discordant or concordant with the lithological banding (often bedding) in the enclosing rocks. We will consider discordant orebodies first. This large class can be subdivided into those orebodies which have an approximately regular shape and those which are thoroughly irregular in their outlines.

Discordant orebodies

REGULARLY SHAPED BODIES

(a) *Tabular orebodies.* These bodies are extensive in two dimensions, but have a restricted development in their third dimension. In this class we have veins (sometimes called fissure-veins) and lodes (Fig. 2.2). In the past, some workers have made a genetic distinction between these terms. Veins were considered to have resulted

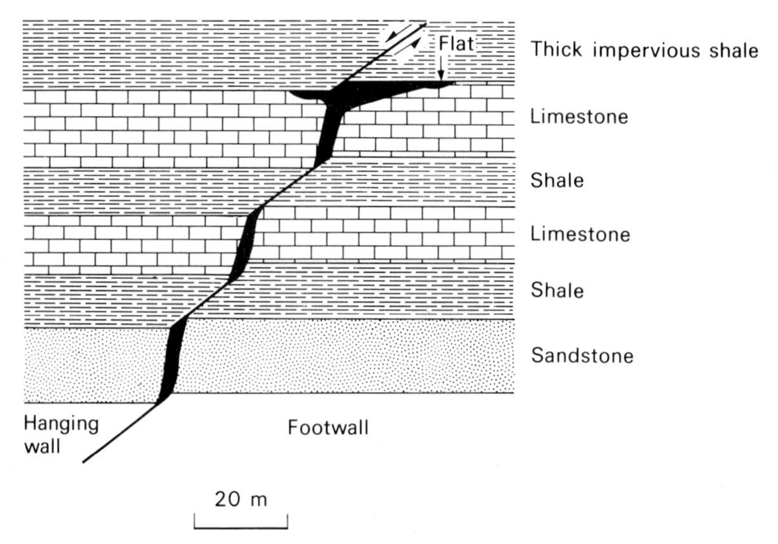

Fig. 2.2. Vein occupying a normal fault and exhibiting pinch-and-swell structure, giving rise to ribbon ore shoots. The development of a flat beneath impervious cover is also shown.

mainly from the infilling of pre-existing open spaces, whilst the formation of lodes was held to involve the extensive replacement of pre-existing host rock. Such a genetic distinction has often proved to be unworkable and the writer advises that all such orebodies be called veins and the term lode be dropped.

Veins are often inclined, and in such cases, as with faults, we can speak of the hanging wall and the footwall. Veins frequently pinch and swell out as they are followed up or down a stratigraphic sequence (Fig. 2.2). This pinch-and-swell structure can create difficulties during both exploration and mining. Often only the swells are workable and if these are imagined in a section at right angles to that in Fig. 2.2 it can be seen that they form ribbon ore shoots. The origin of pinch-and-swell structure is shown in Fig. 2.3. An initial fracture in rocks changes its attitude as it crosses them according to the changes in physical properties of the rocks. These properties are in turn governed by changes in lithology (Fig. 2.3A).

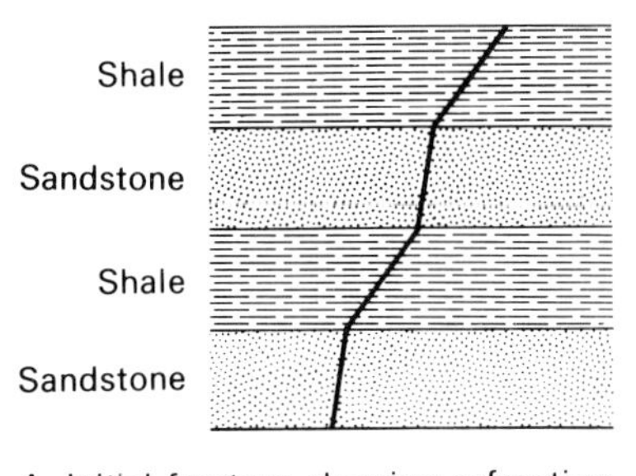

A. Initial fracture showing refraction, as it crosses beds of different competency

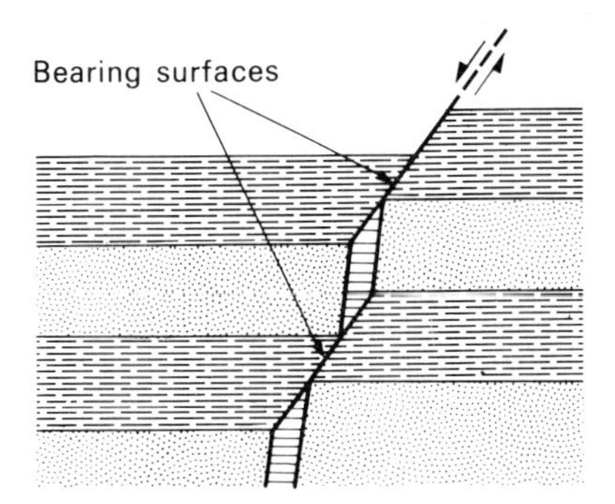

B. Movement along fracture gives rise to open spaces

Fig. 2.3. Formation of pinch-and-swell structure in veins.

When movement occurs producing a normal fault then the less steeply dipping sections are held against each other to become bearing surfaces, and open spaces (dilatant zones) form in the more steeply dipping sections. If minerals are deposited in these cavities then a vein will be formed. If the reader carries out the experiment of reversing the movement on the initial fracture then he will find that the steeper parts of the fault now act as bearing surfaces and the dilatant zones are formed in the less steeply dipping sections. Veins are usually developed in fracture systems and therefore show regularities in their orientation (Figs 2.4 and 14.3).

The infilling of veins may consist of one mineral but more usually it consists of an intergrowth of ore and gangue minerals. The boundaries of vein orebodies may be the vein walls or they can be assay boundaries within the veins.

(b) *Tubular orebodies.* These bodies are relatively short in two dimensions but extensive in the third. When vertical or subvertical they are called pipes or chimneys, when horizontal or subhorizontal mantos. The Spanish word manto is inappropriate in this context for its literal translation is blanket: it is, however, firmly entrenched in the English geological literature. The word has been employed by some workers for flat-lying tabular bodies, but the perfectly acceptable word flat is available for these.

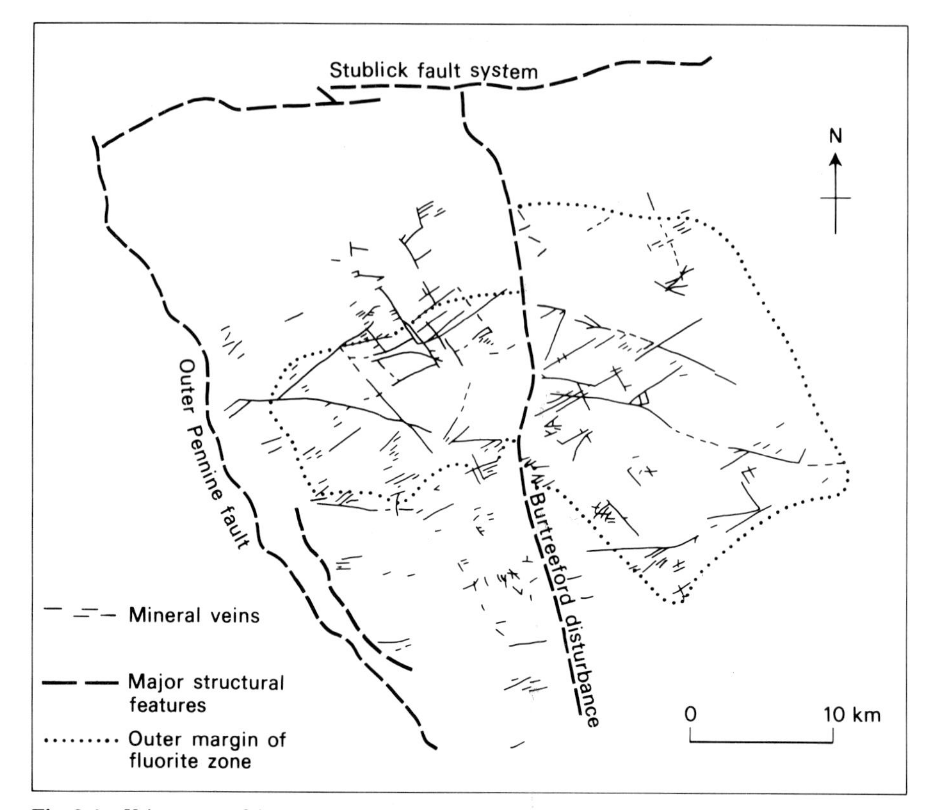

Fig. 2.4. Vein system of the Alston block of the Northern Pennine Orefield, England. Note the three dominant vein directions. (Modified from Dunham 1959.)

Fig. 2.5. Diagram of the Vulcan tin pipe, Herberton, Queensland. The average grade was 4.5% tin. (After Mason 1953.)

In eastern Australia there are hundreds of pipes in and close to granite intrusions. The pipes occur along a 2400 km belt from Queensland to New South Wales. Most have quartz fillings and some are mineralized with bismuth, molybdenum, tungsten and tin. An example is shown in Fig. 2.5. Pipes may be of various types and origins (Mitcham 1974). Infillings of mineralized breccia are particularly common, a good example being the copper-bearing breccia pipes of Messina in South Africa (Jacobsen & McCarthy 1976).

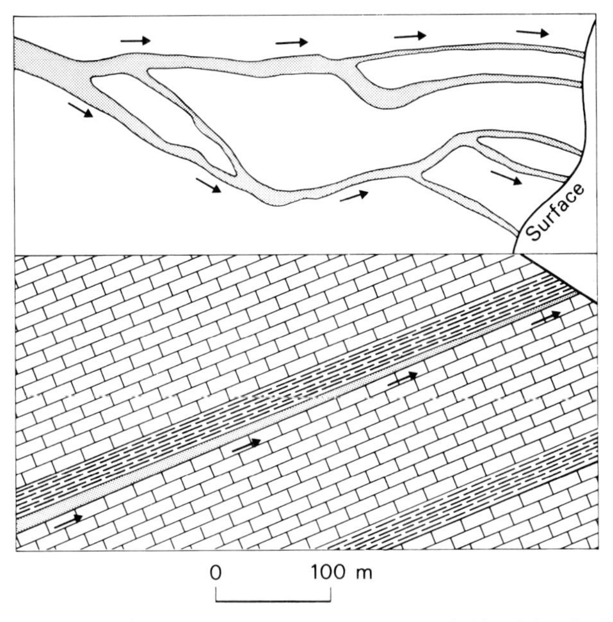

Fig. 2.6. Plan and section of part of the Hidden Treasure manto, Ophir mining district, Utah. (After Gilluly 1932.)

Mantos and pipes may branch and anastomose. An example of a branching manto is given in Fig. 2.6. Mantos and pipes are often found in association, the pipes frequently acting as feeders to the mantos. Sometimes mantos pass upwards from bed to bed by way of pipe connections often branching as they go, an example being the Providencia mine in Mexico where a single pipe at depth feeds into twenty mantos nearer the surface.

IRREGULARLY SHAPED BODIES

(a) *Disseminated deposits.* In these deposits, ore minerals are disseminated through the body of the host rock, a good example being diamonds in kimberlites. In other deposits, the disseminations may be wholly or mainly along close-spaced veinlets cutting the host rock and forming an interlacing network called a stockwork (Fig. 12.1). In some deposits the economic minerals may be disseminated through the host rock and along veinlets (Fig. 12.2). Whatever the mode of occurrence mineralization of this type generally fades gradually outwards into sub-economic mineralization and the boundaries of the orebody are assay limits. They are, therefore, often irregular in form and may cut across geological boundaries. The overall shapes of some are cylindrical (Fig. 12.5) and others are caplike (Fig. 12.13). The mercury-bearing stockworks of Dubník in Slovakia are sometimes pear-shaped.

Stockworks most commonly occur in acid to intermediate plutonic igneous intrusions, but they may cut across the contact (Fig. 2.7) into the country rocks, and a few are wholly or mainly in the country rocks. Disseminated deposits produce most of the world's copper and molybdenum. They are of some importance in the production of tin, silver, mercury and uranium.

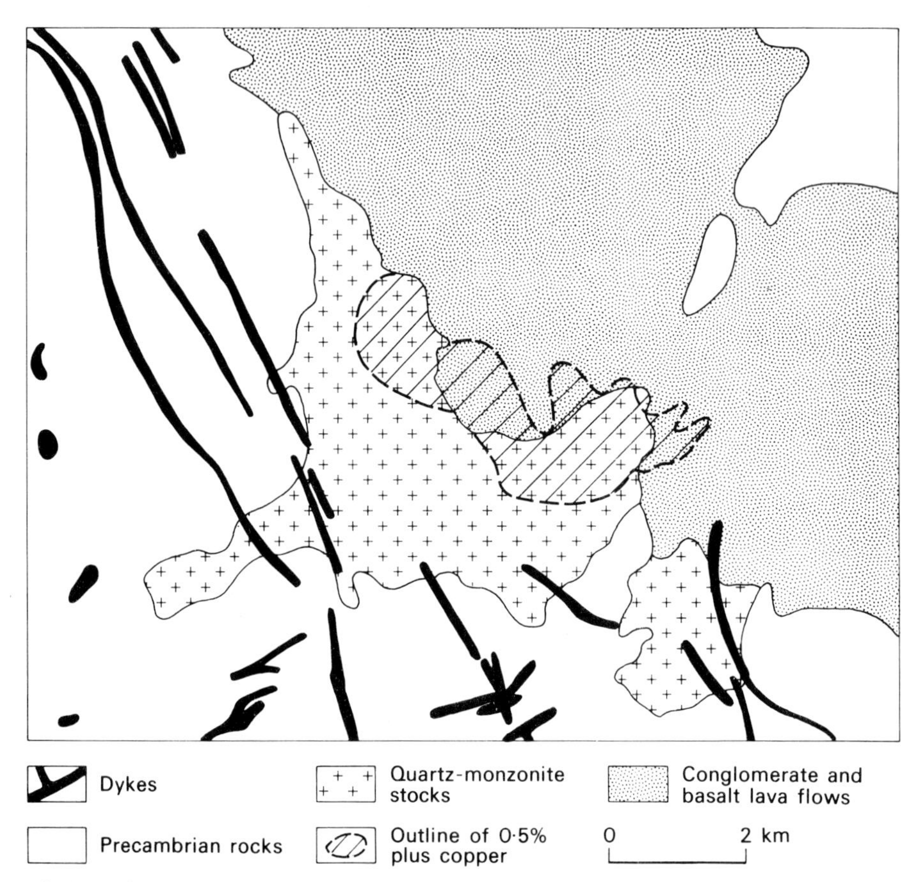

Fig. 2.7. Generalized geological map of the Bagdad mine area, Arizona. (Modified from Anderson 1948.)

(b) *Irregular replacement deposits.* Many ore deposits have been formed by the replacement of pre-existing rocks, particularly carbonate-rich sediments. These replacement processes often occurred at contacts with medium-sized to large igneous intrusions. Such deposits have therefore been called contact metamorphic, however, pyrometasomatic is now the preferred and more popular term. The orebodies are characterized by the development of calc-silicate minerals such as diopside, wollastonite, andradite garnet, actinolite and so on. For this reason another name for these deposits is skarn orebodies. These deposits are extremely irregular in shape (Fig. 2.8), tongues of ore may project along any available planar structure—bedding, joints, faults, etc. The distribution within the contact aureole is often apparently capricious. Structural changes may cause abrupt termination of the orebodies. The principal materials produced from pyrometasomatic deposits are: iron, copper, tungsten, graphite, zinc, lead, molybdenum, tin and uranium.

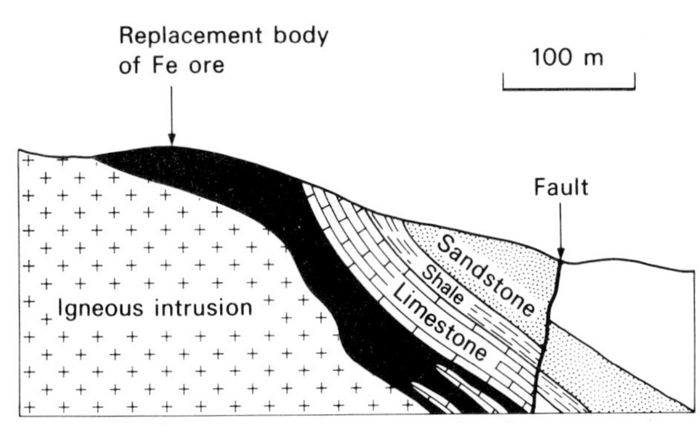

Fig. 2.8. Pyrometasomatic deposit at Iron Springs, Utah. (After Gilluly *et al.* 1959.)

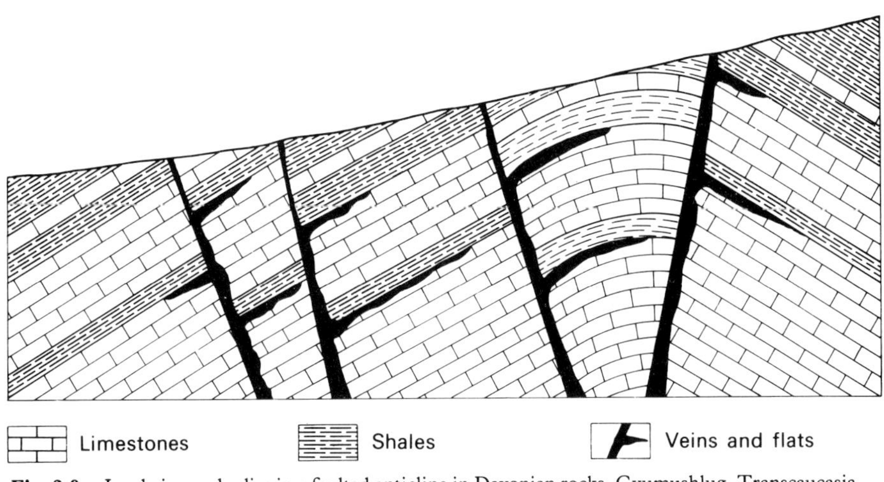

Fig. 2.9. Lead-zinc orebodies in a faulted anticline in Devonian rocks, Gyumushlug, Transcaucasia, USSR. Note the development of veins along the faults with flats branching off them beneath impervious beds of shale. (After Malyutin & Sitkovskiy 1968.)

Other replacement deposits occur which do not belong to the pyrometasomatic class. Examples of these include flats. These are horizontal or subhorizontal bodies of ore which commonly branch out from veins and lie in carbonate host rocks beneath an impervious cover such as shale (Fig. 2.9).

Concordant orebodies

SEDIMENTARY HOST ROCKS

Concordant orebodies in sediments are very important producers of many different metals, they are particularly important for base metals and iron. These orebodies are of course concordant with the bedding. They may be an integral part of the stratigraphic sequence, as is the case with Phanerozoic ironstones, or they may be epigenetic infillings of pore spaces, replacement orebodies or syngenetic ores formed by the exhalation of mineralizing solutions at the sediment-water interface. Usually these orebodies show a considerable development in two dimensions, i.e. parallel to the bedding and a limited development perpendicular to it (Fig. 2.10). Such deposits are referred to as stratiform. This term must not be confused with strata-bound which refers to any type or types of orebody, concordant or discordant, which are restricted to a particular part of the stratigraphic column. Thus the veins, pipes and flats of the Southern Pennine orefield of England can be designated as strata-bound, as they are virtually restricted to the Carboniferous limestone of that region. A number of examples of concordant deposits which occur in different types of sedimentary rocks will be considered.

(a) *Limestone hosts.* Limestones are very common host rocks for base metal sulphide deposits. In a dominantly carbonate sequence ore is often developed in a small number of preferred beds or at certain sedimentary interfaces. These are often zones in which the permeability has been increased by dolomitization or fracturing. When they only form a minor part of the stratigraphical succession, limestones, because of their solubility and reactivity, can become favourable horizons for mineralization. For example, the lead-zinc ores of Bingham, Utah, occur in limestones which make up 10% of a 2300 m succession mainly composed of quartzites.

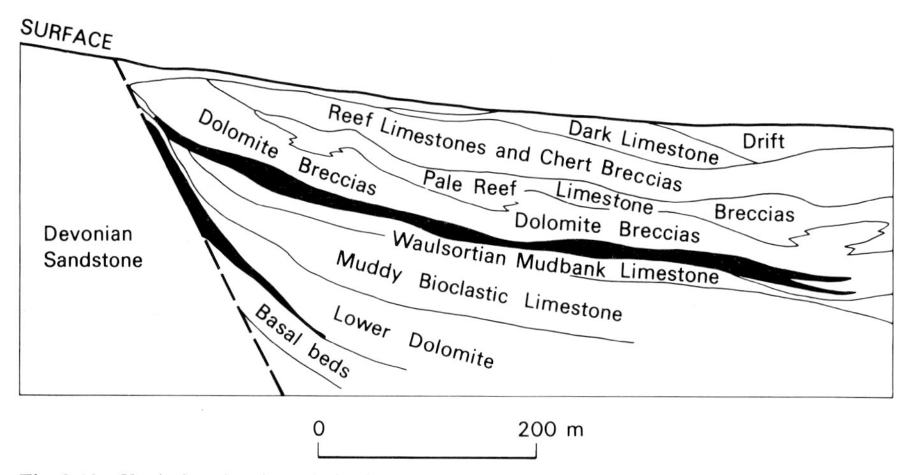

Fig. 2.10. Vertical section through the G zone at Silvermines, Co. Tipperary, Ireland. The orebodies are shown in black. (After Taylor & Andrew 1978.)

At Silvermines in Ireland, lead-zinc mineralization occurs as syngenetic stratiform orebodies in a limestone sequence (Fig. 2.10), as fault-bounded epigenetic strata-bound orebodies in the basal Carboniferous and as structurally controlled vein or breccia zones in the Upper Devonian sandstones (Taylor & Andrew 1978). The larger stratiform orebody shown in Fig. 2.10 occurs in massive, partly brecciated pyrite at the base of a thick sequence of dolomite breccias. The maximum thickness of the ore is 30 m. At the base, there is usually an abrupt change to massive pyrite from a footwall of nodular micrite, shale biomicrite and other limestone units of the Mudbank Limestone. Sometimes the contact is gradational. The upper contact is always sharp. Pyrite and marcasite make up 75% of the ore, sphalerite forms 20% and galena 4%.

(b) *Argillaceous hosts.* Shales, mudstones, argillites and slates are important host rocks for concordant orebodies which are often remarkably continuous and extensive. In Germany, the Kupferschiefer of the Upper Permian is a prime example.

This is a copper-bearing shale a metre or so thick which, at Mansfeld, occurred in orebodies which had plan dimensions of 8, 16, 36 and 130 km^2. Mineralization occurs at exactly the same horizon in Poland, where it is being worked extensively, and across the North Sea in north-eastern England, where it is subeconomic.

The world's largest, single lead-zinc orebody occurs at Sullivan, British Columbia. The host rocks are late Precambrian argillites. Above the main orebody (Fig. 2.11) there are a number of other mineralized horizons with concordant mineralization. This deposit appears to be syngenetic and the lead-zinc minerals form an integral part of the rocks in which they occur. They are affected by sedimentary deformation such as slumping, pull apart structures, load casting, etc., in a manner identical to that in which poorly consolidated sand and mud respond.

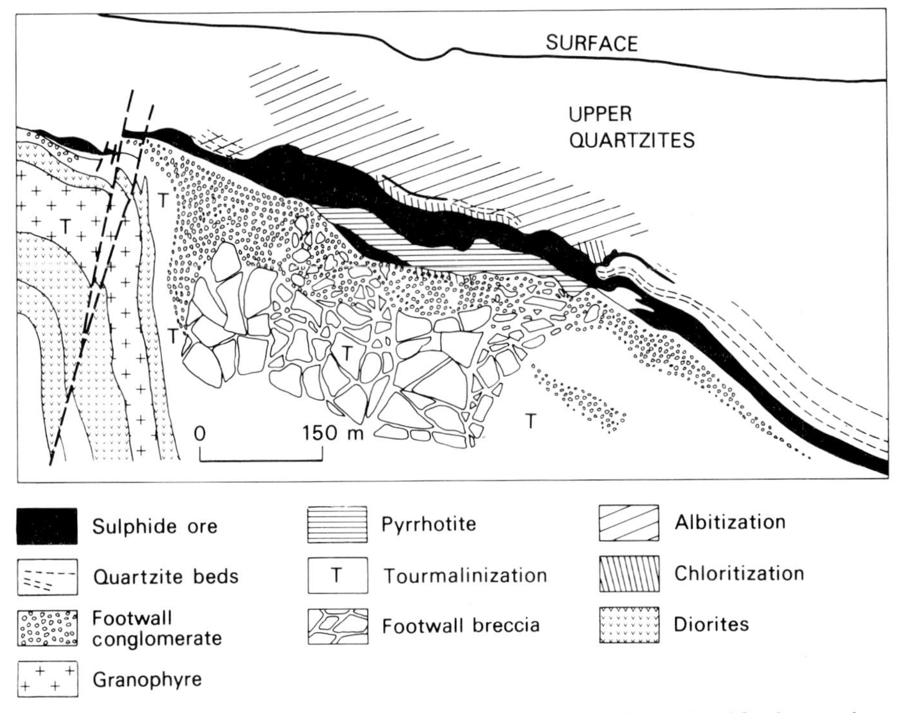

Fig. 2.11. Cross section through the ore zone, Sullivan Mine, British Columbia. (After Sangster & Scott 1976.)

The orebody occurs in a single generally conformable zone between 60 and 90 m thick. It runs 6.6% lead and 5.7% zinc. Other metals recovered are silver, tin, cadmium, antimony, bismuth, copper and gold. Before mining commenced the orebody contained at least 155 million tonnes of ore. The footwall rocks consist of graded impure quartzites and argillites and, in places, conglomerate. The hanging wall rocks are more thickly bedded and arenaceous. The ore zone is a mineralized argillite. The principal sulphide-oxide minerals are pyrrhotite, sphalerite, galena, pyrite and magnetite with minor chalcopyrite, arsenopyrite and cassiterite. Beneath the central part of the orebody there is a roughly funnel-shaped zone of brecciation and tourmalinization which extends downwards for at least 100 m. In places, the matrix of the breccia is heavily mineralized with pyrrhotite and occasionally with galena and sphalerite. This zone may have been a channelway up which solutions moved to debouch on to the seafloor, to precipitate the ore minerals among the accumulating sediment. If this was the case, then Sullivan could be called a sedimentary-exhalative deposit.

Other good examples of concordant deposits in argillaceous rocks, or slightly metamorphosed equivalents, are the lead-zinc deposits of Mount Isa, Queensland, many of the Zambian Copperbelt deposits and the copper shales of the White Pine Mine, Michigan.

(c) *Arenaceous hosts.* Not all the Zambian Copperbelt deposits occur in shales and metashales. Some orebodies occur in altered feldspathic sandstones (Fig. 2.12).

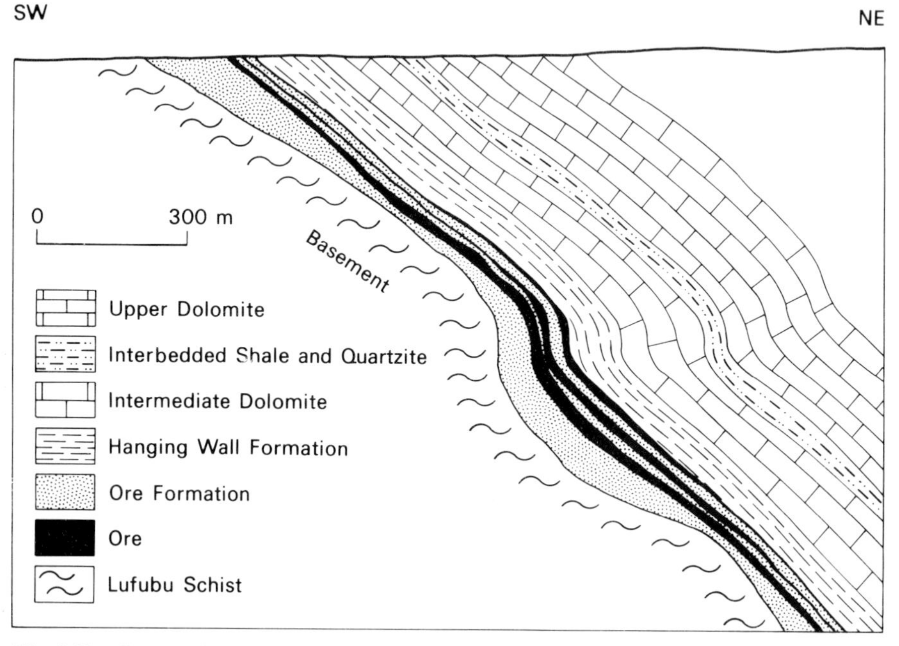

Fig. 2.12. Cross section through the Mufulira orebodies, Zambia. (After Fleischer *et al.* 1976.)

The Mufulira copper deposit occurs in Proterozoic rocks on the eastern side of an anticline (Fleischer *et al.* 1976). It lies just above the unconformity with an older, strongly metamorphosed Precambrian basement. The gross ore reserves in 1974 stood at 282 million tonnes assaying 3.47% copper. The largest orebody stretches for 5800 m along the strike and for several kilometres down dip. Chalcopyrite is the principal sulphide mineral, sometimes being accompanied by significant amounts of bornite. Fluviatile and aeolian arenites form the footwall rocks. The ore zone consists of feldspathic sandstones which, in places, contain carbon-rich lenses with much sericite. The basal portion is coarse-grained and characterized by festoon cross-bedding in which bornite is concentrated along the cross-bedding together with well rounded, obviously detrital zircon. In other parts of the orebody, concentrations of sulphides occur in the hollows of ripplemarks and in desiccation cracks. These features suggest that some of the sulphides are detrital in origin. Mineralization ends abruptly at the hanging wall, suggesting a regression. At this sharp cut-off the facies changes from an arenaceous one to dolomites and shallow-water muds.

Conformable deposits of copper occur in some sandstones which were laid down under desert conditions. As these rocks are frequently red, the deposits are known as Red Bed Coppers. Dune sands are frequently porous and permeable and the copper minerals are generally developed in pore spaces. Examples of such deposits occur in the Permian of the Urals and the Don Basin in the USSR in the form of small sandstone layers 10-40 cm thick running 1.5-1.9% copper. They are also found in the Trias of central England, in Nova Scotia, in Germany and in the south-western USA. At the Naciemento Mine in New Mexico a deposit of 11 million tonnes averaging 0.65% copper is being worked by open pit methods. Like other red bed coppers, this deposit has a high metal/sulphur ratio as the principal mineral is chalcocite. This yields a copper concentrate low in sulphur which is very acceptable to present-day custom smelters faced with stringent anti-pollution legislation.

Copper is not the only base metal which occurs in such deposits. Similar lead ores are known in Germany and silver deposits in Utah. Another important class of pore-filling deposits are the uranium-vanadium deposits of Colorado Plateau or western states-type. These occur mainly in Mesozoic sandstones of continental origin but also in some siltstones and conglomerates. The orebodies are very variable in form and pods and irregularly shaped deposits occur, although large

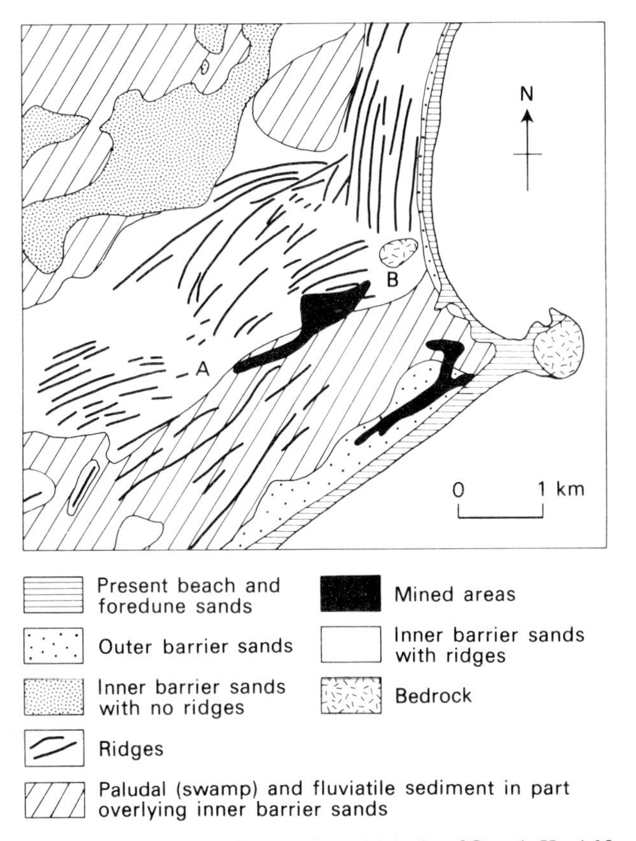

Fig. 2.13. Geology and mining areas of the beach sand deposits of Crowdy Head, New South Wales, Australia. (After Winward 1975.)

concordant sheets up to 3 m thick are also present. The orebodies follow sedimentary structures and depositional features.

Many mechanical accumulations of high density minerals such as magnetite, ilmenite, rutile and zircon occur in arenaceous hosts. These usually take the form of heavy mineral rich layers in Pleistocene and Holocene sands. As the sands are usually unlithified the deposits are easily worked and no crushing of the ore is required. These orebodies belong to the group called placer deposits—beach sand placers are a good example (Fig. 2.13). Beach placers supply much of the world's titanium, zircon, thorium, cerium and yttrium. They occur along present-day beaches or ancient beaches where longshore drift is well developed and frequent storms occur. Economic grades can be very low and sands running as little as 0.6% heavy minerals are worked along Australia's eastern coast. The deposits usually show a topographic control, the shapes of bays and the position of headlands often being very important. Thus in exploring for buried orebodies a reconstruction of the palaeogeography is invaluable.

(d) *Rudaceous hosts.* Alluvial gravels and conglomerates also form important recent and fossil placer deposits. Alluvial gold deposits are often marked by 'white runs' of vein quartz pebbles as in the White Channels of the Yukon, the White Bars of California and the White Leads of Australia. Such deposits form one of the few types of economic placer deposits in fully lithified rocks and indeed the majority of the world's gold is won from Precambrian deposits of this type in South

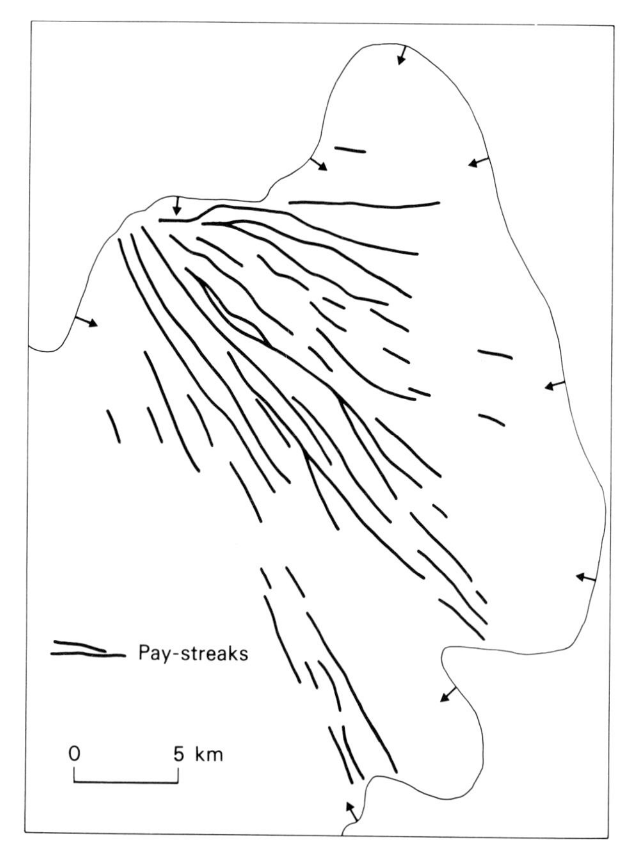

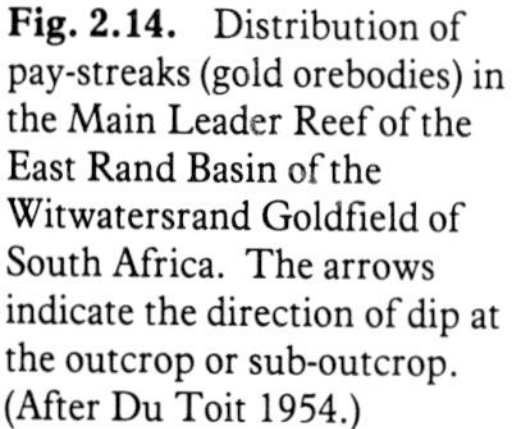

Fig. 2.14. Distribution of pay-streaks (gold orebodies) in the Main Leader Reef of the East Rand Basin of the Witwatersrand Goldfield of South Africa. The arrows indicate the direction of dip at the outcrop or sub-outcrop. (After Du Toit 1954.)

Africa. Fig. 2.14 shows the distribution of the gold orebodies in the East Rand Basin where the vein quartz pebble conglomerates occur in quartzites of the Upper Witwatersrand System. Their fan-shaped distribution strongly suggests that they occupy distributory channels. Uranium is recovered as a by-product of the working of the Witwatersrand goldfields. In the very similar Blind River area of Ontario uranium is the only metal produced. Similar mineralized conglomerates occur elsewhere in the Precambrian.

(e) *Chemical sediments.* Sedimentary iron and manganese formations occur scattered through the stratigraphical column where they form very extensive beds conformable with the stratigraphy. They are described in Chapter 16.

IGNEOUS HOST ROCKS

(a) *Volcanic hosts.* There are two principal types of deposit to be found in volcanic rocks, vesicular filling deposits and volcanic massive sulphide or volcanic-exhalative deposits. The first deposit type is not very important but the second type is a widespread and important producer of base metals often with silver and gold as byproducts.

The first type forms in the permeable vesicular tops of basic lava flows whose permeability may have been increased by autobrecciation. The mineralization is normally in the form of native copper and the best examples occurred in late Precambrian basalts of the Keweenaw Peninsula of northern Michigan. Mining commenced in 1845 and the orebodies are now virtually worked out. They were very large and were mined down to nearly 2750 m. There were six main producing horizons, five in the tops of lava flows and one in a conglomerate. The orebodies averaged 4 m in thickness and 0.8% copper. Occasionally, so-called veins of massive copper were found, one such mass weighed 500 tonnes. Similar deposits occur around the Coppermine River in northern Canada where 3 000 000 t running 3.48% copper have been found. Though the copper content was worth some £100 million at 1979 prices the deposit is unlikely to be worked because of its remote location. Other uneconomic deposits of this type are known in many other countries.

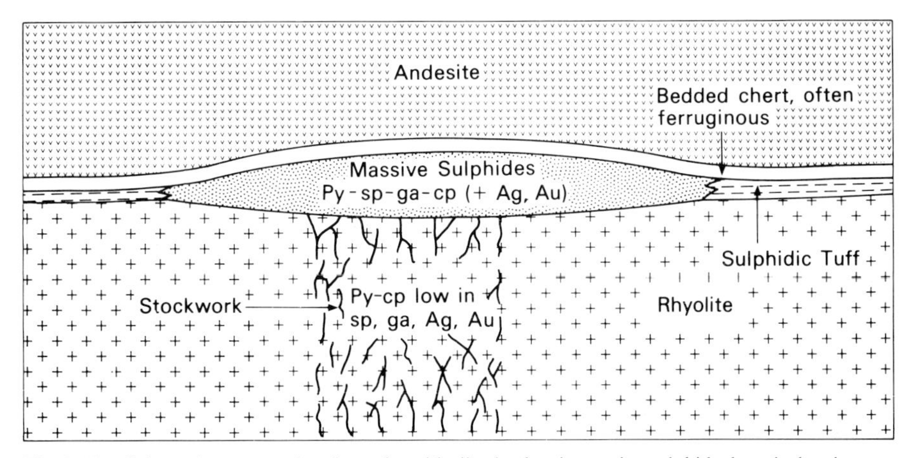

Fig. 2.15. Schematic cross section through an idealized volcanic massive sulphide deposit showing the underlying feeder stockwork and typical mineralogy. Py = pyrite, sp = sphalerite, ga = galena, cp = chalcopyrite.

Volcanic massive sulphide deposits often consist of over 90% iron sulphide usually as pyrite, although pyrrhotite is well developed in some deposits. They are generally stratiform bodies, lenticular to sheetlike (Fig. 2.15), developed at the interfaces between volcanic units or at volcanic-sedimentary interfaces. With increasing magnetite content, these ores grade to massive oxide ores of magnetite and/or hematite such as Savage River in Tasmania, Fosdalen in Norway and Kiruna in Sweden (Solomon 1976). They can be divided into three classes of deposit: (a) zinc-lead-copper, (b) zinc-copper, and (c) copper. Typical tonnages and copper grades are given in Fig. 12.3.

The most important host rock is rhyolite and lead-bearing ores are only associated with this rock-type. The copper class is usually, but not invariably, associated with mafic volcanics. Massive sulphide deposits commonly occur in groups and in any one area they occur at one or a restricted number of horizons within the succession. These horizons may represent changes in composition of the volcanic rocks, a change from volcanism to sedimentation or simply a pause in volcanism. There is a close association with volcaniclastic rocks and many orebodies overlie the explosive products of rhyolite domes. The deposits are usually underlain by a stockwork that may itself be ore grade and which appears to have been the feeder channel up which mineralizing fluids penetrated to form the overlying massive sulphide deposit.

(b) *Plutonic hosts.* Many plutonic igneous intrusions possess rhythmic layering. This is particularly well developed in some basic intrusions. Usually the layering takes the form of alternating bands of mafic and felsic minerals. Sometimes minerals of economic interest such as chromite, magnetite and ilmenite may form discrete mineable seams within such layered complexes (Fig. 8.2). These seams are naturally stratiform and may extend over many kilometres as is the case with the chromite seams in the Bushveld Complex of South Africa (Fig. 8.1).

Another form of orthomagmatic deposit is the nickel-copper sulphide orebody formed by the sinking of an immiscible sulphide liquid to the bottom of a magma chamber containing ultrabasic or basic magma. These are known to the learned as liquation deposits. They may be formed at the base of lava flows as well as in plutonic intrusions. The sulphide usually accumulates in hollows in the base of the igneous body and generally forms sheets or irregular lenses conformable with the overlying silicate rock. From the base upwards, massive sulphide gives way through disseminated sulphides in a silicate gangue to lightly mineralized and then barren rock (Chapter 9).

METAMORPHIC HOST ROCKS

Apart from some deposits of metamorphic origin such as the irregular replacement deposits already described, metamorphic rocks are mainly important for the metamorphosed equivalents of deposits that originated in sedimentary and igneous rocks and which have been discussed above.

RESIDUAL DEPOSITS

These are deposits formed by the removal of non-ore material from protore. For example the leaching of silica and alkalis from a nepheline-syenite may leave behind a surface capping of hydrous aluminium oxides (bauxite). Some residual bauxites occur at the present surface, others have been buried under younger sediments to which they form conformable basal beds.

Other examples of residual deposits include some laterites sufficiently high in iron to be worked and nickeliferous laterites formed by the weathering of peridotites.

SUPERGENE ENRICHMENT

This is a process which may affect most orebodies to some degree. After a deposit has been formed, uplift and erosion may bring it within reach of circulating ground waters which may leach some of the metals out of that section of the orebody above the water table. These dissolved metals may be redeposited in that part of the orebody lying beneath the water table. This can lead to a considerable enrichment in metal values. Supergene processes are discussed in Chapter 17.

3

Textures and Structures of Ore and Gangue Minerals. Fluid Inclusions. Wall Rock Alteration

The study of textures can tell us much about the genesis and subsequent history of orebodies. The textures of orebodies vary according to whether their constituent minerals were formed by deposition in an open space from a silicate or aqueous solution, or by replacement of pre-existing rock or ore minerals. Subsequent metamorphism may drastically alter primary textures. The interpretation of mineral textures is a very large and difficult subject and only a few important points can be touched on here. For further information the reader should consult Edwards (1954), Ramdohr (1969) and Stanton (1972).

Open space filling

PRECIPITATION FROM SILICATE MELTS

Critical factors in this situation are the time of crystallization and the presence or absence of simultaneously crystallizing silicates. Oxide ore minerals such as chromite often crystallize out early and thus may form good euhedral crystals. These may, however, be subsequently modified in various ways. Chromites deposited with interstitial silicate liquid may suffer corrosion and partial resorption to produce atoll textures (Fig. 3.1) and rounded grains, whereas those developed in monomineralic bands (Figs. 3.2, 8.2) may, during the cooling of a large parent intrusion, suffer auto-annealing and develop foam texture (see Chapter 18).

When oxide and silicate minerals crystallize simultaneously, xenomorphic to hypidiomorphic textures similar to those of granitic rocks develop. This is because there has been mutual interference during the growth of the grains of all the minerals. A minor development of micrographic textures involving oxide ore minerals may also occur at this stage.

Fig. 3.1. Chromite grains in anorthosite, Bushveld Complex, South Africa. The chromites are euhedral crystals which have undergone partial resorption producing rounded grains of various shapes, including atoll texture.

Fig. 3.2. Chromite bands in anorthosite, Dwars River Bridge, Bushveld Complex, South Africa. The central band is 1.3 mm thick at the right-hand end.

Sulphides, because of their lower melting points, crystallize after associated silicates and, if they have not segregated from the silicates, they will either be present as rounded grain aggregates representing frozen globules of immiscible sulphide liquid (Fig. 4.1), or as anhedral grains or grain aggregates which have crystallized interstitially to the silicates and whose shapes are governed by those of the enclosing silicate grains.

PRECIPITATION FROM AQUEOUS SOLUTIONS

Open spaces such as dilatant zones along faults, solution channels in areas of karst topography, etc., may be permeated by mineralizing solutions. If the prevailing physico-chemical conditions induce precipitation then crystals will form. These will grow as the result of spontaneous nucleation within the solution, or, more commonly, by nucleation on the enclosing surface. This leads to the precipitation and outward growth of the first formed minerals on vein walls. If the solutions change in composition, there may be a change in mineralogy and crusts of minerals

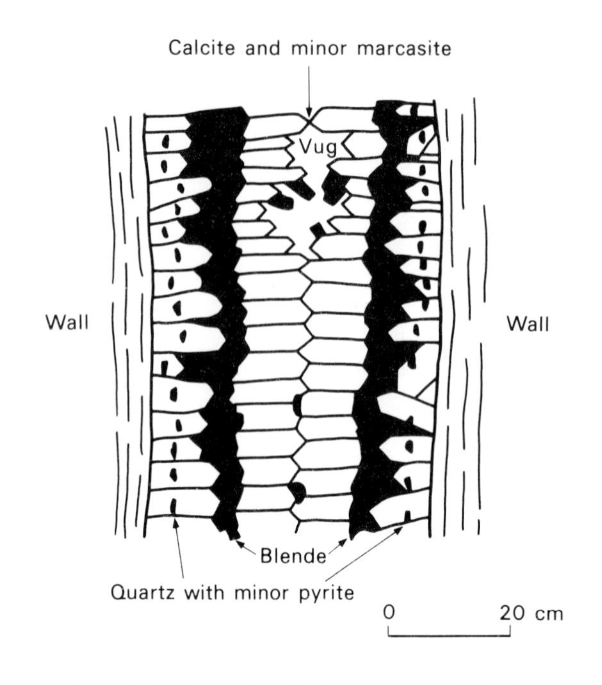

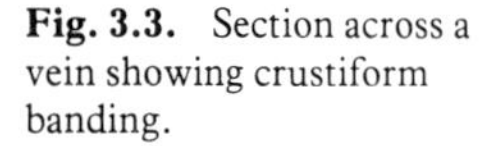
Fig. 3.3. Section across a vein showing crustiform banding.

of different composition may give the vein filling a banded appearance (Fig. 3.3). This is called crustiform banding. Its development in some veins demonstrates that mineralizing solutions may change in composition with time and shows us the order in which the minerals were precipitated. This order is called the paragenetic sequence.

In an example of simple opening and filling of a fissure as depicted in Fig. 3.3 the banding is symmetrical about the centre of the vein. With repeated opening and mineralization this symmetry will of course be disturbed, but it may still be present among the constituents deposited after the last phase of opening.

Open space deposition also occurs at the surface at sediment-water or rock-water interfaces during, for example, the formation of volcanic massive sulphide deposits. Under such situations rapid flocculation of material occurs and a common primary texture which results is colloform banding. This is a very fine scale banding involving one or more sulphides very like the banding in agate. Some workers believe that it results from colloidal deposition and that the banding is analogous to the Liesegang rings formed in some gels. This banding is very susceptible to destruction by recrystallization and may be partially or wholly destroyed by diagenesis or low grade metamorphism, producing a granular ore. The textures of sedimentary iron and manganese ores are discussed in Chapter 16.

Replacement

Edwards (1952) defined replacement as 'the dissolving of one mineral and the simultaneous deposition of another mineral in its place, without the intervening development of appreciable open spaces, and commonly without a change of volume.' Replacement has been an important process in the formation of many ore deposits particularly the pyrometasomatic class. This process involves not only the minerals of the country rocks, but also the ore and gangue minerals.

Nearly all ores, including those developed in open spaces, show some evidence of the occurrence of replacement processes.

The most compelling evidence of replacement is pseudomorphism. Pseudomorphs of cassiterite after orthoclase have been recorded from Cornwall, England, and of pyrrhotite after hornblende from Sullivan, British Columbia. Numerous other examples are known. The preservation of delicate plant cells by marcasite is well known and crinoid ossicles replaced by cassiterite occur in New South Wales. An overall view of ore deposits suggests that there is no limit to the direction of metasomatism. Given the right conditions, any mineral may replace any other mineral, although natural processes often make for unilateral reactions. Secondary (supergene) replacement processes, leading to sulphide enrichment by downward percolating meteoric waters, are sometimes most dramatic and fraught with economic importance. They can be every bit as important as primary (hypogene) replacement brought about by solutions emanating from crustal or deeper sources.

Fluid inclusions

The growth of crystals is never perfect and as a result samples of the fluid in which the crystals grew may be trapped in tiny cavities usually $< 100\ \mu m$ in size. These are called fluid inclusions and can be divided into various types. *Primary* inclusions formed during the growth of crystals provide us with samples of the ore-forming fluid. They also yield crucial geothermometric data and tell us something about the physical state of the fluid, e.g. whether it was boiling at the time of entrapment. They are common in all rocks and veins.

The principal matter in most fluid inclusions is water. Second in abundance is carbon dioxide. The commonest inclusions in ore deposits fall into four groups (Nash 1976). Type I, moderate salinity inclusions, are generally two phase consisting principally of water and a small bubble of water vapour which forms 10-40% of the inclusion (Fig. 3.4). The presence of the bubble indicates trapping at an elevated temperature with formation of the bubble on cooling. Heating on a

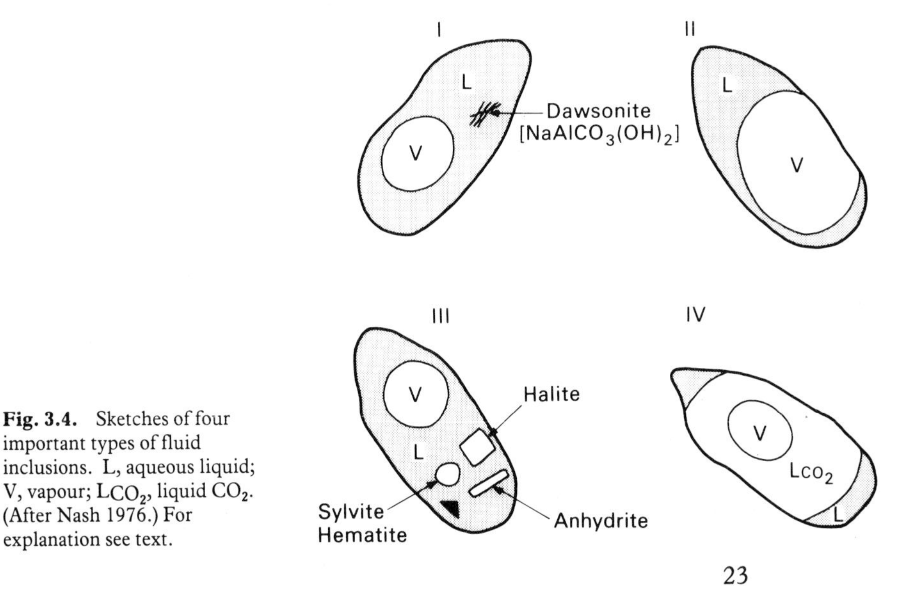

Fig. 3.4. Sketches of four important types of fluid inclusions. L, aqueous liquid; V, vapour; LCO_2, liquid CO_2. (After Nash 1976.) For explanation see text.

microscope stage causes rehomogenization to one liquid phase and the homogenization temperature indicates the temperature of growth of that part of the containing crystal (provided the necessary pressure correction can be made). Sodium, potassium, calcium and chlorine occur in solution and salinities range from 0-23 wt % NaCl equivalent. In some of these inclusions small amounts of daughter salts have been precipitated during cooling. These include carbonates and anhydrite.

Type II, gas-rich inclusions, generally contain more than 60% vapour. Again they are dominantly aqueous but CO_2 may be present in small amounts. They appear to represent trapped steam. The simultaneous presence of gas-rich and gas-poor aqueous inclusions is good evidence that the fluids were boiling at the time of trapping. Type III, halite-bearing inclusions, have salinities ranging up to more than 50%. They contain well-formed, cubic halite crystals and generally several other daughter minerals, particularly sylvite and anhydrite. Type IV, CO_2-rich inclusions, have CO_2:H_2O ratios ranging from 3 to over 30 mol %. They grade into type II inclusions and indeed there is a general gradation in many situations, e.g. porphyry copper deposits (Chivas & Wilkins 1977), between the common types of fluid inclusions.

Perhaps one of the most surprising results of fluid inclusion studies is the evidence of the common occurrence of exceedingly strong brines in nature, brines more concentrated than any now found at the surface (Roedder 1972). These are not only present in mineral deposits but are common in igneous and metamorphic rocks. Many, but not all, strong brine inclusions are secondary features connected with late-stage magmatic and metamorphic phenomena such as the genesis of greisens, pegmatites and ore deposits as well as wall rock alteration processes such as sericitization and chloritization. They are compelling evidence that ore-forming fluids are hot, saline aqueous solutions. They form a link between laboratory and field studies, and it should be noted that there is strong experimental and thermodynamic evidence which shows that chloride in hydrothermal solutions is a potent solvent for metals through the formation of metal-chloride complex ions, and indeed inclusions that carry more than one per cent of precipitated sulphides are known.

Wall rock alteration

Frequently alongside veins or around irregularly shaped orebodies we find alteration of the country rocks. This may take the form of colour, textural, mineralogical or chemical changes, or any combination of these. Alteration is not always present, but when it is, it may vary from minor colour changes to extensive mineralogical transformations and complete recrystallization. Generally speaking, the higher the temperature of deposition of the ore minerals the more intense is the alteration, but it is not necessarily more widespread. This alteration, which shows a spatial and usually a close temporal relationship to ore deposits, is called wall rock alteration.

The areal extent of the alteration can vary considerably. Sometimes it may be limited to a few centimetres on either side of a vein, at other times it may form a thick halo around an orebody and then, since it widens the drilling target, it may be of considerable exploration value. The spatial and temporal relationships suggest that wall rock alteration is due to reactions caused by the mineralizing fluid permeating parts of the wall rocks. Many alteration haloes show a zonation of mineral assemblages resulting from the changing nature of the hydrothermal solution as it passes through the wall rocks. The associated orebodies in certain deposits (e.g. porphyry coppers, Chapter 12) may show a special spatial relationship to the zoning, knowledge of which may be invaluable in probing for the deposit with a diamond drill.

The different wall rock alteration mineral assemblages can be compared with metamorphic facies as, like these, they are formed in response to various pressure, temperature and compositional changes. They are not, however, generally referred to as alteration facies but as types of wall rock alteration. It can be very difficult in some areas to distinguish wall rock alteration, such as chloritization, from the effects of low-grade regional or contact metamorphism. It is, however, essential that the two effects are separately identified, otherwise a considerable degree of exploration effort may be expended in vain.

There are two main divisions of wall rock alteration: hypogene and supergene. Hypogene alteration is caused by ascending hydrothermal solutions, and supergene alteration by descending meteoric water reacting with previously mineralized ground. A third mechanism giving rise to the formation of wall rock alteration is the metamorphism of sulphide orebodies. In this chapter we will be concerned mainly with hypogene alteration.

The study of hydrothermal fluids has shown that they are commonly weakly acidic, but may become neutral or slightly alkaline by reaction with wall rocks (or by mixing with other waters, e.g. ground water). The solutions contain dissolved ions that are important in ion exchange reactions, and the composition of a particular hydrothermal solution will have an important bearing on the nature of the wall rock alteration it may give rise to. Since the chemistry of wall rocks can also vary greatly according to their petrography, it is clear that predictions as to the course of wall rock alteration reactions are fraught with difficulties. Nevertheless, there is, despite a variety of controls, a considerable uniformity in the types of wall rock alteration which facilitates their study and classification. The controls of wall rock alteration fall into two groups governed respectively by the nature of the host rocks and the nature of the ore-forming solution. Besides the chemistry of the host rocks, other factors of importance are their grain size, physical state (e.g. sheared or unsheared) and permeability, and for the hydrothermal solution important properties are the pressure, temperature, chemistry, pH and Eh.

Although most rock-forming minerals are susceptible to attack by acid solution, carbonates, zeolites, feldspathoids and calcic plagioclase are least resistant; pyroxenes, amphiboles and biotite are moderately resistant, and sodic plagioclase, potash feldspar and muscovite are strongly resistant. Quartz is often entirely unaffected.

Studies of wall rock alteration are important because they (a) contribute to our knowledge of the nature and evolution of ore-forming solutions, (b) are often valuable in exploration, and (c) produce minerals such as phyllosilicates which can be used to obtain radiometric dates on the wall rock alteration and, by inference, on the associated mineralization.

TYPES OF WALL ROCK ALTERATION

These have been extensively described by Meyer & Hemley (1967) from whose work much of the following is drawn.

(a) *Advanced argillic alteration.* This alteration is characterized by dickite, kaolinite [both $Al_2Si_2O_5(OH)_4$], pyrophyllite [$Al_2Si_4O_{10}(OH)_2$] and quartz. Sericite is usually present and frequently alunite, pyrite, tourmaline, topaz, zunyite and amorphous clay minerals. This is one of the more intense forms of alteration, often present as an inner zone adjoining many base metal vein or pipe deposits associated with acid plutonic stocks as at Butte, Montana, and Cerro de Pasco in Peru. It is also found in hot spring environments and in telescoped shallow precious metal deposits. The associated sulphides of the orebodies are generally sulphur-rich; covellite, digenite, pyrite and enargite are most common.

This alteration involves extreme leaching of bases (alkalies and calcium) from all aluminous phases such as feldspars and micas, but is only present if aluminium is not appreciably mobilized. When aluminium is also removed it grades into silicification and, with increasing sericite, it grades outwards into sericitization.

(b) *Sericitization.* In orefields the world over this is one of the commonest types of alteration in aluminium-rich rocks such as slates, granites, etc. The dominant minerals are sericite and quartz; pyrite often accompanies them. Care should be taken to ensure that the sericite is muscovite. Illite, paragonite, phlogopite, talc and pyrophyllite can be mistaken for sericite. Muscovite is stable over a wide pressure-temperature field and this accounts for its common occurrence as an alteration mineral. If potassium is introduced into the wall rocks then rocks low in this element, such as diorites, can be sericitized. The reader must not assume that during this and other wall rock alteration processes the wall rocks necessarily become solid sericite or clay minerals, as the case may be. What we see is the appearance in significant amounts, or an increase in quantity, of the mineral or minerals concerned. Sometimes the new mineral(s) may be developed to the exclusion of all other minerals, but this is not necessarily the case. During the sericitization of granite, the feldspars and micas may be transformed to sericite, with secondary quartz as a reaction by-product, but the primary quartz may be largely unaffected except for the development of secondary fluid inclusions. Wall rock alteration is progressive, with some minerals reacting and being altered more rapidly than others. If shearing accompanies this, or any of the other types of alteration carrying phyllosilicates, then a schistose rock may result, otherwise hornfelsic-like textures develop.

With the appearance of secondary potash feldspar and/or secondary biotite, sericitization grades into potassium silicate alteration which is very common in the central deeper portions of porphyry copper deposits as at Bingham Canyon, Utah. In fluorine-rich environments, topaz together with zunyite and quartz may accompany the sericite to form greisen. Outside the sericitization zone lower grade intermediate argillic alteration may occur. Thus sericitization may grade into three types of higher grade alteration and one of lower grade.

(c) *Intermediate argillic alteration.* The principal minerals are now kaolin- and montmorillonite-group minerals occurring mainly as alteration products of plagioclase. The intermediate argillic zone may itself be zoned with montmorillonite minerals dominant near the outer fringe of alteration and kaolin minerals nearer the sericitic zone. Sulphides are generally unimportant. Outwards from the intermediate argillic alteration zone propylitic alteration may be present before fresh rock is reached.

(d) *Propylitic alteration.* This is a complex alteration generally characterized by chlorite, epidote, albite and carbonate (calcite, dolomite or ankerite). Minor sericite, pyrite and magnetite may be present, less commonly zeolites and montmorillonites. The term propylitic alteration was first used by Becker in 1882 for the alteration of diorite and andesite beside the Comstock Lode, Nevada (a big gold-silver producer in the boom days of the last century). Here the main alteration products are epidote, chlorite and albite. The propylitic alteration zone is often very wide and therefore, when present, is a useful guide in mineral exploration. For example at Telluride, Colorado, narrow sericite zones along the veins are succeeded outwards by a wide zone of propylitization.

With the intense development of one of the main propylitic minerals we have what are sometimes considered as subdivisions of propylitization: chloritization, albitization and carbonatization. Albitization will be dealt with under feldspathization.

(e) *Chloritization.* Chlorite may be present alone or with quartz or tourmaline in very simple assemblages. However other propylitic minerals are usually present, and anhydrite may also be in evidence. Hydrothermal chlorites often show a change in their Fe/Mg ratio with distance from the orebody. Usually they are richer in iron adjacent to the sulphides, but the reverse has been reported. The change in this ratio can be recorded by simple refractive index measurements so that this offers the possibility of a cheap exploration tool.

The development of secondary chlorite may result from the alteration of mafic minerals already present in the country rocks or from the introduction of magnesium and iron. Of course both processes can occur together, as at Ajo, Arizona. Chloritization is common alongside tin veins in Cornwall where progressive alteration of the country rocks occurs:

Unaltered porphyritic granite	Pinking of feldspars	Chlorite in groundmass developed from biotite and small feldspars	Rim, core and cleavage replacement of phenocrysts	Quartz-chlorite rock carrying tin values	Quartz vein with tin.

(f) *Carbonatization.* Dolomitization is a common accompaniment of low to medium temperature ore deposition in limestones, and dolomite is probably the commonest of the carbonates formed by hydrothermal activity. Dolomitization is most commonly associated with low temperature lead-zinc deposits of 'Mississippi Valley type'. These can have wall rocks of pure dolomite. Usually this rock is coarser and lighter in colour than the surrounding limestone. Dolomitization generally appears to have preceded sulphide deposition, but the relationship and timing have been much debated in the various fields where dolomitization occurs. Nevertheless, dolomitization often appears to have preceded mineralization, to have increased the permeability of the host limestones and thus to have prepared them for mineralization. Of course not all dolomitized limestone contains ore and sometimes in such fields ore may occur in unaltered limestone. Often the limestone is then recrystallized to a coarse white calcite rock, as at Bisbee, Arizona.

Other carbonates may be developed in silicate rocks, especially where iron is available. Ankerite may then be common particularly in the calcium-iron environment of carbonatized basic igneous rocks and volcaniclastics. This is particularly the case with many Precambrian and Phanerozoic vein gold deposits, for example the Mother Lode in California where ankerite, sericite, albite, quartz, pyrite and arsenopyrite are well developed in the altered wall rocks. At Larder Lake in Ontario, dolomite and ankerite have replaced large masses of greenstone. As with the chlorites, there may be a chemical variation in the Fe/Mg ratio with proximity to ore.

Most of the alteration types we have dealt with so far have involved hydrolysis, i.e. the introduction of hydrogen ion for the formation of hydroxyl-bearing minerals such as micas and chlorite, this is often accompanied by the removal of bases (K^+,Na^+,Ca^{2+}). Thus the H^+/OH^- ratio of the mineralizing solutions will decrease and concomitantly they will be enriched in bases. The chemical significance of carbonatization may take two different forms. Dolomitization of limestone involves very extensive magnesium metasomatism, but only as a base exchange process. This is called cation metasomatism. In this case the mineralizing (or pre-mineralization) solutions must have carried abundant magnesium. Carbonatization of silicate rocks involves *inter alia* anion metasomatism with the introduction of CO_3^{2-} rather than bases.

(g) *Potassium silicate alteration.* Secondary potash feldspar and/or biotite are the essential minerals of this alteration. Clay minerals are absent but minor chlorite may be present. Anhydrite is often important especially in porphyry copper deposits, e.g. El Salvador, Chile, and it can form up to 15% of the altered rock. It is, however, easily hydrated and removed in solution, even at depths of up to 1000 m, so that it has probably been removed from many deposits by groundwater solution. Magnetite and hematite may be present and the common sulphides are pyrite, molybdenite and chalcopyrite, i.e. there is an intermediate sulphur/metal ratio.

(h) *Silicification.* This involves an increase in the proportion of quartz or cryptocrystalline silica (i.e. cherty or opaline silica) in the altered rock. The silica may be introduced from the hydrothermal solutions, as in the case of chertified limestones associated with lead-zinc-fluorite-baryte deposits, or it may be the by-product of the alteration of feldspars and other minerals during the leaching of bases. Silicification is often a good guide to ore, e.g. the Black Hills, Dakota. At the Climax porphyry molybdenum deposit in Colorado, intensive and widespread silicification accompanied the mineralization.

The silication of carbonate rocks leading to the development of skarn is dealt with under pyrometasomatic deposits (see Chapters 4 and 11).

(i) *Feldspathization.* This leads to the development of either potash feldspar or albite. Secondary orthoclase or microcline results from the introduction of potassium as in the deeper zones or porphyry copper deposits. Albitization, on the other hand, may result from the introduction of sodium or from removal of calcium from plagioclase-bearing rocks. Albitization is found adjacent to some gold deposits, often replacing potash feldspar, e.g. Treadwell, Alaska.

(j) *Tourmalinization.* This is associated with medium to high temperature deposits, e.g. many tin and some gold veins have a strong development of tourmaline in the wall rocks and often actually in the veins as well. The Sigma Gold Mine in Quebec has veins which in places are massive tourmaline and this mineral is well developed in the adjacent wall rock. This is also the case in the granodiorite wall rock of the Siscoe Mine, Quebec. At Llallagua, Bolivia, the world's largest primary tin mine, the porphyry host is altered to a quartz-sericite-tourmaline rock.

If the altered country rocks are lime-rich then axinite rather than tourmaline may be formed.

(k) *Other alteration types.* There are many other types of alteration, among these may be mentioned *alunitization,* which may be of either hypogene or supergene origin; *pyritization,* due to the introduction of sulphur which may attack both iron oxides and mafic minerals; *hematitization,* an alteration type often associated with uranium (particularly pitchblende deposits); *bleaching,* due in many cases to the reduction of hematite; *greisenization,* a frequent form of alteration alongside tin-tungsten and beryllium deposits in granitic rocks or gneisses; *fenitization,* which is associated with carbonatite hosted deposits and which is characterized by the development of nepheline, aegirine, sodic amphiboles and alkali feldspars in the aureoles of the carbonatite masses; *serpentinization* and the allied development of talc can occur in both ultrabasic rocks and limestones, it is associated with some gold and nickel deposits—where serpentine and talc are developed in limestones there is generally an introduction of SiO_2 and H_2O and frequently some Mg; finally *zeolitization* is marked by the development of stilbite, natrolite, heulandite, etc., and often accompanies native copper mineralization in amygdaloidal basalts—calcite, prehnite, pectolite, apophyllite and datolite are generally also present.

INFLUENCE OF ORIGINAL ROCK-TYPES

A survey of world literature shows that wall rock alteration exhibits a certain regularity with respect to the nature of the host rock (Boyle 1970). There are of course exceptions, but certain generalizations can be made. Thus, for example, the most prevalent types of alteration in acidic rocks are sericitization, argillization, silicification and pyritization. Intermediate and basic rocks generally show chloritization, carbonatization, sericitization, pyritization and propylitization. In carbonate rocks the principal high temperature alteration is skarnification, whereas normal shales, slates and schists are frequently characterized by tourmalinization, especially when hosting tin and tungsten deposits. A more detailed discussion of this subject can be found in Boyle's work.

CORRELATION WITH TYPE OF MINERALIZATION

Boyle has also shown that certain types of mineralization tend to be accompanied by characteristic types of alteration. Space only permits the mention here of a few examples. Red bed uranium, vanadium, copper, lead and silver deposits are generally accompanied by bleaching. Vein deposits of native silver are usually characterized by carbonatization and chloritization and molybdenum-bearing veins by silicification and sericitization. Other examples have been given above.

TIMING OF WALL ROCK ALTERATION

The problem of disentangling age relationships between different assemblages of wall rock alteration minerals in a given deposit can be extremely difficult. Their presence, together with evidence such as crustiform banding, has suggested to many workers that the mineralizing solutions came in pulses of different compositions (polyascendant solutions). In some deposits, however, such as Butte, Montana, there is a considerable body of data suggesting a single long-lasting phase of mineralization and accompanying alteration. Both mechanisms have probably taken place and some deposits may have been formed from monoascendant solutions whilst others are the result of polyascendant mineralization. This important aspect is discussed at length by Meyer & Hemley (1967).

THE NATURE OF ORE-FORMING SOLUTIONS AS DEDUCED FROM WALL ROCK ALTERATION

Studies of wall rock alteration indicate that aqueous solutions played a large part in the formation of epigenetic deposits. Clearly, in some cases these solutions carried other volatiles such as CO_2, S, B and F. The pH of ore-forming solutions is difficult to assess from wall rock alteration studies, but in cases of hydrogen metasomatism the pH value must have been low; it would have increased in value during reactions with the wall rocks, so that in some cases solutions may have become neutral or even slightly alkaline. A very instructive study of the use of wall rock alteration studies in throwing a light on the nature of ore-forming solutions is to be found in Fournier (1967).

4

Some Major Theories of Ore Genesis

Some indication has been given in Chapter 2 of the very varied nature of ore deposits and their occurrence. This variety of form has given rise over the last hundred years or more to an equally great variety of hypotheses of ore genesis. The history of the evolution of these ideas is an interesting study in itself and one which has been well documented by Stanton (1972). There is no room for such a discussion here and the reader is referred to Stanton's work and to references given by him. This chapter will, therefore, be concerned only with major theories of ore genesis current at the present time. These will be divided for the sake of convenience into internal and surface processes. The reader should be warned, however, that very often several processes contribute to the formation of an orebody. Thus, where we have rising hot aqueous solutions forming an epigenetic stockwork deposit just below the surface and passing on upwards through it to form a contiguous syngenetic deposit under, say, marine conditions, even the above simple classification is in difficulties. This is the reason why ore geologists besides producing a plethora of ore genesis theories have also created a plethora of orebody classifications! A summary of the principal theories of ore genesis is given in Table 4.1.

Table 4.1. Simple classification of ore genesis theories.

Theory	Nature of process	Typical deposits
ORIGIN DUE TO INTERNAL PROCESSES		
Magmatic segregation	Separation of ore minerals by fractional crystallization and related processes during magmatic differentiation	Chromite layers in the Great Dyke of Rhodesia and the Bushveld lopolith, South Africa
	Liquation, liquid immiscibility. Settling out from magmas of sulphide, sulphide-oxide or oxide melts which accumulated beneath the silicates or were injected into wall rocks or in rare cases erupted on the surface	Copper-nickel orebodies of Sudbury, Canada; Pechenga, USSR and the Yilgarn Block, Western Australia
Pegmatitic deposition	Crystallization as disseminated grains or segregations in pegmatites	Lithium-tin-caesium pegmatites of Bikita, Rhodesia. Uranium pegmatites of Bancroft, Canada
Hydrothermal	Deposition from hot aqueous solutions which may have had a magmatic, metamorphic, surface or other source	Tin-tungsten-copper veins and stockworks of Cornwall, UK. Molybdenum stockworks of Climax, USA. Porphyry copper deposits of Panguna PNG and Bingham, USA

Table 4.1. (contd.)

Theory	Nature of process	Typical deposits
Lateral secretion	Diffusion of ore- and gangue-forming materials from the country rocks into faults and other structures	Yellowknife gold deposits, Canada. Mother Lode, USA
Metamorphic processes	Pyrometasomatic deposits formed by replacement of wall rocks adjacent to an intrusion	Copper deposits of Mackay, USA and Craigmont, Canada. Magnetite bodies of Iron Springs, USA
	Initial or further concentration of ore elements by metamorphic processes, e.g. granitization, alteration processes	Some gold veins, and disseminated nickel deposits in ultramafic dykes
ORIGIN DUE TO SURFACE PROCESSES		
Mechanical accumulation	Concentration of heavy durable minerals into placer deposits	Rutile-zircon sands of New South Wales, Australia, and Trail Ridge, USA. Tin placers of Malaysia. Gold placers of the Yukon, Canada
Sedimentary precipitates	Precipitation of particular elements in suitable sedimentary environments, with or without the intervention of biological organisms	Banded iron formations of the Precambrian shields. Manganese deposits of Chiaturi, USSR
Residual processes	Leaching from rocks of soluble elements leaving concentrations of insoluble elements in the remaining material	Nickel laterites of New Caledonia. Bauxites of Hungary, France, Jamaica and Arkansas, USA
Secondary or supergene enrichment	Leaching of valuable elements from the upper parts of mineral deposits and their precipitation at depth to produce higher concentrations	Many gold and silver bonanzas. The upper parts of a number of porphyry copper deposits
Volcanic-exhalative (=sedimentary exhalative)	Exhalations of hydrothermal solutions at the surface, usually under marine conditions and generally producing stratiform orebodies	Meggan, Germany; Sullivan, Canada; Mount Isa, Australia; Rio Tinto, Spain; Kuroko deposits of Japan

Origin due to internal processes

MAGMATIC SEGREGATION

The terms magmatic segregation deposit or orthomagmatic deposit are used for those ore deposits, apart from pegmatites, that have crystallized direct from a magma. Those formed by fractional crystallization are usually found in plutonic igneous rocks; those produced by liquation (separation into immiscible liquids) may be found associated with both plutonic and volcanic rocks. Magmatic segregation deposits may consist of layers within or beneath the rock mass (chromite layers, subjacent copper-nickel sulphide ores).

(a) *Fractional crystallization.* This includes any mechanical process by which early-formed crystals are prevented from equilibrating with the melt from which they grew. The important processes are gravity fractionation, flowage differentiation, filter pressing and dilatation (Carmichael *et al.* 1974), the most important of these from the ore forming point of view is gravity fractionation.

Gravity fractionation occurs during the earlier stages of magmatic crystallization when the magma is still dominantly liquid and crystals can sink through it to accumulate at the base of the magma chamber or interlayered with other early-formed minerals. The rocks so formed are called cumulates and they often display conspicuous lithological alternations called rhythmic layering, since it is frequently repeated many times over in vertical sections of the plutonic bodies in which it occurs. Usually it leads to the formation of olivine-, pyroxene- or plagioclase-rich layers; but when oxides such as chromite are precipitated, layers of this mineral may develop as in the Bushveld Complex of South Africa. This enormous lopolith is characterized by cumulus magnetite in the upper zone. The chromite layers have been mined for decades, the magnetite will shortly be exploited, the main prize here being its high vanadium content. Another mineral which may be concentrated in this way is ilmenite. Whilst chromite accumulations are nearly all in ultrabasic rocks and to a lesser extent in gabbroic or noritic rocks, ilmenite accumulations show an association with anorthosites or anorthositic gabbros. These striking rock associations are strong evidence for the magmatic origin of the minerals.

(b) *Liquation.* A different form of segregation results from liquid immiscibility. In exactly the same way that oil and water will not mix but form immiscible globules of one within the other, so in a mixed sulphide-silicate magma the two liquids will tend to segregate. Sulphide droplets separate out and coalesce to form globules which, being denser than the magma, sink through it to accumulate at the base of the intrusion or lava flow (Fig. 4.1). Iron sulphide is the principal constituent of these droplets which are associated with basic and ultrabasic rocks, because sulphur and iron are both more abundant in these rocks than in acid or intermediate rocks. Chalcophile elements such as copper and nickel also enter these droplets and sometimes the platinum group metals.

The accumulation of Fe-Ni-Cu sulphide droplets beneath the silicate fraction can produce massive sulphide orebodies. These are overlain by a zone with subordinate silicates enclosed in a network of sulphides—net-textured ore sometimes called disseminated ore. This zone is, in turn, overlain by one of weak mineraliza-

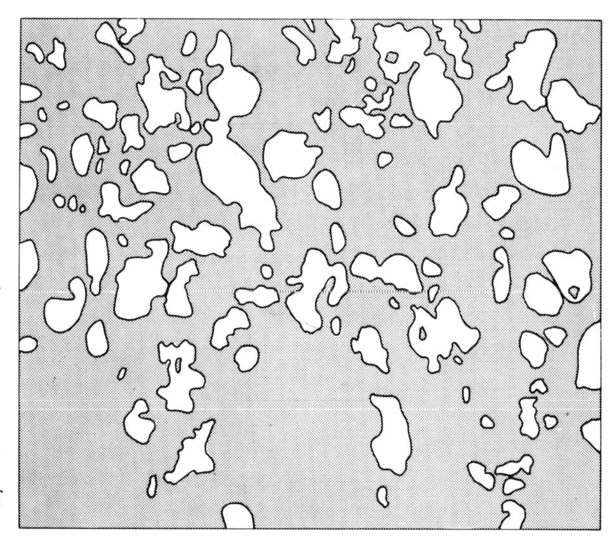

Fig. 4.1. Tracing of an ore specimen from Sudbury, Ontario. Sulphides, mainly pyrrhotite with minor pentlandite and chalcopyrite, are shown white; surrounding silicates are shown grey. Note the rounded discontinuous nature of the sulphide globules. They appear to have formed as a result of liquid immiscibility from a silicate-sulphide melt. Note especially the rounded silicate blebs within the sulphide bodies and that many of the sulphide globules appear to have formed from the coalescence of smaller bodies of sulphide liquid.

tion which grades up into overlying peridotite, gabbro or komatiite, depending on the nature of the associated silicate fraction. To explain the mechanism of formation of these zones Naldrett (1973) proposed his 'billiard ball' model (Fig. 4.2).

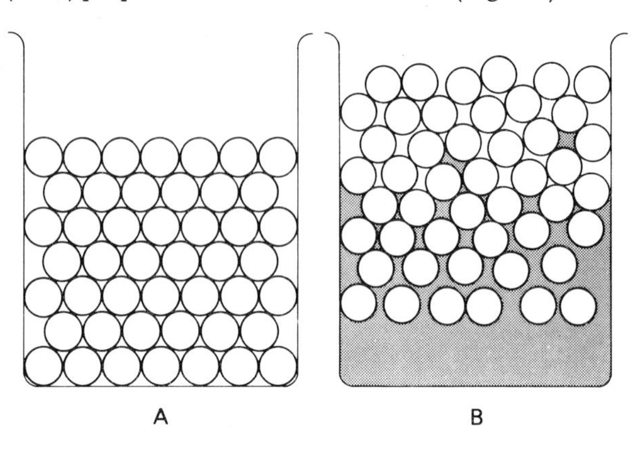

Fig. 4.2. The billiard ball model illustrating the formation of Fe-Ni-Cu sulphide deposits by liquation. For explanation see text. (After Naldrett 1973.)

Imagine a large beaker partly filled with billiard balls and water (Fig. 4.2A). These represent olivine grains and interstitial silicate liquid. Then consider the effect of adding mercury to represent the immiscible sulphide liquid. This will sink to the bottom and the balls will tend to float on it. The lower balls will be forced down into the mercury by the weight of the overlying ones. If the contents of the beaker are frozen before all the mercury has had a chance to percolate to the bottom then the situation shown in Fig. 4.2B will exist. There is an obvious analogy between the massive mercury and the zone of massive sulphides, between the overlying zone of balls immersed in mercury and net-textured ore, and between the zone of scattered globules of mercury and the zone of weak mineralization.

PEGMATITIC DEPOSITION

Pegmatites are very coarse-grained igneous or metamorphic rocks. They commonly form dyke-like or lenticular bodies usually ranging from a few metres to tens of metres in size, though occasional large dykes over 2 km long have been found. The vast bulk of pegmatites of economic importance are granitic in composition. Granitic magmas are richer than basic magmas in water and other volatiles but the principal minerals of granite are anhydrous, therefore as crystallization proceeds the water content of the remaining magma increases. This can produce a residual melt rich in volatiles and elements such as Li, Be, Nb, Ta, Sn and U which cannot be accommodated in the crystal lattices of the quartz and feldspar of the granite. If this melt is injected into fractures in the now mainly solid granite or its country rocks then pegmatites will crystallize from it. Thus they are late-stage products of granitic activity associated with plutonic intrusions from which the volatiles could not readily escape. Formational temperatures are low being about 700-250°C (Jahns 1955).

Pegmatites are often divided into simple and complex. Simple pegmatites have simple mineralogy and no well-developed zoning, complex pegmatites may have a complex mineralogy with many rare minerals such as pollucite [$CsSi_2AlO_6$] and amblygonite [$LiAlPO_4(F,OH)$], but their marked feature is the arrangement of their minerals in a zonal sequence from the contact inwards. An example of a complex

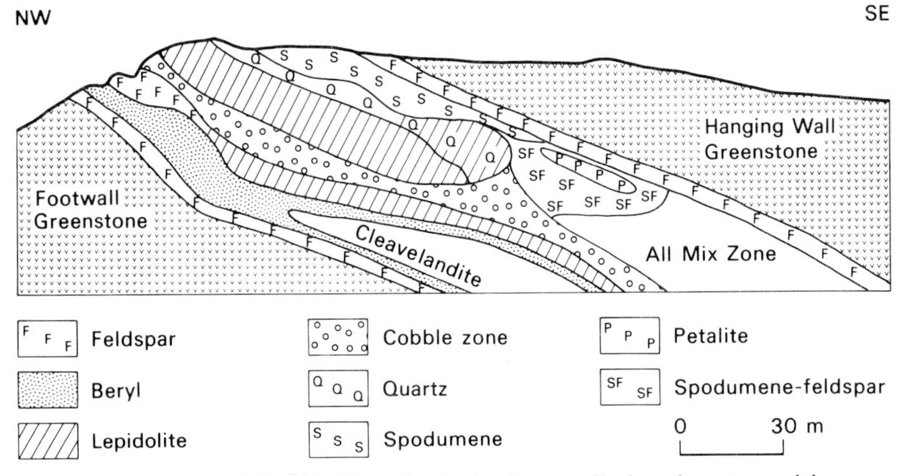

Fig. 4.3. Section through the Bikita Pegmatite showing the generalized zonal structure and the important minerals of each zone. (After Symons 1961.) (Cleavelandite is a lamellar variety of white albite.)

Table 4.2. Zoning in the Bikita Pegmatite, Rhodesia. (From Symons 1961).

	Hanging wall greenstone
Border zone:	Selvage of fine-grained albite, quartz, muscovite.
Wall zones:	Mica band. Coarse muscovite some quartz. Hanging wall feldspar zone. Large microcline crystals.
Intermediate zones:	Petalite-feldspar zone.
	Spodumene zone (a) massive, (b) mixed spodumene, quartz, plagioclase and lepidolite.
	Pollucite zone. Massive pollucite with 40% of quartz.
	Feldspar-quartz zone. Virtually devoid of lithium minerals.
	'All mix' zone. Microcline, lepidolite, quartz.
Core zones:	Massive lepidolite (a) high grade core, nearly pure lepidolite, (b) lepidolite-quartz subzone.
	Lepidolite-quartz shell.
Intermediate zones:	'Cobble' zone. Rounded masses of lepidolite in an albite matrix.
	Feldspathic lepidolite zone.
Wall zone:	Beryl zone. Albite, lepidolite, beryl.
	Footwall feldspar. Albite, muscovite, quartz.
	Footwall greenstone

zoned pegmatite is given in Fig. 4.3 and Table 4.2. The crystals in complex pegmatites can be very large. In the Bikita Pegmatite, for example, the spodumene crystals are commonly 3 m long. Contacts between different zones may be sharp or gradational. Inner zones may cut across or replace outer zones, but not vice versa. There is a world-wide similarity of zonal sequences. All these features must be explained by any proposed genetic theory.

Broadly speaking, three hypotheses have been put forward to account for such zoning. The first is that of fractional crystallization under non-equilibrium conditions leading to a steady change in the composition of the melt with time. The second is of deposition along open channels from solutions of changing composition.

The third is a two-stage model: (a) crystallization of a simple pegmatite with (b) partial or complete replacement of the pegmatite as hot aqueous solutions pass through it. At the present time, most workers prefer the first or third hypothesis as the majority of the evidence, such as complete enclosure in many pegmatites of the interior zones, does not favour the existence of open channels during crystallization. As the third theory encounters difficulties in explaining the world-wide similarity of zonal sequences, the first theory finds most favour.

HYDROTHERMAL PROCESSES

Hot aqueous solutions have played a part in the formation of many different types of ore deposit, for example veins, stockworks of various types, volcanic-exhalative deposits and others. Such fluids are usually called hydrothermal solutions. Many lines of evidence attest to their important role as mineralizers. The evidence from wall rock alteration and fluid inclusions has been discussed in Chapter 3. Homogenization of fluid inclusions in minerals from hydrothermal deposits and other geothermometers has shown that the depositional range for all types of deposit is approximately 50-650°C. Analysis of the fluid has shown water to be the common phase and usually it has salinities far higher than that of sea water. Hydrothermal solutions are believed to be capable of carrying a wide variety of materials and of depositing these to form minerals as diverse as gold and muscovite. The physical chemistry of such solutions is therefore complex and very difficult to imitate in the laboratory. Our knowledge of their properties and behaviour is still somewhat hazy and there are many ideas about the origin of such solutions and the materials they carry (see relevant chapters in Barnes 1967a).

The principal problems are then the source and nature of the solutions, the sources of the metals and sulphur in the solutions, the means of transport of these substances, and the mechanisms of deposition.

(a) *Sources of the solutions and their contents.* Because of the spatial relationship that exists between many hydrothermal deposits and igneous rocks, a strong school of thought holds that consolidating magmas are the source of many, if not all, hydrothermal solutions. The solutions are considered to be low temperature residual fluids left over after pegmatite crystallization and containing the base metals and other elements which could not be accommodated in the crystal lattices of the silicate minerals precipitated by the freezing magma. This model derives not only the metals and other elements from a hot body of igneous rock, but also the heat to drive the mineralization system. The solutions are assumed to move upwards along fractures and other channelways to cooler parts of the crust where deposition of minerals occurs.

In many orefields, however, such as the Northern Pennine Orefield of England, there are no acid or intermediate plutonic intrusions which might be the source of the ores. Some workers have therefore postulated a more remote magmatic source such as the lower crust or, more frequently, magmatic processes in the mantle, whilst an important body of opinion has favoured deposition from connate solutions—that is water which was trapped in sediments during deposition and which has been driven out by the rise in temperature and pressure caused by deep burial. Such burial might occur in sedimentary basins, and solutions from this source are often called basinal brines. With a geothermal gradient of 1°C per 30 m, temperatures around 300°C would be reached at a depth of 9 km. Hot solutions from this source are believed to leach metals, but not necessarily sulphur, from the rocks through which they pass, ultimately precipitating them near the surface far from any igneous intrusion.

(b) *Means of transport.* Sulphides and other minerals have such low solubilities in pure water that it is now generally believed that the metals were transported either as sulphide or bisulphide complexes or as halide complexes. A few simple figures will illustrate this. The amount of zinc in a saturated zinc sulphide solution at a pH of 5 and a temperature of 100°C (possible mineralizing conditions) is about 1×10^{-5} g l^{-1}. A small orebody containing 1 million tonnes of zinc could have been formed from a solution of this strength (assuming all the zinc was precipitated) provided 10^{17} l of solution passed through the orebody. This is equal to the volume of a tank having an area of 10 000 km² and sides 10 km high—an impossible quantity of solution. This difficulty is further illustrated in Fig. 4.4 where the

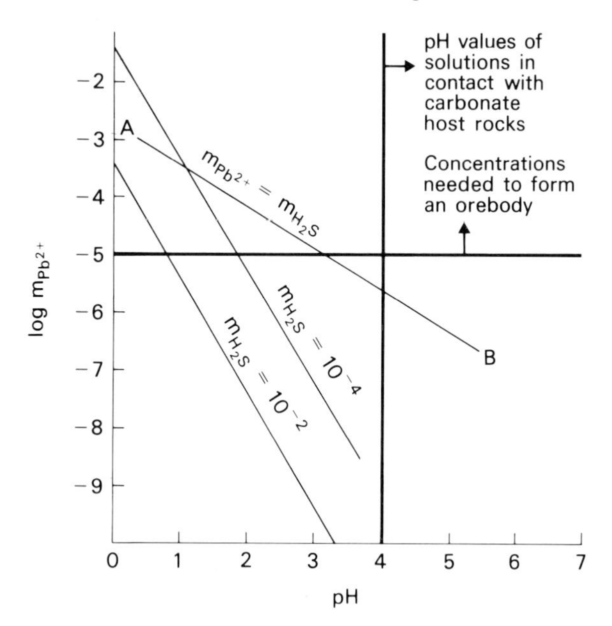

Fig. 4.4. Calculated lead ion concentrations in water in equilibrium with galena at 80°C—line AB. Two lines showing H_2S concentrations of 10^{-2} and 10^{-4}m are plotted. (After Anderson 1977.)

calculated lead ion (line AB) and H_2S concentrations in water in equilibrium with galena at 80°C are shown. This indicates that (in the absence of ion complexing) concentrations of lead and H_2S high enough to form an orebody can only be achieved in very acid solutions, about pH = 0.3 (Anderson 1977). These pH values are most unlikely to hold for hydrothermal solutions except those in contact with a relatively insoluble rock such as quartzite. With other rocks, hydrogen ion would be consumed by wall rock alteration reactions until a pH nearer neutral was achieved. Probable values of pH for hydrothermal solutions lie to the right of the thick vertical line, particularly when the solutions are passing through limestones. So we can rule out the possibility that pure acidified water could be the transporting medium. What we require is a mechanism of transportation which will operate in the upper right-hand portion of such a diagram.

One suggestion to overcome this difficulty is that ore solutions carry an excess of S^{2-} and the metals stay in solution as complex sulphide ions like HgS_2^{2-}. Such solutions would be decidedly alkaline, a condition which is geologically unlikely.

On the other hand, bisulphide complexes can exist stably in near neutral solutions containing abundant H_2S. Ions such as $PbS(HS)^-$ are formed and these have much higher solubilities than pure ionic solutions. The main objection to what is a very useful and promising hypothesis is the high concentration of H_2S and HS^- required to keep the complexes stable, a concentration much higher than that usually found in hot springs and fluid inclusions. For this and other reasons many workers favour the idea of metal transport in chloride complexes. Silver chloride in a solution of sodium chloride forms the complex $AgCl_2^-$. Ag_2S may be prevented from forming by the reaction

$$Ag_2S + 4Cl^- = 2AgCl_2^- + S^{2-}.$$

Nevertheless, the bisulphide transportation hypothesis has much to recommend it and Barnes & Czamanske (1967) have argued with reference to ZnS that bisulphide complexing is important and chloride complexing relatively insignificant. Helgeson (1964) contended that chloride complexing increases sphalerite solubility in chloride solutions when sulphur is low, whereas Barnes (1967b) has proposed that transportation as chloride complexes requires a solution to be virtually sulphur-free. Barnes & Czamanske showed that in alkaline solutions the solubility of lead can be increased by ten times, probably due to the formation of $Pb(HS)_3^-$. Another possibility is the formation of thio-complexes which Nriagu (1971) has demonstrated to be powerful transporting agents in the system $PbS\text{-}NaCl\text{-}H_2S\text{-}H_2O$. They can exist even when the total reduced sulphur is low. One particular advantage of the bisulphide hypothesis is that it gives us an explanation of the zonal distribution of minerals in epigenetic deposits as will be discussed later.

The origin and transport of sulphur to the site of deposition is also a difficult problem. Anderson (1977) has shown that, if the molality of H_2S is 10^{-5}, a 3 molal NaCl solution will transport almost 10^{-5} m (about 2 ppm) Pb at a pH of 4 at 80°C. At temperatures up to 150°C and otherwise similar conditions a little more than 10^{-5} m Pb would dissolve. This means that ore transport and deposition in equilibrium with carbonate rocks could occur at these temperatures. pH 4 is at the acid end of the possible pH range. With pH values in the more probable range 5-6, chloride solutions at up to 150°C cannot carry more than 10^{-5} m of Pb and H_2S at the same time. This has caused Anderson to agree with the many workers who conclude that in the formation of low temperature lead-zinc ores in limestones the metals are supplied by a hot brine solution, whilst H_2S is supplied at the site of sulphide precipitation. Support for this school of thought comes from the discovery of natural *sulphur-poor* hot brine solutions in various parts of the crust. The first was discovered in 1961-2 during drilling for geothermal power near the Salton Sea, California. One borehole tapped a concentrated brine at 1600 metres. Its temperature was 300-350°C and it carried about 8 ppm copper, 102 ppm lead and 540 ppm zinc. S (as H_2S) was 16 ppm—a level low in comparison to the amounts of dissolved base metals. Similar sulphur-poor brines have since been reported from deeps along the median rift zone of the Red Sea and Cheleken (USSR). On the other hand the evidence of sulphur isotope investigations suggests that, in some cases at least, the sulphur was transported by the mineralizing fluid, e.g. Silvermines, Ireland (Rye & Ohmoto 1974). It is the presence of too much *reduced* sulphur which would cause precipitation. Further sulphur could be carried as sulphate in the metal-bearing brines. This would have to be reduced by organic material, or in some other way, for sulphide formation. Sulphate values in fluid inclusions are usually comparable with base metal concentrations.

(c) *Sulphide deposition.* As has already been implied, the provision of abundant H_2S by the reduction of sulphate in carbonate host rocks would cause precipitation of sulphides from metal-bearing chloride brines. In the case where metal and reduced sulphur were transported in the same solution, precipitation could be caused by (a) cooling, (b) dilution, or (c) neutralization. Much of the material in the foregoing sections can be illustrated by the work of Anderson (1975). In Fig. 4.5, in which oxygen fugacity is plotted against pH, the line AB is one of a family

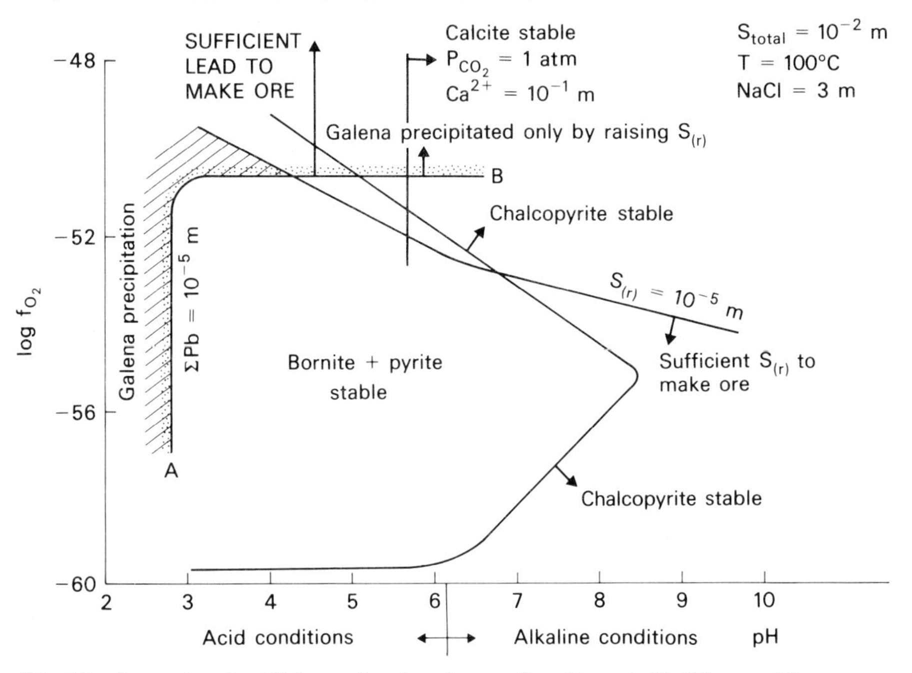

Fig. 4.5. Some mineral stabilities as a function of oxygen fugacities and pH. Other conditions as indicated. The stippled line is the 10^{-5}m contour for total lead in equilibrium with galena. Larger solubilities lie to the left and above this line. (Adapted from Anderson 1975.) $S_{(r)}$ = reduced sulphur. For further discussion see text.

of contours of lead concentrations of solutions in equilibrium with galena. Only solutions *above* or *to the left* of this line can be considered to be carrying enough lead to form orebodies. The line labelled $S_{(r)}$ = 10^{-5} m is one of a family of reduced sulphur contours *below which* the solution contains enough reduced sulphur to produce ore, and above which further sulphur must be added to precipitate all the lead. The intersection of these two contours outlines a potential ore-forming area in the left of the diagram. 10^{-5} m of Pb (2 ppm) is, however, the lowest possible concentration for ore formation and a pH of 4 is at the very acid end of the possible pH range. Other constraints are that, in general, as we have seen, calcite must be stable during transportation and at the site of deposition and that copper, in the form of chalcopyrite is often present in lead-zinc deposits. If chalcopyrite was precipitated in equilibrium with other ore constituents then we are restricted to areas lying to the right of or below the chalcopyrite + bornite + pyrite contour. Thus, the constraint of carbonate stability and the presence of chalcopyrite suggest we are dealing with deposition under conditions depicted by the top right-hand part of the figure—that is a region to which reduced sulphur must be added for full precipitation of the lead and formation of an orebody. It therefore appears likely, as has been suggested above, that reduced sulphur does not become available until the time of deposition, and its presence is the cause of deposition.

In the foregoing, some ideas concerning the genesis and nature of hydrothermal solutions have been dealt with briefly by particular reference to lead-zinc deposits in limestone host rocks. The subject is, however, vast and the above discussion must only be considered as a short introduction to some of the principles involved.

LATERAL SECRETION

It has been accepted for many years that quartz lenses and veins in metamorphic rocks commonly result from the infilling of dilatational zones and open fractures by silica which has migrated out of the enclosing rocks, and that this silica may be accompanied by other constituents of the wall rocks including metallic components and sulphur. This derivation of materials from the immediate neighbourhood of the vein is called lateral secretion. A very interesting example of deposits formed in this way has been described by Boyle (1959) from the Yellowknife Goldfield of the North-west Territories of Canada; but, before discussing it we must consider the probable behaviour of element levels in rocks adjacent to veins forming under different conditions. In Fig. 4.6A we have a vein forming from an uprising hydrothermal solution supersaturated in silica. Some of this diffuses into the wall rocks

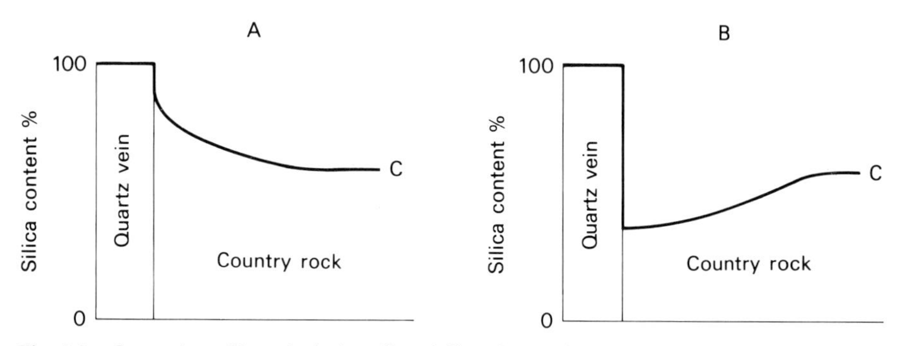

Fig. 4.6. Comparison of hypothetical profiles of silica. In case A, silica is added to the wall rocks from the hydrothermal solution which is depositing quartz in the vein. In B, silica is abstracted from the wall rocks and deposited as quartz in the vein. C indicates the normal level of silica in the country rocks.

and causes some silicification. The curve showing the level of silica decreases away from the source (i.e. the vein). In Fig. 4.6B we have the opposite situation where silica is being supplied to the vein from the wall rocks. The curve now climbs as it leaves the vein, indicating a zone of silica depletion in the rocks next to the vein. Clearly silica has been abstracted from the wall rocks and has presumably accumulated in the vein.

The principal economic deposits of the Yellowknife field occur in quartz-carbonate lenses in extensive chloritic shear zones cutting amphibolites (metabasites). The deposits represent concentrations of silica, carbon dioxide, sulphur, water, gold, silver and other metallic elements. The principal minerals are quartz, carbonates, sericite, pyrite, arsenopyrite, stibnite, chalcopyrite, sphalerite, pyrrhotite, various sulphosalts, galena, scheelite, gold and aurostibnite. The regional metamorphism of the host rocks varies from amphibolite to greenschist facies. Alteration haloes of carbonate-sericite schist and chlorite- and chlorite-carbonate-schist occur in the host rocks adjoining the deposits.

It is very instructive to remember that the dominant mineral of the veins is quartz. The profile of silica alongside the lenses is shown in Fig. 4.7. This demonstrates that a very substantial amount of silica has been subtracted from both the

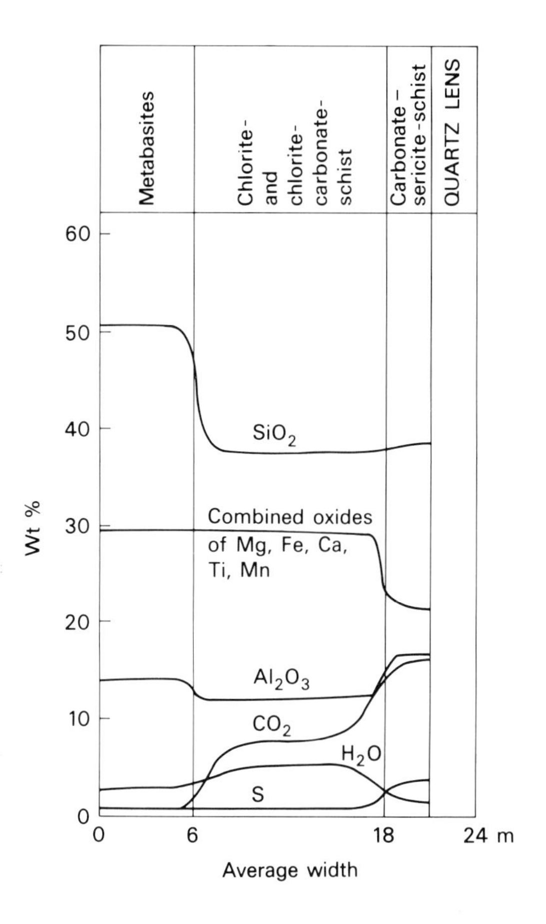

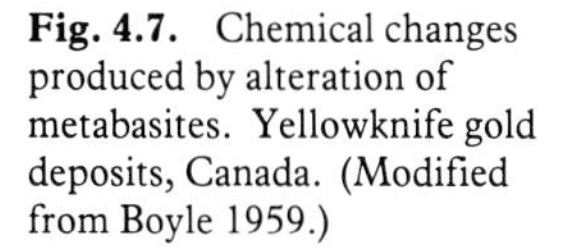
Fig. 4.7. Chemical changes produced by alteration of metabasites. Yellowknife gold deposits, Canada. (Modified from Boyle 1959.)

alteration zones and this has occurred of course *on both sides* of the vein. Clearly more silica has been subtracted from the wall rocks than is present in the lenses and the problem is not, where has the silica in the lenses come from, but where has the surplus silica gone to? Some subtraction of magnesia, iron oxides, lime, titania and manganese oxide has also occurred and doubtless this is the source of iron in such minerals as pyrite, pyrrhotite and chalcopyrite in the lenses. Alumina shows a depletion in the outer zone of alteration and a concentration in the inner zone where it has collected for the formation of sericite. As a result of the extensive development of carbonates in the alteration zones carbon dioxide develops a much higher level than in the unaltered country rocks. Water shows a similar but not identical behaviour. Boyle produced good evidence that these two oxides were passing through the rocks in considerable quantity, being mobilized by regional metamorphism and migrating down the metamorphic facies. They passed into the shear zones to form the chlorite, carbonate and sericite.

It appears highly probable that the major constituents of the shear zones resulted from rearrangement and introduction of material from the country rocks; the remaining question is whether the metabasites could have been the source of the sulphur and metallic elements in the deposits. The metabasites consist of metamorphosed basic volcanic lavas and tuffs. These rocks are richer in elements such as gold, silver, arsenic, copper, etc., than other igneous rocks, and for the unsheared metabasites of the Yellowknife area Boyle obtained the following values: (all in ppm) S = 1500; As = 12; Sb = 1; Cu = 50; Zn = 50; Au = 0.01; Ag = 1. For purposes of calculation the rock system was taken to be: length = 16 km; width = 152 m; depth = 4.8 km. The amount of ore in the system was assumed to be 6×10^6 t with average grades of S = 2.34%; As = 1.35%; Sb = 0.15%; Cu = 0.07%; Zn = 0.28%; Au = 0.654 oz/ton and Ag = 0.139 oz/ton. The total contents of these elements in the shear system prior to shearing and alteration, and in the

Table 4.3. Contents of chalcophile elements in shear zones and deposits, Yellowknife gold deposit, Canada.

Element	Total content in shear system before shearing and alteration (millions of tons)	Total content in deposits (millions of tons)
S	62	0.14
As	0.5	0.081
Sb	0.04	0.009
Cu	2.0	0.004
Zn	2.0	0.017
Au	12.2×10^6oz	3.9×10^6oz
Ag	1219×10^6oz	0.834×10^6oz

deposits is shown in Table 4.3. It is apparent from these figures that all the elements considered could have been derived solely from the sheared rock of the shear zone and there is no need to postulate another source. Indeed, there is such a difference between the values in columns two and three that it may well be that significant quantities of chalcophile elements accompanied the surplus silica to higher zones in the crust to form deposits which have now been eroded away.

METAMORPHIC PROCESSES

(a) *Pyrometasomatic deposits.* These deposits have been termed hydrothermal metamorphic, igneous metamorphic, contact metamorphic, skarn and pyrometasomatic. The last term is gaining general acceptance, particularly because it neatly summarizes the origin of these deposits which are formed at high temperatures (pyro) with the addition and subtraction of material (metasomatic). Their general morphology and nature have been summarized in Chapter 2. They are developed most often, but not invariably, at the contact of intrusive plutons and carbonate country rocks. The latter are converted to marbles, calc-silicate hornfelses and/or skarns by contact metamorphic effects.

The calc-silicate minerals such as diopside, andradite and wollastonite, which are often the principal minerals in these ore-bearing skarns, attest to the high temperatures involved. The pressures at the time of formation were very variable as the depths of formation were probably from one to several kilometres. Some of the classic pyrometasomatic deposits of the United States are associated with porphyry copper intrusions, indicating a relatively shallow depth of emplacement.

During the genesis of these deposits, the first stage was that of recrystallization of the country rocks. (In any carbonate rocks impurities led to the formation of calc-silicate hornfelses.) The next stage was the introduction of calc-silicate-forming materials such as silica, iron, magnesia and alumina. This was the stage of large-scale metasomatism and Lindgren (1924) showed that at Morenci, Arizona, vast quantities of material had been added *and subtracted.* He pointed out that if all the CaO in 1 cm^3 of calcite is converted into andradite [$Ca_3Fe_2(SiO_4)_3$] skarn then the volume would increase to 1.4 cm^3. All the evidence at Morenci, as in similar deposits, suggests a volume for volume replacement process with no concomitant expansion. In that case, for every m^3 of altered limestone 460 kg of CaO and 1190 kg of CO_2 were removed, and 1330 kg of SiO_2 and 1180 kg Fe_2O_3 were added.

The next stage in the development of many deposits was the formation of volatile-bearing minerals such as amphiboles, epidote, idocrase, fluorite, tourmaline and axinite. The introduction of ore minerals may have overlapped stages two and three. Oxide minerals appear to have crystallized before sulphides.

The origin of all this introduced material is much debated. In the past, it was held that the pluton responsible for the contact metamorphism was also the source of the metasomatizing solutions. Whilst it is conceivable that a granitic pluton might supply much silica, it is unlikely that it could have supplied the amount of iron which is present in some deposits. On the other hand, where the pluton concerned is basic, the supply of iron does not present such great problems. These difficulties do become insurmountable, however, for the small class of pyrometasomatic deposits such as the Ausable Magnetite District, New York State, which have no associated intrusions. Perhaps the main function of the intrusion is that of a heat engine. For Ausable, Hagner & Collins (1967) suggested migration of iron from accessory magnetite in granite-gneiss, with concentration in shear zones to form magnetite-rich bodies, together with the release of iron during the recrystallization of clinopyroxene- and hornblende-gneisses.

The experimental replacement of marble by sulphides has been achieved by Howd & Barnes (1975) who found that ore-bearing bisulphide solutions at 400-450°C and 0.5 kb, when oxidized, produced acid solutions which dissolved the marble and provided sites for sulphide deposition. These experiments and the proximity of many pyrometasomatic deposits to porphyry copper intrusions suggest that circulating hydrothermal solutions may have played a part in the ore genesis. The introduction of sulphur may have led to the formation of iron oxides by the following and similar reactions:

$$\underset{\text{hedenbergite}}{4CaFeSi_2O_6} + 2S \rightleftharpoons FeS_2 + Fe_3O_4 + \underset{\text{wollastonite}}{4CaSiO_3} + SiO_2.$$

(b) *The role of other metamorphic processes in ore formation.* Some examples of lateral secretion are clearly the result of metamorphism. This subject has, however, already been covered above and will not be discussed further. In this section we are concerned with those metamorphic changes which involve recrystallization and redistribution of materials by ionic diffusion in the solid state or through the medium of volatiles, especially water. Under such conditions relatively mobile ore constituents may be transported to sites of lower pressure such as shear zones, fractures or the crests of folds. In this way, the occurrence of quartz-chalcopyrite-pyrite veins in amphibolites and schists may have come about.

The behaviour of trace amounts of ore minerals in large volumes of rock undergoing regional metamorphism is uncertain and is a field for more extensive research. It might be thought that with the progressive expulsion of large volumes of water and other volatiles during prograde metamorphism, natural nydrothermal systems might evolve which would carry away elements such as copper, zinc or uranium which are enriched in trace amounts in pelites. Shaw (1954), however, in a study of the progressive metamorphism of pelitic sediments from clays through to gneisses showed that such changes are but slight. Taylor (1955) in a study of greywackes reported similar results. On the other hand, De Vore (1955) calculated that during the transformation of one cubic mile of epidote-amphibolite facies hornblendite into the granulite facies there may be a release of 8 million tons of Cr_2O_3, 4 million tons of NiO and 800 000 tons of CuO. Similarly, retrograde metamorphis ı can release large quantities of zinc, lead and manganese. In most cases, the liberated elements are probably dispersed rather than concentrated, but with the diversion of hydrothermal fluids also expelled by the metamorphism into suitable structural situations, ore concentration may occur. Fyfe & Henley (1973) have suggested just such a mechanism.

They envisage a situation where a volcanic-sedimentary pile is being metamorphosed under amphibolite facies conditions. It would be losing about 2% water, and if salt is present and oxygen buffered by magnetite-ferrous silicate assemblages, then gold solubilities of the order of 0.1 ppm at 500°C would be achieved. This gold would either be dispersed through greenschist facies rocks or concentrated into a favourable structure. This could happen if the solution flow was focused into a large vein or pipe system as appears to have happened at Morro Velho, Brazil (Fig. 4.8). Morro Velho is a quartz vein that has been mined down to a depth of 3 km. Fyfe & Henley show that a source region of 30 km^3 could have provided all the gold and silica plus the water to carry them. Their figures are as follows. The orebody of auriferous quartz occupies 0.01 km^3 and contains about 3×10^8 g of gold. With an average crustal gold content of about 3 parts per billion (ppb), approximately 30 km^3 of volcanics or sediments are required to form the source region. About 2×10^{15} g of water would be released which at 0.1 ppm could transport 2×10^8 g of gold. At 5 kb and 500°C this volume of water could also transport 2×10^{13} g silica (solubility under these conditions 10 g kg^{-1}). This silica is about 0.01 km^3 in volume, i.e. that of the orebody. This model of gold mineralization is similar to that proposed by Boyle (see earlier), but now the major constituents are derived from a deep source region and not by lateral secretion.

The processes of granitization and migmatization have been put forward by a number of authors as possible ore-concentrating processes. Sullivan (1948) suggested that during granitization ore elements may be segregated into ore deposits, the valuable elements being 'concentrated in inverse ratio to the extent to which they are incorporated by isomorphous substitution in the common rock-forming minerals.' Thus tin may accumulate locally in granites because its ionic radius is incompatible with substitution in quartz and feldspar lattices, whereas

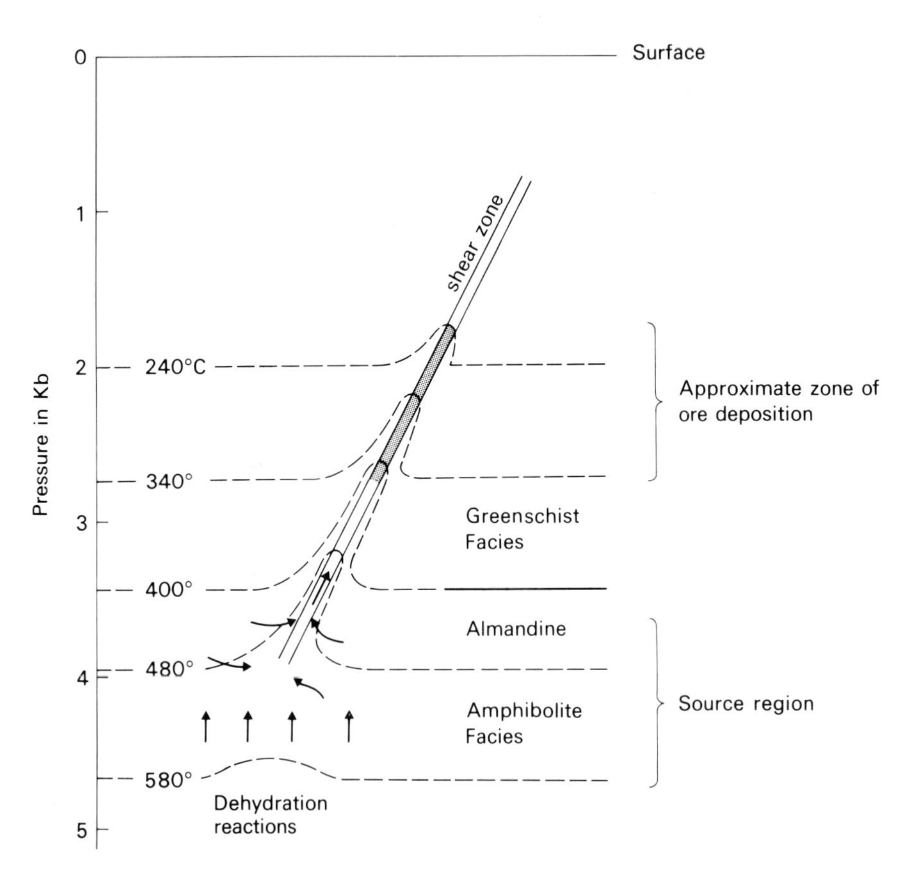

Fig. 4.8. Diagram of a shear zone where metamorphic water from a large volume of rock is rising to higher levels. (After Fyfe & Henley 1973.)

stanniferous deposits are unlikely to occur in rocks carrying abundant ferromagnesian minerals capable of accommodating small amounts of tin in their lattices. Sullivan also suggested that gold may be driven outwards from greenstones (metabasites) by advancing fronts of granitization, and initial gold concentration in fields such as the Kalgoorlie region of Western Australia may have been due to advancing migmatizing fronts of sodic metasomatism driving the gold before them. This could come about due to the fact that gold (ionic radius 1.37 Å) can be accommodated in significant trace amounts in potash feldspar (ionic radius of potassium = 1.33 Å) but not in albite (ionic radius sodium = 0.97 Å).

Origin due to surface processes

Processes involving mechanical and chemical sedimentation will not be dealt with here. The reader is referred to texts on sedimentology for the general principles involved, e.g. Selley (1976). However, certain aspects will be touched upon in Chapter 16. Residual processes and supergene enrichment will be dealt with in Chapter 17. The remaining space in this chapter will be devoted to a consideration of exhalative processes. These, it should be noted, are a surface expression of the activity of hydrothermal solutions.

VOLCANIC-EXHALATIVE (SEDIMENTARY-EXHALATIVE) PROCESSES

We are concerned here with a group of deposits often referred to as massive sulphide ores. Some of the characteristics of these ores have been dealt with on pp. 17-18. They frequently show a close spatial relationship to volcanic rocks but this is not the case with all the deposits, e.g. Sullivan, Canada (Fig. 2.11). They are conformable and frequently banded, the principal constituent is usually pyrite with varying amounts of copper, lead, zinc and baryte; precious metals together with other minerals may be present. For many decades they were considered to be epigenetic hydrothermal replacement orebodies (Bateman 1950). In the 1950s, however, they were recognized as being syngenetic, submarine-exhalative, sedimentary orebodies. They are now often referred to by one or other of the two terms in this section heading, or as volcanic massive sulphide deposits.

The ores with a volcanic affiliation show a progression of types. Associated with basic volcanics, usually in the form of ophiolites and presumably formed at oceanic or back arc spreading ridges, we find the Cyprus types. These are essentially cupriferous pyrite bodies. They are exemplified by the deposits of the Troodos Massif in Cyprus and the Ordovician Bay of Islands Complex in Newfoundland. Associated with the early part of the main calc-alkaline stage of island arc formation are the Besshi-type deposits. These occur in successions of mafic volcanics in complex structural settings characterized by thick greywacke sequences. They commonly carry zinc as well as copper and are exemplified by the Palaeozoic Sanbagawa deposits of Japan, and the Ordovician deposits of Folldal in Norway. The more felsic volcanics, developed at a later stage in island arc evolution, have a more varied metal association. They are copper-zinc-lead ores often carrying gold and silver. Large amounts of baryte, quartz and gypsum may be associated with them. They are called Kuroko deposits after the Miocene ores of that name in Japan. Older examples are the Ordovician bodies of Avoca, Ireland, and the Precambrian ones of Noranda, Canada. All these different types are normally underlain in part by a stockwork up which the generating hydrothermal solutions appear to have passed (Fig. 2.15).

There is today wide agreement that these deposits are submarine-hydrothermal in origin, but there is a divergence of opinion as to whether the solutions responsible for their formation are magmatic in origin or whether they represent circulating sea water. Here, there is only room to deal briefly with this controversy.

The general correlation of ore types with basic to acid volcanics favours a magmatic origin. In the case of the Kuroko deposits, Sato (1977) has argued strongly for a magmatic origin. In particular he notes the close correlation of the Kuroko deposits with rhyolite domes rather than with the preceding andesitic and dacitic vulcanism, together with supporting isotopic and fluid inclusion evidence. The lead isotopic ratios are almost identical with the rock leads of the Neogene volcanics which, in turn, show a uniform variation across the island arc. This suggests a common direct magmatic source for the volcanics and the ore lead, because the volcanics beneath the deposits are too thin for circulating brines to have leached the lead from them.

The circulating sea water model has been put forward mainly by workers on stable isotopes who argue that sea water is the main source of the water and sulphur, and that the metals were leached out of the rocks through which the fluids flowed. The four major types of water—magmatic, sea water and/or connate, meteoric and metamorphic—have characteristic hydrogen (D/H) and oxygen ($^{18}O/^{16}O$) isotopic ratios (Sheppard 1977). Using these ratios it has been shown that the first three types of water were involved in general mineralization processes. Magmatic fluids were dominant in some cases, and in others initiated the mineralization and wall rock alteration only to be swamped by convective meteoric water set in motion by hot intrusions or other heat sources. In massive sulphide deposits, the isotopic evidence favours sea water as the principal or only fluid. The possibility must be borne in mind, however, (Sato 1977) that the sea water was a late addition to the hydrothermal system and that it has overprinted the pre-existing magmatic values. Let us examine the evidence a little more deeply, as isotopic studies are now being applied to all types of mineral deposits and they will be mentioned again in later chapters.

Variations in the isotopic ratios of hydrogen and oxygen are given in the δ notation in parts per thousand (per mil,‰) where

$$\delta_x = \frac{R_{sample}}{R_{standard}} - 1 \times 1000$$

In the above formula for hydrogen $\delta_x = \delta D$ and $R = D/H$; for oxygen, $\delta_x = \delta^{18}O$ and $R = {}^{18}O/{}^{16}O$.

The standard for both hydrogen and oxygen is standard mean ocean water or SMOW. In nature, D/H is about 1/7000 and $^{18}O/^{16}O$ is about 1/500. These values are measured directly on natural substances such as thermal waters, connate waters in sediments and fluid inclusions, or they are determined indirectly using minerals after removal of all the adsorbed water. In the latter case the isotopic composition of the mineral is not that of the fluid with which it was in contact at the time of crystallization or recrystallization. The δ values for the fluid have to be calculated from the mineral values using equilibrium fractionation factors determined by laboratory experiment or from studies of active geothermal systems. A temperature fractionation effect also occurs, so the temperature must be known

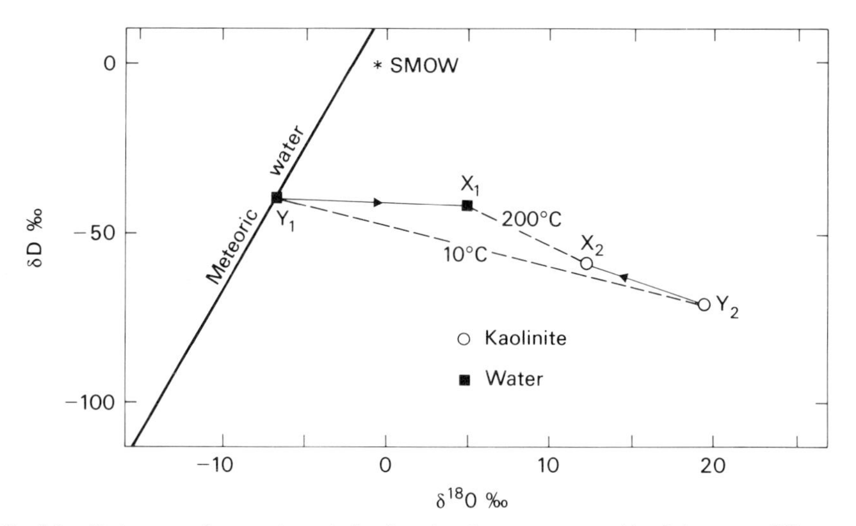

Fig. 4.9. Hydrogen and oxygen isotopic fractionations between water and kaolinite at two different temperatures. From a knowledge of the system kaolinite-water and given the values X_2 and 200°C, the value of X_1 can be calculated. (Modified from Sheppard 1977.)

(from fluid inclusion studies, etc.) to determine the isotopic composition of the water in equilibrium with the mineral. For example, in Fig. 4.9 raising the temperature from 10 to 200°C gives rise to isotope exchange and re-equilibration such that the isotopic composition of the water changes from Y_1 to X_1 whilst that of the coexisting kaolinite changes from Y_2 to X_2. In other words rock-water reactions cause a shift in the delta values of both the circulating meteoric water and the rock with which it is in contact, with the result that the water is enriched in ^{18}O as the temperature rises.

The isotopic compositions of the various types of water show useful differences. Sea water in general plots very close to SMOW (Fig. 4.10) and shows very little variation. Meteoric water varies fairly systematically with latitude along the line shown in Fig. 4.10. Values for metamorphic and magmatic waters have been

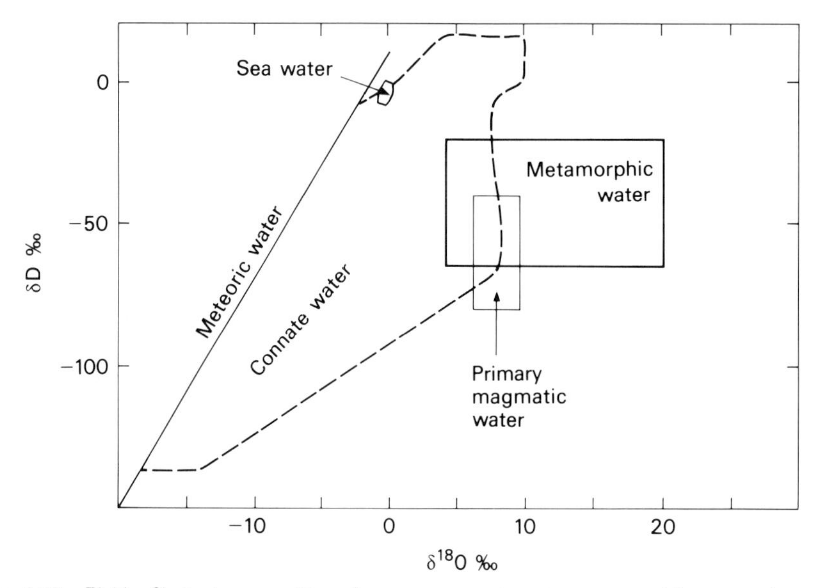

Fig. 4.10. Fields of isotopic composition of sea water, connate water, metamorphic water and magmatic water. (Modified from Sheppard 1977.)

deduced from measurements on minerals. Connate (formational) water can be measured directly and plots as shown. Since many of the connate waters are richer in ^{18}O than SMOW they cannot have resulted from simple mixing of meteoric and sea water. There must have been isotopic exchange with the sediments at elevated temperatures (shown by Fig. 4.9 to result in ^{18}O enrichment of the water, e.g. the change from Y_1 to X_1), addition of rising metamorphic water or some other process. We are now in a position to compare the results obtained from a study of the Cyprus and Kuroko deposits.

These results are plotted in Fig. 4.11. Those for the Cyprus stockwork deposits coincide exactly with the values for sea water, and Heaton & Sheppard (1977) suggest a model involving deeply circulating sea water as sketched in Fig. 4.12. They also present evidence that the associated country rocks were thoroughly permeated by sea water during their metamorphism into the greenschist and zeolitic facies. The Kuroko fluids show $\delta^{18}O$ values commensurate with a sea water origin but δD is depleted by 11-26‰ relative to sea water. Ohmoto & Rye (1974) concluded

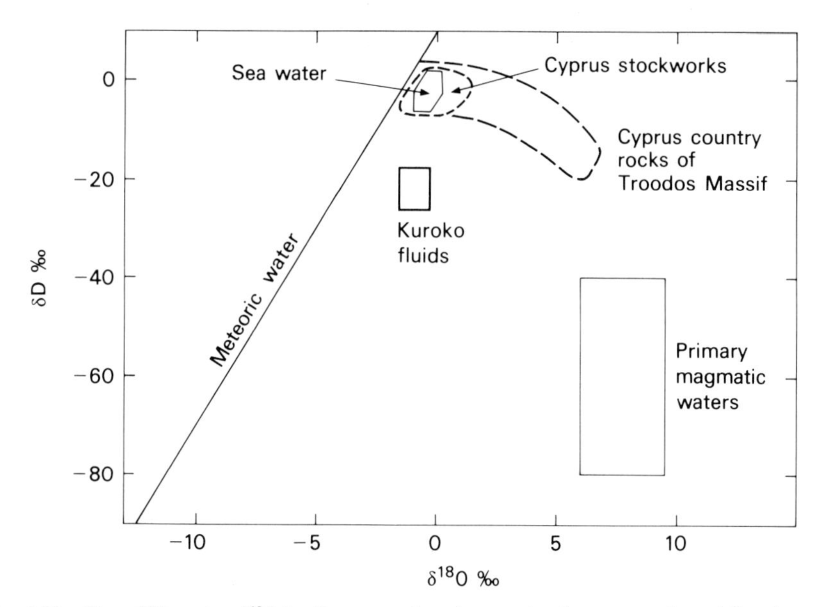

Fig. 4.11. Plot of δD against $\delta^{18}O$ for Cyprus stockworks, associated country rocks and Kuroko fluids. (Modified from Sheppard 1977.)

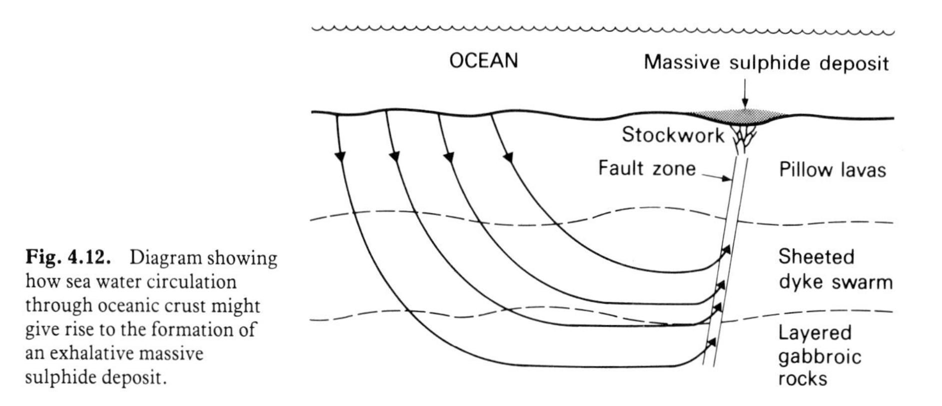

Fig. 4.12. Diagram showing how sea water circulation through oceanic crust might give rise to the formation of an exhalative massive sulphide deposit.

that sea water was the dominant source of the hydrothermal fluid but that it contained a small meteoric and/or magmatic contribution. As Kuroko deposits belong to the Island Arc environment, meteoric water could be involved. This would suggest a model, similar to that in Fig. 4.12, of circulating sea water becoming a concentrated brine at depth, dissolving copper and other metals from the rocks it traversed and carrying these up to the surface where they were precipitated as sulphides with sulphur derived from the sea water. This leads us into another field of fruitful research in isotopic studies—sulphur.

Sulphur isotopic data for the Cyprus ores have been reviewed by Spooner (1977) who has pointed out that $\delta^{34}S$ for the pyrite is higher than that of the country rocks from which the sulphur may have been partly derived. This suggests an additional source of isotopically heavy sulphur. Spooner feels that this was probably the circulating Cretaceous sea water which would have had a value of $\delta^{34}S = +16‰$. Spooner *et al.* (1977) also showed that the mineralized zones are enriched in ^{87}Sr

relative to the initial magmatic $^{87}Sr/^{86}Sr$ of the igneous country rocks. The values obtained range up to the value for Upper Cretaceous sea water but no higher.

Sato's point (Sato 1977) that the sea water circulation can overprint an earlier magmatic-hydrothermal mineralization event must, however, be borne in mind. One of the difficulties concerning the isotope studies outlined above is that they have been largely confined to mineralized areas. One non-mineralized area where such studies have been carried out in depth is Mull in Scotland. The isotopic evidence from this district shows without doubt that vast quantities of meteoric water can be flushed through igneous intrusive and extrusive rocks without the genesis of any significant mineralization. This fact should be remembered when considering the evidence of the circulation of meteoric water through some porphyry copper deposits (Chapter 12).

The most important factor in controlling the behaviour of exhalative hydrothermal solutions when they reach the sea floor is probably the density difference between the ore solution and sea water. When the hydrothermal solution has a high salinity and a low temperature it will be heavier than sea water and will collect in depressions on the sea floor. As a result, massive ores may not necessarily be underlain by a feeder vent or stockwork. The location of ore deposition is, however, mainly controlled by the site of discharge of the hydrothermal solutions and this may lead to deposition in gravitationally unstable positions. The unconsolidated ores may therefore be reworked by submarine sliding, leading perhaps to turbidity current transportation and deposition. The common occurrence of graded fragmental ore is a characteristic feature of many Kuroko deposits (Sato 1977), and Thurlow (1977) and Badham (1978) have described the results of transportation by slumping and sliding at Buchans, Newfoundland, and Avoca, Ireland.

5

Geothermometry, Geobarometry, Paragenetic Sequence, Zoning and Dating of Ore Deposits

Geothermometry and geobarometry

Ores are deposited at temperatures and pressures ranging from very high, at deep crustal levels, to atmospheric, at the surface. Some pegmatites and magmatic segregation deposits have formed at temperatures around 1000°C and under many kilometres of overlying rock. Placer deposits and sedimentary ores have formed under surface conditions. Most orebodies were deposited between these two extremes. Clearly, knowledge of the temperatures and pressures obtaining during the precipitation of the various minerals will be invaluable in assessing their probable mode of genesis. Such knowledge will also be of great value in formulating exploration programmes. In this small volume it is only possible to touch on a few of the methods that can be used.

FLUID INCLUSIONS

The nature of fluid inclusions and the principal of this method have been outlined in Chapter 3. Clearly, primary inclusions are those which must be examined. Secondary inclusions produced after the mineral was deposited and commonly formed by the healing of fractures will not give us data on the mineral depositional conditions, but their study can be very important in the investigation of certain deposits, e.g. porphyry coppers (Chapter 12).

The filling (homogenization) temperatures of aqueous inclusions (i.e. the temperature at which the inclusion becomes a single phase fluid) indicate the depositional temperature of the enclosing mineral if a correction can be made for the confining pressure and salinity of the fluid. The salinity, in terms of equivalent weight per cent NaCl, can be determined by studying the depression of the freezing point using a freezing stage. Frequently, confining pressures have to be estimated by reconstructing the stratigraphical and structural succession above the point of mineral depositional. This clearly leads to a degree of uncertainty. Pressure corrections increase the homogenization temperatures obtained in the laboratory, so these are still of great value even when uncorrected as they record minimum temperatures of deposition. Where CO_2-rich and H_2O-rich inclusions coexist, the pressure can be estimated from the filling temperature of the aqueous inclusions and the density of the CO_2 inclusions (Groves & Solomon 1969). If the fluid inclusion assemblage indicates boiling at the time of trapping then the depth of formation can be estimated (Haas 1971). However, if boiling did not occur, then the confining pressure must have exceeded the vapour pressure of the fluid, which can be calculated from the salinity and temperature data, giving us a minimum pressure value.

INVERSION POINTS

Some natural substances exist in various mineral forms (polymorphs) and some use of their inversion temperatures can be made in geothermometry. For example, β-quartz inverts to α-quartz with falling temperature at 573°C. We can determine whether quartz originally crystallized as β-quartz by etching with hydrofluoric acid, and thus decide whether it was deposited above or below 573°C. Examples among ore minerals include:

$$\underset{\text{monoclinic}}{\text{acanthite}} \overset{177°C}{\rightleftharpoons} \underset{\text{cubic}}{\text{argentite}}$$

$$\text{orthorhombic chalcocite} \overset{104°C}{\rightleftharpoons} \text{hexagonal chalcocite.}$$

This second low inversion temperature is important in distinguishing hypogene chalcocite from low temperature near-surface supergene chalcocite, a distinction which can be of great economic importance in evaluating many copper deposits, for if the near surface ore is only just of economic grade and much of the mineralization is supergene, then the ore below the zone of supergene enrichment may be uneconomic. Fortunately, the distinction can be made, for relict cleavage remains after chalcocite has inverted from the high temperature hexagonal form, and this cleavage is revealed by etching polished sections with nitric acid.

EXSOLUTION TEXTURES

As a result of restricted solid solution at lower temperatures between various pairs of oxide and sulphide minerals, exsolution bodies of the minor phase segregate from the host solid solution on cooling (Fig. 5.1). Their presence indicates the former existence of a solid solution of the two minerals which was deposited at an elevated temperature. An idea of that temperature can be obtained by reference to laboratory work on the sulphide system concerned (Edwards 1960), or by resolution of natural exsolution bodies in the host grain by heating samples in the laboratory. An example of the latter approach was reported by Edwards & Lyon (1957) who performed resolution experiments on samples from the Aberfoyle tin mine, Tasmania. After the geologically short annealing time of one week they obtained the following temperatures for the onset of resolution:

chalcopyrite and stannite in sphalerite - 550°C,
sphalerite in chalcopyrite - 400°C,
stannite in chalcopyrite - 475°C,
sphalerite in stannite - 325°C,
chalcopyrite in stannite - 400-475°C.

These results, they suggested, indicated a temperature of about 600°C for the formation of the original solid solutions. There is much data in the literature which suggests that with longer annealing times lower temperatures would have been obtained and the original solid solutions might have been deposited at temperatures in the range 400-500°C. This range would agree better with the homogenization temperatures of around 400°C obtained by Groves *et al.* (1970) on fluid inclusions in cassiterites from this deposit.

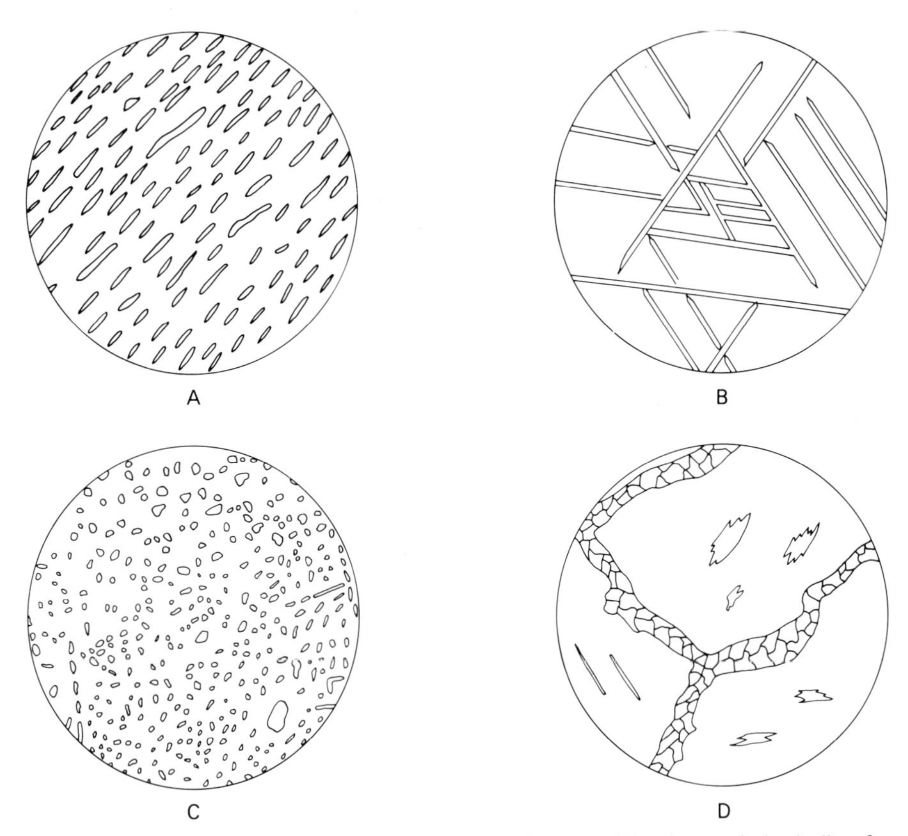

Fig. 5.1. Exsolution textures in oxide and sulphide mineral systems: (A) seriate exsolution bodies of hematite-rich material in an ilmenite-rich base, × 300, Tellnes, Norway; (B) exsolution lamellae of ilmenite in magnetite, × 134, Sudbury, Ontario; (C) emulsoid exsolution bodies of chalcopyrite in sphalerite, × 80, Geevor mine, Cornwall; (D) rim or net exsolution texture formed by the exsolution of pentlandite from pyrrhotite, × 536, Sudbury, Ontario.

SULPHIDE SYSTEMS

The initial studies of these systems were carried out to discover those which would provide data on the temperature and pressure of ore deposition. The first results, e.g. the sphalerite geothermometer, were very promising. Further study, however, revealed that the formation of mineral assemblages and variations in mineral composition depend on many factors. Thus, Kullerud (1953) suggested that the FeS content of sphalerite gave a direct measurement of its temperature of deposition. Later, Barton & Toulmin (1963) showed that the fugacity of sulphur was also an important control.

More research into this system, however, has produced some rewards. Although the work of Scott & Barnes (1971) indicated that the composition of sphalerite coexisting with pyrrhotite and pyrite over the range 525-250°C is essentially constant (Fig. 5.2) their microprobe studies did reveal a possible geothermometer. They discovered metastable iron-rich patches within sphalerites whose composition, compared with their matrix, appeared to be constant and temperature dependent. This applies only to coexisting sphalerite-pyrrhotite-pyrite assemblages. Iron-rich patches have been found in natural sphalerites from veins but they do not

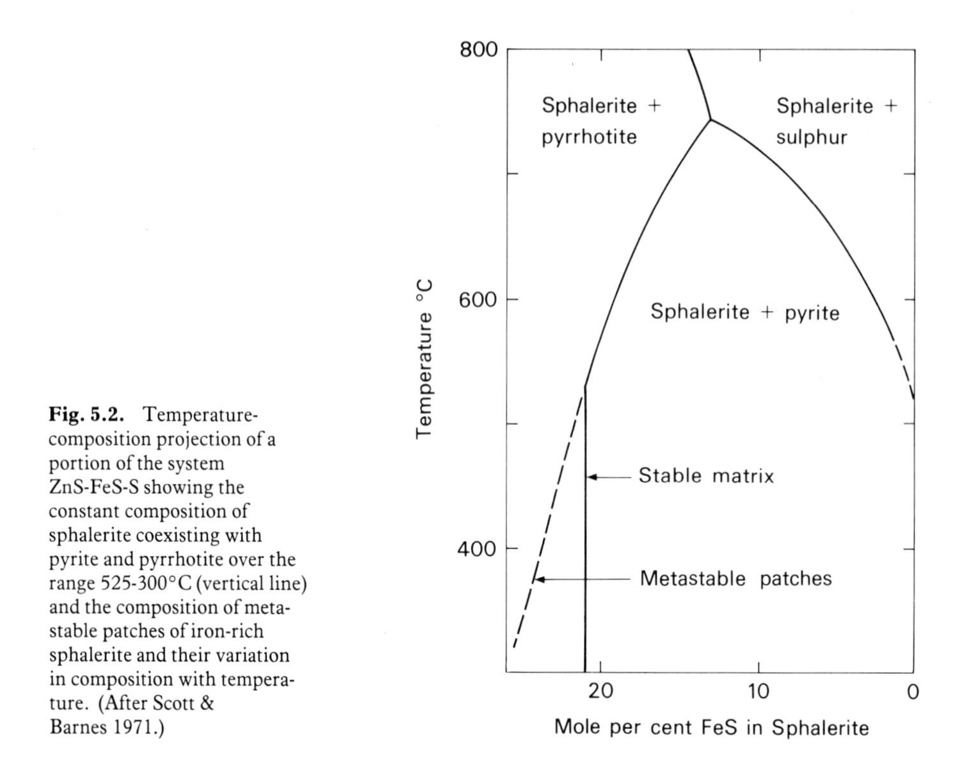

Fig. 5.2. Temperature-composition projection of a portion of the system ZnS-FeS-S showing the constant composition of sphalerite coexisting with pyrite and pyrrhotite over the range 525-300°C (vertical line) and the composition of metastable patches of iron-rich sphalerite and their variation in composition with temperature. (After Scott & Barnes 1971.)

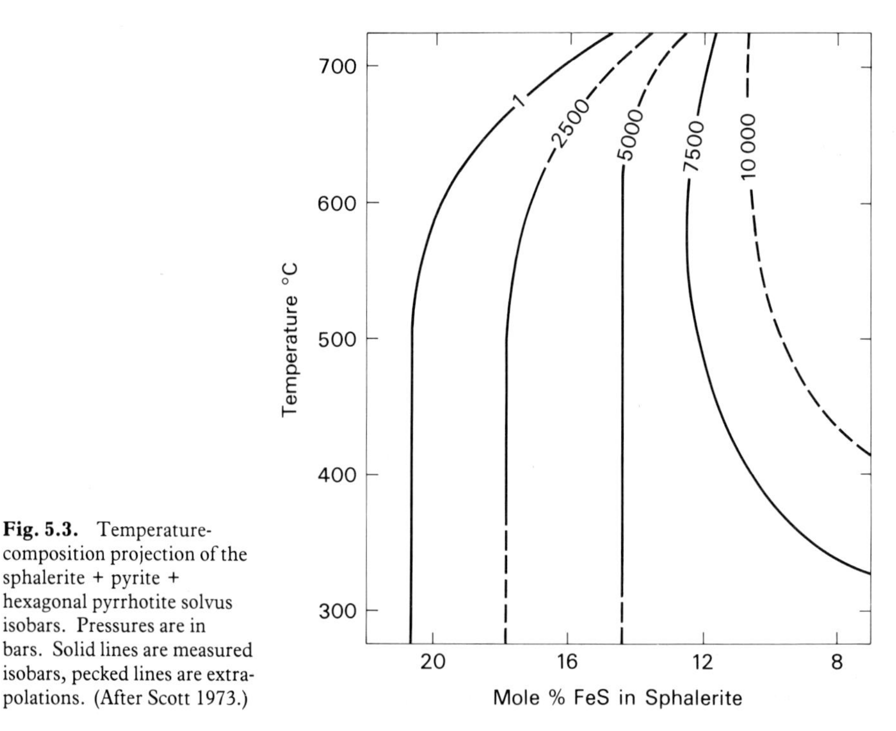

Fig. 5.3. Temperature-composition projection of the sphalerite + pyrite + hexagonal pyrrhotite solvus isobars. Pressures are in bars. Solid lines are measured isobars, pecked lines are extrapolations. (After Scott 1973.)

appear to be present in metamorphosed sphalerite because of annealing and re-equilibration.

Scott & Barnes (1971) and Scott (1973) indicated that the iron content of sphalerite *in equilibrium* with pyrite and pyrrhotite, although not temperature dependent below 525°C (Fig. 5.2), is strongly controlled by pressure above 300°C (Fig. 5.3). Thus, sphalerite can be used as a geobarometer provided equilibrium has been reached. This is most likely to be the case in metamorphosed ores and Lusk *et al.* (1975) have described its use in just such a case.

This brief discussion must suffice to illustrate the use of sulphide systems in geothermometry and geobarometry. Excellent summaries of experimental techniques and problems are provided by Barton & Skinner (1967) and Scott (1974). A list of the sulphide systems studied experimentally up to 1974 is given by Craig & Scott (1974). There is also considerable experimental work on non-sulphide systems which is applicable to orebodies, particularly their gangue minerals (Levin *et al.* 1969). For carbonate composition variation with temperature see Goldsmith & Newton (1969).

STABLE ISOTOPE STUDIES

In Chapter 4, when discussing the uses of hydrogen and oxygen isotopes to determine the origin of waters which had reacted with minerals, it was pointed out that if we know the temperature of reaction we can determine the isotopic composition of the water. Similarly, given the isotopic compositions of cogenetic minerals and water we can determine the temperature. This is the principal of the use of oxygen isotopes on gangue minerals as a geothermometer. A much more extensively used method for ore deposits has been the sulphur isotope geothermometer.

Sulphur has four isotopes, but the major geological interest is in the variation $^{34}S/^{32}S$ because this varies significantly in the minerals of hydrothermal deposits for a number of reasons:

(i) the fractionation of the isotopes between individual sulphide and sulphate minerals in the ore varies with the temperature of deposition,

(ii) the initial isotopic ratio is controlled by the source of the sulphur, e.g. mantle, crust, sea water, etc.,

(iii) the variable proportions of oxidized and reduced sulphur species in solution—in its simplest form the H_2S/SO_4^{2-} ratio. As this ratio depends on temperature, pH and fO_2 we can also evaluate the variation of sulphur isotopic values in terms of T, pH and fO_2, Rye & Ohmoto (1974).

Variations in $^{34}S/^{32}S$ are expressed in delta notation ($\delta^{34}S$) where:

$$\delta^{34}S = \frac{(^{34}S/^{32}S)_{sample} - (^{34}S/^{32}S)_{standard}}{(^{34}S/^{32}S)_{standard}}$$

Point (i) above implies that sulphides and sulphates in hydrothermal deposits will show a variation (fractionation) in $\delta^{34}S$ values from mineral to mineral which is dependent on their temperature of deposition, provided the minerals crystallized in equilibrium with each other. This fractionation is found in nature and its extent is illustrated in Fig. 5.4. The curves in this figure are based on experimental and theoretical data. They show that sulphur isotopic fractionation at 200°C between pyrite and galena is about 4.6‰ and between sphalerite and galena is a little over 3‰. Thus, these and other mineral pairs can be used in sulphur isotope geothermometry.

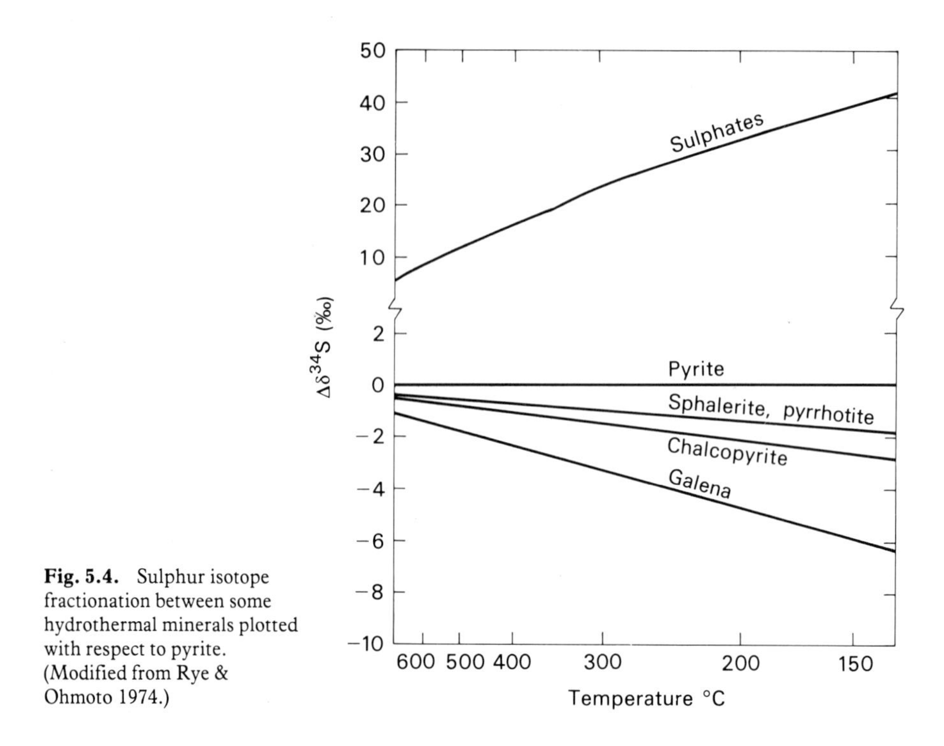

Fig. 5.4. Sulphur isotope fractionation between some hydrothermal minerals plotted with respect to pyrite. (Modified from Rye & Ohmoto 1974.)

Depositional temperatures obtained by fluid inclusion methods on transparent minerals, sulphur isotope methods on sulphides intergrown with them and oxygen isotopic compositions of coexisting oxides generally give results in close agreement, as in the work on the Echo Bay uranium-nickel-silver-copper deposits by Robinson & Ohmoto (1973). This paper is a good illustration of the use of these methods in obtaining temperatures of mineral deposition.

Paragenetic sequence and zoning

The time sequence of deposition of minerals in a rock or mineral deposit is known as its paragenetic sequence. If the minerals show a spatial distribution then this is known as zoning. The paragenetic sequence is determined from studying such structures in deposits as crustiform banding and from the microscopic observation of textures in polished sections.

PARAGENETIC SEQUENCE

Abundant evidence has been accumulated from world-wide studies of epigenetic-hydrothermal deposits indicating that there is a general order of deposition of minerals in these deposits. Exceptions and reversals are known but not in sufficient number to suggest that anything other than a common order of deposition is generally the case. A simplified, general paragenetic sequence is as follows:

(1) silicates;
(2) magnetite, ilmenite, hematite;
(3) cassiterite, wolframite, molybdenite;
(4) pyrrhotite, löllingite, arsenopyrite, pyrite, cobalt and nickel arsenides;
(5) chalcopyrite, bornite, sphalerite;

(6) galena, tetrahedrite, lead sulphosalts, tellurides, cinnabar.
Of course, not all these minerals are necessarily present in any one deposit and the above list has been drawn up from evidence from a great number of orebodies.

ZONING

Zones may be defined by changes in the mineralogy of ore or gangue minerals or both, by changes in the percentage of metals present, or by more subtle changes from place to place in an orebody or mineralized district of the ratios between certain elements or even the isotopic ratios within one element. Zoning was first described from epigenetic vein deposits but it is also present in other types of deposit. For example, syngenetic deposits may show zoning parallel to a former shore line as is the case with the iron ores of the Mesabi Range, Minnesota; alluvial deposits may show zoning along the course of a river leading from the source area; some exhalative syngenetic sulphide deposits show a marked zonation of their metals and pyrometasomatic deposits often show a zoning running parallel to the igneous-sedimentary contact. In this discussion, attention will be focused on the zoning of epigenetic-hydrothermal, exhalative syngenetic and sedimentary syngenetic sulphide deposits.

(a) *Epigenetic hydrothermal zoning.* Zoning of this type can be divided into three intergradational classes, these are regional, district and orebody zoning (Park & MacDiarmid 1975). Regional zoning occurs on a very large scale often corresponding to large sections of orogenic belts and their foreland (Fig. 5.5). A number of examples of zoning on this scale are described from the circum-Pacific orogenic belts by Radkevich (1972). Some regional zoning of this type, e.g. the Andes,

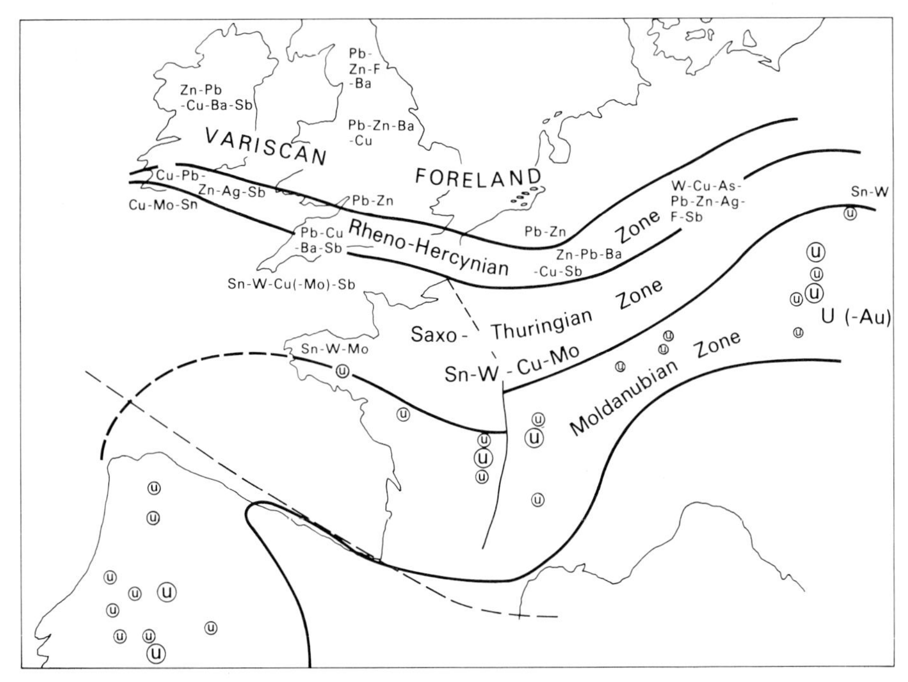

Fig. 5.5. Regional metal zonation in the Variscan Metallogenic Province of north-west Europe. (After Evans 1976a and Cuney 1978.) Sizes of symbols in Moldanubian Zone and Spain indicate relative sizes of uranium deposits.

Table 5.1. A generalized mineral paragenesis of the mineral deposits of south-west England. (After El Shazly *et al.* 1957.)

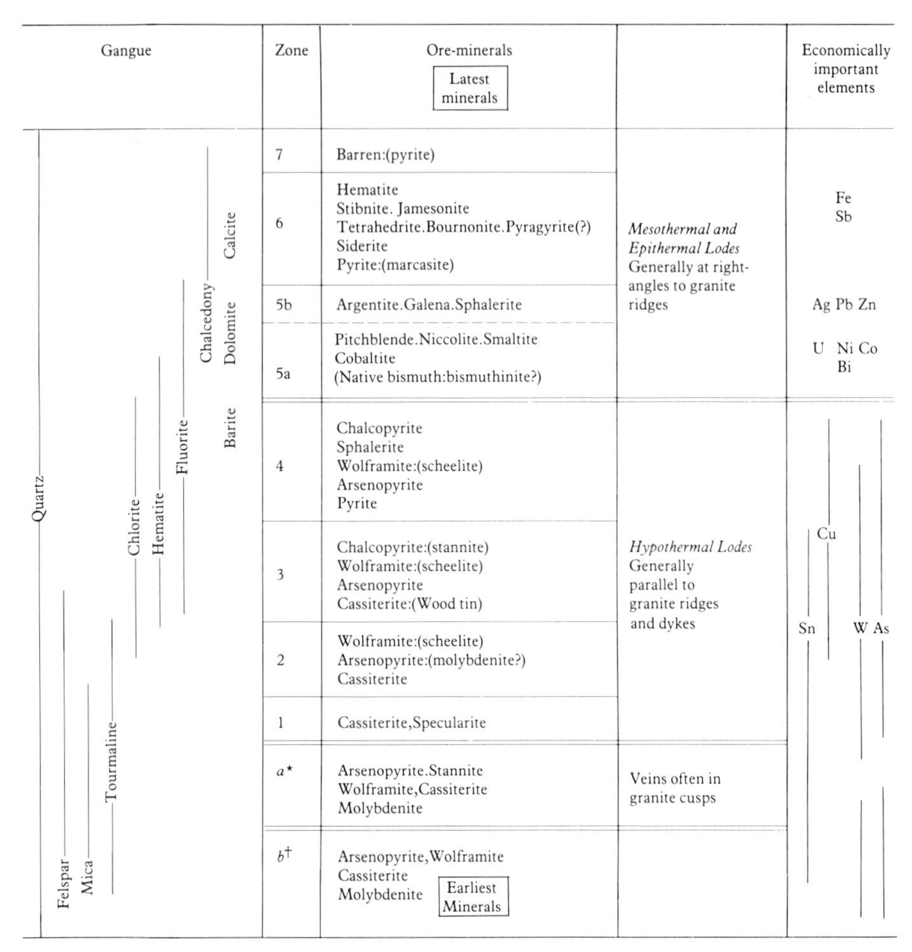

Gangue: Quartz, Felspar, Mica, Tourmaline, Chlorite, Hematite, Fluorite, Chalcedony, Barite, Dolomite, Calcite

Zone	Ore-minerals (Latest minerals at top)		Economically important elements
7	Barren:(pyrite)	Mesothermal and Epithermal Lodes Generally at right-angles to granite ridges	
6	Hematite Stibnite. Jamesonite Tetrahedrite.Bournonite.Pyragyrite(?) Siderite Pyrite:(marcasite)		Fe Sb
5b	Argentite.Galena.Sphalerite		Ag Pb Zn
5a	Pitchblende.Niccolite.Smaltite Cobaltite (Native bismuth:bismuthinite?)		U Ni Co Bi
4	Chalcopyrite Sphalerite Wolframite:(scheelite) Arsenopyrite Pyrite	Hypothermal Lodes Generally parallel to granite ridges and dykes	Cu
3	Chalcopyrite:(stannite) Wolframite:(scheelite) Arsenopyrite Cassiterite:(Wood tin)		Sn W As
2	Wolframite:(scheelite) Arsenopyrite:(molybdenite?) Cassiterite		
1	Cassiterite,Specularite		
a*	Arsenopyrite.Stannite Wolframite,Cassiterite Molybdenite	Veins often in granite cusps	
b†	Arsenopyrite,Wolframite Cassiterite Molybdenite (Earliest Minerals)		

a* = Greisen bordered veins
b† = Pegmatites

appears to be related to the depth of the underlying Benioff Zone which suggests a deep level origin for the metals as well as the associated magmas (Chapter 19). District zoning is the zoning seen in individual orefields such as Cornwall, England (Fig. 5.6 and Table 5.1) and Flat River, Missouri (Fig. 5.7). Zoning of this type is most clearly displayed where the mineralization is of considerable vertical extent and was formed at depth where changes in the pressure and temperature gradients were very gradual. If deposition took place near to the surface, then steep temperature gradients may have caused superimposition of what would, at deeper levels, be distinct zones, thus giving rise to the effect known as telescoping. Orebody zoning takes the form of changes in the mineralization within a single orebody. A good example occurs in the Emperor Gold Mine, Fiji, where vertical zoning of gold-silver tellurides in one of the main ore shoots gives rise to an increase in the Ag/Au

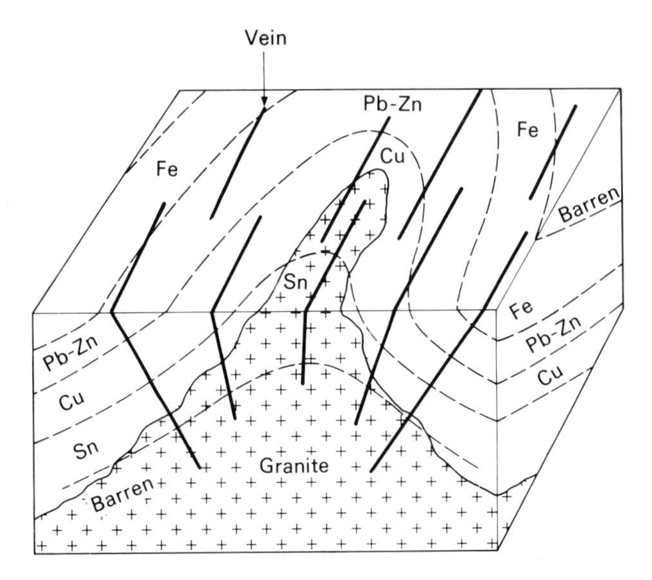

Fig. 5.6. Diagrammatic illustration of district zoning in the orefield of south-west England showing the relationship of the zonal boundaries to the granite-metasediment contact. (After Hosking 1951.)

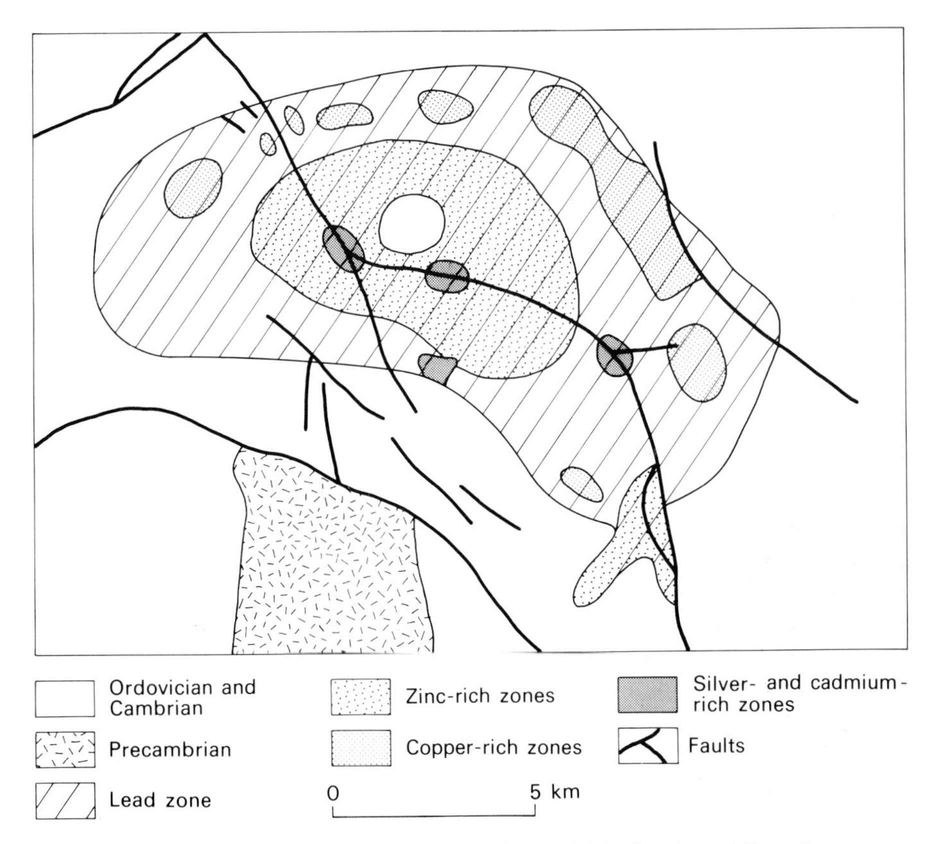

Fig. 5.7. District zoning in the Flat River area of the Old Lead Belt of south-east Missouri. (After Heyl 1969.)

ratio with depth (Forsythe 1971).

(b) *Syngenetic hydrothermal zoning.* This is the zoning found in stratiform sulphide bodies principally of volcanic affiliation (Chapter 2). These deposits are frequently underlain by stockwork deposits and many of them appear to have been formed from hydrothermal solutions which reached the sea floor. As Barnes (1975) has pointed out, the zonal sequence is not always clearly seen in this type of ore deposit, but where it has been established, it is identical to that of epigenetic hydrothermal ores. This zonal sequence is well exemplified by the Precambrian deposits of the Canadian shield, by the Devonian Rammelsberg and Meggen deposits of Germany and the Kuroko deposits of Japan. In these deposits there is a general sequence upwards and outwards through the orebodies of Fe→(Sn)→Cu→Zn→Pb→Ag→Ba which should be compared with the similar zoning shown in Table 5.1. As in vein deposits, the zonal boundaries are normally gradational with frequent overlap of zones.

(c) *Sedimentary syngenetic sulphide zoning.* This zoning is found in stratiform sulphide deposits of sedimentary affiliation usually of wide regional development such as the Permian Kupferschiefer of Germany and Poland, and the Zambian Copperbelt deposits. Underlying stockwork feeder channelways are not known beneath these deposits and they have usually formed in euxinic environments. The zoning appears to show a relationship to the palaeogeography, and proceeding basinwards through a deposit it takes the form of Cu+Ag→Pb→Zn. In the Zambian Copperbelt, however, the zoning is principally one of copper minerals and pyrite, as lead, zinc and silver are virtually absent.

SULPHIDE PRECIPITATION, PARAGENETIC SEQUENCE AND ZONING

In Chapter 4 it was pointed out that there are two schools of thought concerning the transportation of metal ions in hydrothermal solutions. The first favours transportation as bisulphide complexes, and the second as complex chloride ions in chloride-rich, sulphide-poor brines. Barnes (1975) has shown how the first hypothesis can account for the zoning seen in epigenetic and syngenetic hydrothermal deposits. Clearly, the relative stabilities of complex metal bisulphide ions will control their relative times of precipitation and hence both the resulting paragenetic sequence and any zoning which may be developed. Barnes has calculated these stabilities which are shown in Table 5.2. The data in this table suggest that iron and tin would be precipitated early in the paragenetic sequence and would be present in the lowest zone of a zoned deposit, whilst silver and mercury would be late precipitates which would travel furthest from the source of the mineralizing solutions (cf. Table 5.1). Precipitation itself would be occasioned by the interplay of a number of factors including changes in pH due to wall rock alteration reactions, decrease in temperature with distance from source, reaction with carbonaceous materials in wall rocks, mixing with meteoric waters, and so on.

Syngenetic sedimentary sulphide deposits may have been precipitated from chloride-rich brines, such as those found near the Salton Sea (Chapter 4), if these brines encountered H_2S or HS^- supplied by a large volume of marine, euxinic water. On the other hand, they may have been introduced into a basin of deposition in ionic solution by rivers. It is, therefore, pertinent to look at the relative stabilities

Table 5.2. Predicted sequence of stabilities of bisulphide complexes (in kilocalories).

← Least soluble	....			Most soluble →			
Fe	Ni	Sn	Zn	Cu	Pb	Ag	Hg
79	84	126	132	135	153	157	226

Table 5.3. Relative stabilities of sulphides in chloride solutions (as expressed by the equilibrium constant for the reaction: $MeCl_2(aq) + HS^- \rightarrow MeS(s) + H^+ + Cl^-$).

	← Least soluble		Most soluble →
	CuS	PbS	ZnS
At 25°C	38.4	28.2	27.7

Table 5.4. Relative stabilities of sulphides in ionic solutions (as expressed by the equilibrium constant for the reaction: $Me^{2+} + HS^- \rightarrow MeS(s) + H^+$).

← Least soluble		Most soluble →
$CuFeS_2$	PbS	ZnS
23.7	20.7	19.2

of metal sulphides which show a zonal distribution in such deposits in the presence and absence of chloride ions. Some relative solubilities are shown in Tables 5.3 and 5.4 (drawn from data in Barnes 1975).

From these tables it can be seen that there is excellent agreement between the zonal arrangement of metals in syngenetic sedimentary deposits and the relative solubilities shown. Barnes felt that this suggests that the metals travelled in ionic solution or as chloride complexes. It must be noted, however, that studies of the transport of base metals in surface streams at the present day show that only a small proportion of the base metal contents are carried in ionic solution, which means that the above picture is a somewhat simplified version of the true situation. In addition, faunal (including bacterial) and facies factors can influence zonation significantly. Zoning, however, cannot be used to distinguish between the products of these two types of solution, because chloride complexing up to at least 100°C is not sufficiently strong to change significantly the normal order of precipitation of sulphides from that found for simple metallic ions. It does appear though from the work of Barnes that the nature of the zoning in hydrothermal deposits points to deposition from sulphide complexes, and that chloride complexes or ionic solutions may play a significant role as metal transporting agents in the formation of sedimentary sulphide deposits.

The dating of ore deposits

When orebodies form part of a stratigraphical succession like the Mesozoic ironstones of north-western Europe their age is not in dispute. Similarly the ages of orthomagmatic deposits may be almost as certainly fixed if their parent pluton can be well dated. On the other hand, epigenetic deposits may be very difficult to date, especially as there is now abundant evidence that many of them may have resulted from polyphase mineralization, with epochs of mineralization being separated by intervals in excess of 100 Ma. There are three main lines of evidence which can be used: the field data, radiometric and palaeomagnetic age determinations. The field evidence may not be very exact. It may take the form of an unconformity cutting a vein and thus giving a minimum age for the mineralization. Similarly, if mineralization occurs in rocks reaching up to Cretaceous in age we know that some of it is at least as young as the Cretaceous, but we cannot tell when the first mineralization occurred or how young some of it may be.

RADIOMETRIC DATING

Sometimes a mineral clearly related to a mineralization episode will permit a direct dating, e.g. uraninite in a vein, but more commonly it is necessary to use wall rock alteration products such as micas, feldspar and clay minerals. In the latter case, the underlying assumption is that the wall rock alteration is coeval with some of or all the mineralization. Holmes & Holmes (1978) give a simple summary of the theory of how ages are deduced from the radioactive decay of an unstable parent isotope to a more stable isotope of a different element (daughter) and it is assumed here that the reader is familiar with the principles of the theory. Time is but part of the story. Almost as a by-product, the isochron method also reveals the isotopic composition of the lead and strontium which were present when a rock or orebody formed; information which may point to a specific source for the element concerned. Table 5.5 summarizes the parent-daughter relationships which are of major interest for dating rocks and ores. Perusal of this table will show that in dating mineralization by these methods we are heavily dependent on the products of wall rock alteration and the assumption of its contemporaneity with the associated mineralization.

It is also important to note that our radiometric clocks can be reset. Broken Hill, New South Wales, gives us a dramatic example of this phenomenon. The lead-zinc-silver orebodies of Broken Hill are generally held to be regionally metamorphosed syngenetic deposits. Rb-Sr and K-Ar ages of biotites (450-500 Ma), however, markedly post-date the 1700 Ma of the dominant metamorphism in the country rocks (Richards & Pidgeon 1963, Shaw 1968). The heating event reflected by the mica ages may have been associated with formation of the nearby Thackeringa-type orebodies.

Porphyry coppers in many parts of the world have been dated using Rb-Sr and K-Ar methods. The results of this work have shown (a) that it is usually the youngest intrusions in any particular area that carry mineralization, and (b) that there is a close temporal relationship between porphyry copper mineralization and associated calc-alkaline magmatism. Thus at Panguna, Papua New Guinea, the mineralization dates at 3.4 ± 0.3 Ma, compared with intrusives 4-5 Ma old. At Ok Tedi, north-west Papua, alteration as young as 1.1-1.2 Ma has been found

Table 5.5. Major isotopic dating methods.

Method	Generalized decay scheme	Half-life of radioactive isotope	Materials that can be dated
U-Pb	$^{238}U \rightarrow {}^{206}Pb$	4.5×10^9y	Uraninite, pitchblende, zircon, sphene, apatite, epidote, whole rock
Rb-Sr	$^{87}Rb \rightarrow {}^{87}Sr$	5.0×10^{10}y	Mica, feldspar, amphibole, whole rock
K-Ar	$^{40}K \rightarrow {}^{40}Ar$	1.31×10^9y	Mica, feldspar, amphibole, pyroxene, whole rock

(Page & McDougall 1972). This contrasts with the oldest mineralization yet found, a banded iron formation with associated stratabound copper sulphides at Isua, West Greenland, dated at 3760±70 Ma (Moorbath *et al.* 1973, Appel 1979).

Finally, mention should be made of the model lead ages obtained on lead mineralization (normally from galenas). These, the reader should remember, record the time when the radiometric clock stopped, not the time when it was started or restarted as with the U-Pb, Rb-Sr and K-Ar methods. The Pb-Pb method is dependent on assumptions concerning the evolution of the lead isotopes in the source material, and for this mathematical models have to be deduced. The method is regarded with suspicion by many workers but if the basis of the method is borne in mind, its results can be of considerable value, not only for dating purposes but also in throwing light on the origin of the lead.

PALAEOMAGNETIC DATING

This method deserves to be more widely used than it has been. Krs & Šťovíčková (1966) applied it to veins of the Jáchymov (Joachimsthal) region of the Czechoslovakian section of the Erzgebirge, where they demonstrated the presence of Hercynian (late Carboniferous to early Permian) and Saxonian (middle Triassic to Jurassic) mineralization. Further they showed that the Ag-Co-Ni-U-Bi mineralization was not associated temporally with the Hercynian granites but with the later Saxonian to Tertiary basaltic magmatism. Similar work by Krs in the Freiberg region of East Germany showed an excellent correlation with radiometric data on the veins he investigated (Baumann & Krs 1967).

The successful palaeomagnetic dating of ore deposits depends on a number of factors (Evans & Evans 1977). Some of the most important are:

(a) the development of magnetic minerals in a deposit or its wall rocks during one of the principal phases of mineralization;

(b) a lack of complete oxidation or alteration, which may be accompanied by overprinting with a later period of magnetization;

(c) the availability of an accurate polar wandering curve for the continent or plate in which the deposit occurs.

As is well known, magnetite and hematite are the two principal carriers of magnetization in rocks; this is also true for ore deposits. In general, magnetization carried wholly or in the main by magnetite is, with present palaeomagnetic techniques, more easily measured and interpreted. This mineral is, however, by no means common in epigenetic ore deposits and hematite-mineralized specimens often have to be used. Using specimens mineralized with hematite, Evans & Evans (1977) dated as Saxonian the primary hematite mineralization in the base metal veins of the Mendip Orefield, England.

6

Metallogenic Provinces and Epochs

It has long been recognized that specific regions of the world possess a notable concentration of deposits of a certain metal or metals, these regions are known as metallogenic provinces. Such provinces can be delineated by reference to a single metal (Figs 6.1, 6.2 and 6.3) or to several metals or metal associations. In the latter case, the metallogenic province may show a zonal distribution of the various metallic deposits (Fig. 5.5). The recognition of metallogenic provinces has usually been by reference to epigenetic hydrothermal deposits, but there is no reason why the concept should not be used to describe the regional development of other types of deposit provided they show a geochemical similarity. For example, the volcanic-exhalative antimony-tungsten-mercury deposits in the Lower Palaeozoic inliers of the Eastern Alps form a metallogenic province stretching from eastern Switzerland through Austria to the Hungarian border.

Within a metallogenic province there may have been periods of time during which the deposition of a metal or a certain group of metals was most pronounced. These periods are called metallogenic epochs. Some epochs are close in time to orogenic maxima, others may occur later. Thus in the Variscan orogenic belt of north-western Europe and its northern foreland, which form the metallogenic province shown in Fig. 5.5, the principal epochs of epigenetic mineralization were Hercynian (end Carboniferous to early Permian) and Saxonian (middle Triassic to Jurassic). The orogenic events in this belt culminated about the end of the Carboniferous and the Saxonian mineralization and associated vulcanicity is post-orogenic.

Metallogenic provinces and epochs of tin mineralization

Tin deposits are an excellent example of an element restricted almost entirely from the economic point of view to a few metallogenic provinces. Those outside the USSR are shown in Figs 6.1, 6.2 and 6.3. Even more striking is the fact that most tin mineralization is post-Precambrian and confined to certain well-marked epochs. Equally striking is the strong association of these deposits with post-tectonic granites. Among tin deposits of the whole world, 63.1% are associated with Mesozoic granites, 18.1% with Hercynian (late Palaeozoic) granites, 6.6% with Caledonian (mid Palaeozoic) granites and 3.3% with Precambrian granites. In some belts, tin mineralization of different types and ages occurs. For example, in the Erzgebirge of East Germany Lower Palaeozoic or Precambrian stratiform deposits of tin are developed which are probably volcanic-exhalative in origin (Baumann 1965) together with epigenetic deposits associated with Hercynian granites. It is possible that these granites represent anatectic material from the Lower Palaeozoic or Precambrian of the region and that some of the stratiform tin was remobilized when the granitic magmas were formed by partial melting.

Fig. 6.1. Tin belts on continents around the Atlantic Ocean. Dotted areas indicate concentrations of workable deposits. (Modified from Schuiling 1967.)

The figures quoted above suggest that increasing amounts of tin have been added to the crust with decreasing antiquity. The puzzling feature in this connection is that there does not appear to be any concomitant increase in the level of trace element tin with decrease in geological age in granites inside or outside tin belts.

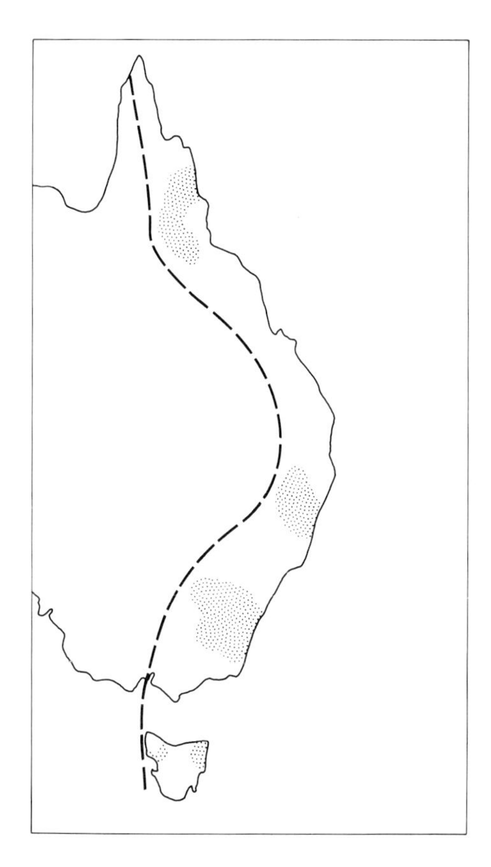

Fig. 6.2. The Palaeozoic tin belt of eastern Australia. The principal fields are shown with dotted ornamentation. (Modified from Hills 1953.)

As epigenetic tin deposits are developed in the cupolas and ridges on the tops of granitic batholiths, it has been suggested that the paucity of such deposits in geologically older terrains is a function of their deeper levels of erosion. If this is the case, one might ask why no Precambrian tin placers are known containing cassiterite derived from such deposits since Precambrian gold and uranium placers are present in a number of continents. The ultimate source of tin, as for other metals, must have been the mantle and this is shown by the recent tin mineralization discovered along the Mid-Atlantic Ridge. The more immediate source, however, must have been the crust or upper mantle, as Schuiling (1967) has argued from a consideration of the south-west African-Nigerian belt where pre-continental drift Precambrian mineralization, syn-drift Jurassic mineralization and post-drift Eocene mineralization occur. If the source of the tin belonging to these different epochs had been below the African plate, then the tin deposits of different ages should occur in parallel belts. That they are not so distributed suggests derivation of the tin from within the crust or upper mantle of the African plate. The existence of tin metallogenic provinces suggests that parts of the upper mantle were (and are?) relatively richer in tin, and that this tin has been progressively added to the crust where it has been recycled and further concentrated by magmatic and hydrothermal processes. The recognition of the development of tin provinces is of fundamental importance to the mineral exploration geologist searching for this metal. It is probable that any further discoveries of important deposits of this metal will be

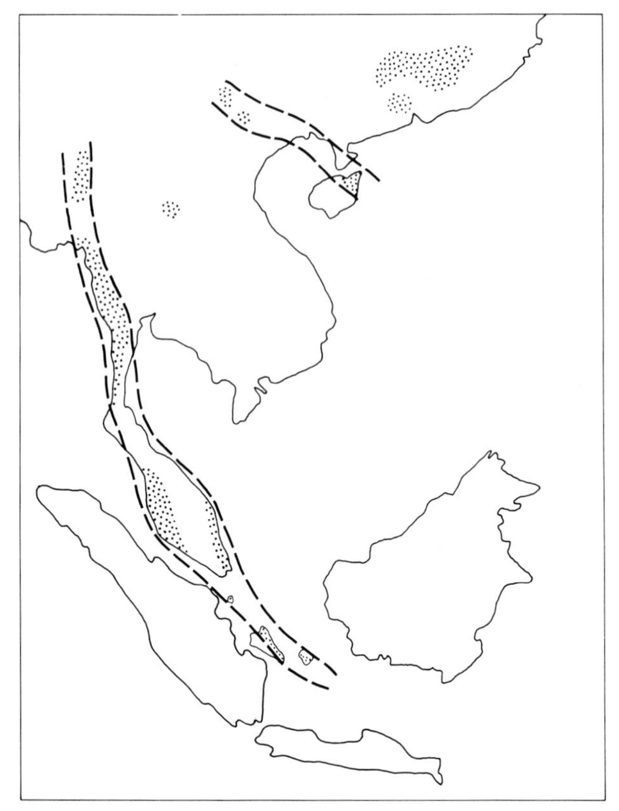

Fig. 6.3. Tin belts and fields of south-eastern Asia. These form the so-called tin girdle of south-eastern Asia, but it should be noted that if lines were drawn to join all the fields together, then the lines would cross tectonic trends at a high angle.

made within these provinces or their continuations. Continental drift reconstructions suggest that a continuation of the eastern Australian tin belt may be present in Antarctica.

Some other examples of metallogenic epochs and provinces

BANDED IRON FORMATION

These rocks, of which the commonest facies is a rock consisting of alternating quartz- and hematite-rich layers, are virtually restricted to the Precambrian. They occur in the oldest (> 3760 Ma old) Isua sediments of western Greenland and in most Archaean greenstone belts. Their best development took place in the interval 2600-1800 Ma ago, in early Proterozoic basins, after which they failed to form and no banded iron formations are found in middle to late Proterozoic or Phanerozoic successions. Instead, later ferruginous sediments are of the hematite-chamosite-siderite type. The latter also tend to be grouped into narrow stratigraphical intervals, for example the Minette ores of this class which are well developed in the Jurassic of western Europe.

NICKEL SULPHIDE DEPOSITS

These deposits are almost entirely Precambrian in age and are mainly restricted to a few Archaean greenstone belts in Ontario, Manitoba and Ungava, Canada, the Western Australian Shield, the northern part of the Baltic Shield and in

Zimbabwe-Rhodesia. They show, therefore, a good development of metallogenic provinces, but there is considerable evidence that those with mantle-derived sulphur are confined to the Archaean and early Proterozoic (see Chapter 9).

TITANIUM OXIDE ORES OF THE ANORTHOSITE ASSOCIATION

The mid-Proterozoic mobile belts (1800-1000 Ma old) of Laurasia and Gondwanaland are characterized by post-tectonic andesine-labradorite anorthosites containing titanium-iron oxide deposits. Examples occur at Bergen, Egersund and Lofoten, Norway; St. Urbain and Allard Lake, Quebec; Iron Mountain, Wyoming, and Sanford Lake, New York. According to Windley (1977), an average age of emplacement of the anorthosites is about 1600 ± 200 Ma which indicates an important metallogenic epoch. These anorthosites are confined to two linear belts in the northern and southern hemispheres when plotted on a pre-Permian continental drift reconstruction and thus they also outline two metallogenic provinces for titanium.

TRACE ELEMENT DELINEATED PROVINCES

Burnham (1959) showed that the major ore deposits of the south-western USA lie in provinces outlined by greater than average trace element content in the crystal lattices of chalcopyrite and sphalerite. The most useful trace elements were found to be cobalt, gallium, germanium, indium, nickel, silver and tin. These exhibit well-defined geographical distributions (Fig. 6.4). Burnham suggested that the variations in trace element content are probably due to variations in the compositions of the fluids during crystal growth. The metallogenic belts appear to be independent of time, wall rock type or intrusions, and Burnham consequently suggested that they are of deep-seated origin and related to deep-seated compositional heterogeneities in the mantle.

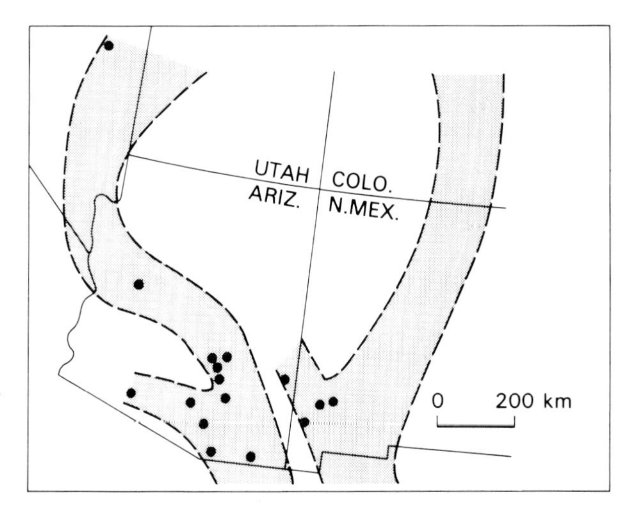

Fig. 6.4. Trace element delineated metallogenic provinces in the south-western USA. The positions of some major porphyry copper deposits are indicated. (Modified from Burnham 1959.)

Part II

Examples of the More Important Types of Ore Deposit

7

Classification of Ore Deposits

The classification of objects in the natural world has always been one of the prime interests of the physical and biological sciences. In geology and biology it is particularly important as it provides a shorthand method of referring to groups of objects having common properties. Without the use of classifications, the comparison of fossils from an evolutionary or palaeogeographical point of view would be practically impossible. Similarly, for the discussion of magmatic processes we must classify igneous rocks. The best classifications are generally those which have no reference to the origin of geological material. Once genetic factors are brought into a classification, difficulties arise. The student may have already noted that most igneous rock classifications in use today are essentially free of genetic parameters whereas classifications of pyroclastic (volcaniclastic) rocks have many genetic

Table 7.1. General characteristics of hypothermal deposits. (After Lindgren 1933.)

Depth of formation	3000 - 15 000 m.
Temperature of formation	≈ 300 - 600°C
Occurrence	In or near deep-seated acid plutonic rocks in deeply eroded areas. Usually found in Precambrian terranes, rarely in young rocks. Often found in reverse faults
Nature of ore zones	Fracture-filling and replacement bodies with the latter phenomenon often more prevalent leading to irregular-shaped orebodies; nevertheless these are frequently broadly tabular. Sheeted zones common, also bedding plane deposits and short, irregular veins. Boundaries usually sharp with limited amount of ore disseminated in walls. Good persistence in depth
Ores of	Au, Sn, Mo, W, Cu, Pb, Zn, As
Ore minerals	*Magnetite,* specularite, *pyrrhotite, cassiterite, arsenopyrite,* molybdenite, bornite, chalcopyrite, Ag-poor gold, *wolframite, scheelite,* pyrite, galena. *Fe-rich sphalerite* (marmetite)
Gangue minerals	*Garnet,* plagioclase, *biotite, muscovite, topaz, tourmaline,* epidote, quartz (often originally *high quartz*), *chlorite* (*high Fe variety*), carbonates
Wall-rock alteration	*Albitization, tourmalinization, rutile development,* sericitization in siliceous rocks, chloritization. Wall rocks are often crisp and sparkling
Textures and structures	Often very coarse-grained, frequently banded, fluid inclusions present in quartz
Zoning	Textural and mineralogical changes with increasing depth are very gradual over thousands of metres. Gold tellurides may give rise to spectacular bonanzas
Examples	Au of Kirkland Lake, Ontario; Kolar, Mysore; Kalgoorlie, W. Australia; Homestake, Dakota; Cu-Au of Rouyn area, Quebec; Sn of Cornwall

overtones. These can affect the usefulness of the classification as our ideas on genesis evolve.

In older classifications of ore deposits much emphasis was placed on the mode of origins of deposits, with the result that as ideas concerning these changed, many classifications became obsolete. A good example of this occurs with the massive volcanic sulphide deposits which, twenty-five years ago, were generally held to be formed by replacement at considerable depths within the crust, but are now thought to be the product of deposition in open spaces at the volcanic- or sediment-sea water interface. They have, therefore, moved from the class of hydrothermal-replacement deposits to that of volcanic-exhalative deposits, and who will be so intrepid as to suggest that future generations of geologists may not postulate new theories concerning the genesis of this and other classes of deposit! If a classification is to be of any value it must be capable of including all known ore deposits so that it will provide a framework and a terminology for discussion and so be of use to the mining geologist, the prospector and the exploration geologist.

Classifications of ore deposits have been based upon commodity (copper deposits, iron deposits, etc.) morphology, environment and origin. Commodity and morphological classifications may be of value to economists and mining engineers, but they lump too many fundamentally different deposit types together to be of much use to geologists. In the past, ore geologists have been inclined to favour genetic classifications but in recent years there has been a swing away from such ideas towards environmental-rock association classifications. Good examples of this trend are to be found in Stanton (1972) and Dixon (1979).

Table 7.2. General characteristics of mesothermal deposits. (After Lindgren 1933.)

Depth of formation	1200 - 4500 m
Temperature of formation	200 - 300°C
Occurrence	Generally in or near intrusive igneous rocks. May be associated with regional tectonic fractures. Common in both normal and reversed faults
Nature of ore zones	Extensive replacement deposits or fracture-fillings. Boundaries of orebodies often gradational from massive to disseminated ore. Tabular bodies, sheeted zones, stockworks, pipes, saddle-reefs, bedding-surface deposits. Fissures fairly regular in dip and strike
Ores of	Au, Ag, Cu, As, Pb, Zn, Ni, Co, W, Mo, U, etc.
Ore minerals	*Native Au, chalcopyrite, bornite,* pyrite, *sphalerite, galena, enargite, chalcocite, bournonite,* argentite, *pitchblende, niccolite, cobaltite, tetrahedrite,* sulphosalts
Gangue minerals	Lack high temperature minerals (garnet, tourmaline, topaz, etc.), albite, *quartz, sericite, chlorite, carbonates, siderite,* epidote, montmorillonite
Wall-rock alteration	Intense chloritization, carbonitization or sericitization. Walls often dull
Textures and structures	Less coarse than hypothermal ores, pyrite, when present, is often very fine-grained. Veins are often banded, large lenses usually massive
Zoning	Gradual but definite change in mineralization with depth, e.g. Butte. Good vertical range, many deposits not bottomed after 1800 m of mining
Examples	Au of Bendigo, Australia; Ag of Cobalt, Ontario; Ag-Pb of Coeur d'Alene, Idaho and Leadville, Colorado; Cu of Butte, Montana

Table 7.3. General characteristics of epithermal deposits. (After Lindgren 1933.)

Depth of formation	Near surface to 1500 m
Temperature of formation	50 - 200°C
Occurrence	In sedimentary or igneous rocks, especially in or associated with extrusive or near-surface intrusive rocks, usually in post-Precambrian rocks not deeply eroded since ore formation. Often occupy normal fault systems, joints, etc.
Nature of ore zones	Simple veins—some irregular with development of ore chambers—also commonly in pipes and stockworks. Rarely formed along bedding surfaces. Little replacement phenomena
Ores of	Pb, Zn, Au, Ag, Hg, Sb, Cu, Se, Bi, U
Ore minerals	*Native Au now often Ag-rich,* native Ag, *Cu,* Bi. Pyrite, *marcasite, sphalerite, galena,* chalcopyrite, *cinnabar,* jamesonite, *stibnite, realgar, orpiment, ruby silvers, argentite, selenides,* tellurides
Gangue minerals	SiO_2 as *chert, chalcedony* or crystalline quartz—often amethystine, (sericite), low Fe chlorite, epidote, carbonates, fluorite, baryte, *andularia, alunite, dickite,* rhodochrosite, *zeolites*
Wall-rock alteration	Often lacking, otherwise chertification, kaolinization, pyritization, dolomitization, chloritization
Textures and structures	Crustification (banding) very common, often with development of fine banding, cockade ore, vugs and brecciation of veins. Grain size very variable
Zoning	Type of mineralization may vary abruptly with depth, often having only a small vertical range (telescoping) mostly bottom at 300 - 900 m. Grade variable with occurrence of bonanzas within low-grade ore
Examples	Au of Cripple Creek, Colorado; Comstock, Nevada; Keweenawan Coppers; Sb of China

Table 7.4. General characteristics of telethermal deposits. These are believed to have formed a long way from the parent magma at low temperatures and high in the crust. Temperatures are still higher than in ground waters.

Depth of formation	Near surface
Temperature	± 100°C
Occurrence	In sedimentary rocks or lava flows, often in areas where plutonic rocks are apparently absent
Nature of ore zones	In open fractures, cavities, joints, fissures, caverns, etc. No replacement phenomena
Ores of	Pb, Zn, Cd, Ge
Ore minerals	*Galena* (poor in Ag), *sphalerite* (poor in Fe, may be rich in Cd), *marcasite* in excess over pyrite, cinnabar, etc. Viz: similar to epithermal mineralogy (Table 7.3)
Gangue minerals	Calcite, low-Fe dolomite, etc.
Wall-rock alteration	Dolomitization and chertification
Textures and structures	As epithermal
Examples	Tri-State, USA, Pb-Zn ores etc.; many Hg deposits

In 1913, Lindgren put forward the most influential classification which has ever been proposed. It is still used by many geologists particularly when discussing epigenetic-hydrothermal deposits, and all ore geologists should be aware of the descriptive terms used by Lindgren and added to by L. C. Graton in later modifications. In this part of Lindgren's classification, epigenetic-hydrothermal deposits are classified according to their depth and temperature of formation—hypothermal deposits being deep-seated high temperature deposits; mesothermal deposits those formed at low temperatures and medium depths, and, thirdly, epithermal near-surface deposits. Later terms include leptothermal to cover deposits gradational from mesothermal to epithermal, and telethermal for very low-temperature deposits formed at great distances from the source of the hydrothermal solutions which gave rise to them. In the field, these various deposit types have to be recognized from their mineral assemblages, type of wall rock alteration, etc. The features used for this purpose and general features of these deposits are given in Tables 7.1 - 7.4. One of the uses of this and other classifications is that if we can classify a particular deposit then we can compare it with others of the same class and make predictions as to its behaviour in depth. Thus, recognition that a deposit is hypothermal suggests that it will have great continuity in depth, whereas epithermal and telethermal deposits may bottom very quickly and be of limited vertical development. Again, recognition that a deposit is of massive volcanic-exhalative type should stimulate the mining geologist into searching for a possible underlying feeder stockwork which may be exploitable.

A glance at the contents pages of this book will show that an environmental-rock association classification is favoured although a whiff of genesis and morphology is included!

8

Orthomagmatic Deposits of Chromium, Platinum, Titanium and Iron Associated with Basic and Ultrabasic Rocks

Orthomagmatic ores of these metals are found almost exclusively in association with basic and ultrabasic plutonic igneous rocks—some platinum being found in nickel-copper deposits associated with extrusive komatites.

Chromium

Chromium is won only from chromite [$FeCr_2O_4$]. This spinel mineral can show a considerable variation in composition with magnesium substituting for the ferrous iron and aluminium, and/or ferric iron substituting for the chromium. It may also be so intimately intergrown with silicate minerals that these too act as an ore dilutant. Because of these variations there are three grades of chromite ore. The specifications for these are somewhat variable, typical figures are given in Table 8.1.

Table 8.1. Chromite ore grades and specifications.

	Cr_2O_3	$^{Cr}/_{Fe}$	$Cr_2O_3+Al_2O_3$	Fe	SiO_2
Metallurgical grade	$>48\%$	>2.8	-	-	-
Refractory grade	$>30\%$	not critical	$>57\%$	$\ngtr 10\%$	$\ngtr 5\%$
Chemical grade	$>45\%$	-	-	-	$\ngtr 8\%$

OCCURRENCE

Three-quarters of the world's chromium reserves are in the Republic of South Africa and 23% in Zimbabwe-Rhodesia, most deposits outside these two countries being small. The only other countries with appreciable reserves are the USSR and Turkey. Lower grade deposits were found recently in the Fiskenæsset Complex of Greenland and there are large subeconomic deposits in Canada.

All economic deposits of chromite are in ultrabasic and anorthositic plutonic rocks. There are two major types: stratiform and podiform (often referred to respectively as Bushveld- and Alpine-types).

STRATIFORM DEPOSITS

This type contains over 90% of the world's chromite resources. These deposits consist of layers usually formed in the lower parts of stratified igneous complexes of the Bushveld and Stillwater class. The immediate country rocks of the complexes are ultrabasic differentiates of the parent gabbroic magma—dunites, peridotites and pyroxenites. The layers of these rocks have great lateral extent, uniformity and consistent positions within the complexes. Chromite in these deposits is usually iron-rich but an outstanding exception is the Great Dyke of Zimbabwe-Rhodesia with its high chromium ores.

BUSHVELD COMPLEX

This is an enormous differentiated lopolith in South Africa (Fig. 8.1). The chromite occurs in the western and eastern outcrops of ultrabasic rocks, with layers a few centimetres to two metres thick (Fig. 8.2). They make up an enormous tonnage (van Gruenewaldt 1977). Assuming a maximum vertical mining depth of only 300 m gives a figure of 2.3×10^9 t, a figure which can be multiplied by ten if lower grade deposits are included and the vertical mining depth is increased to 1200 m. Potentially the largest orebodies are the LG3 and LG4 chromite layers present only in the western Bushveld. In these layers the chromite grades 50% Cr_2O_3, $^{Cr}/_{Fe}=2.0$, the strike length is 63 km, the thickness 50 cm and the resources (300 m vertical depth) are 156×10^6 t. The ore grades about 45% Cr_2O_3.

The Bushveld chromite zone as a whole contains as many as 29 chromite layers or groups of layers. Above this zone is the platinum-bearing Merensky Reef (Figs 8.1 and 8.2), and near the top of the basic part of the complex vanadiferous magnetite layers occur.

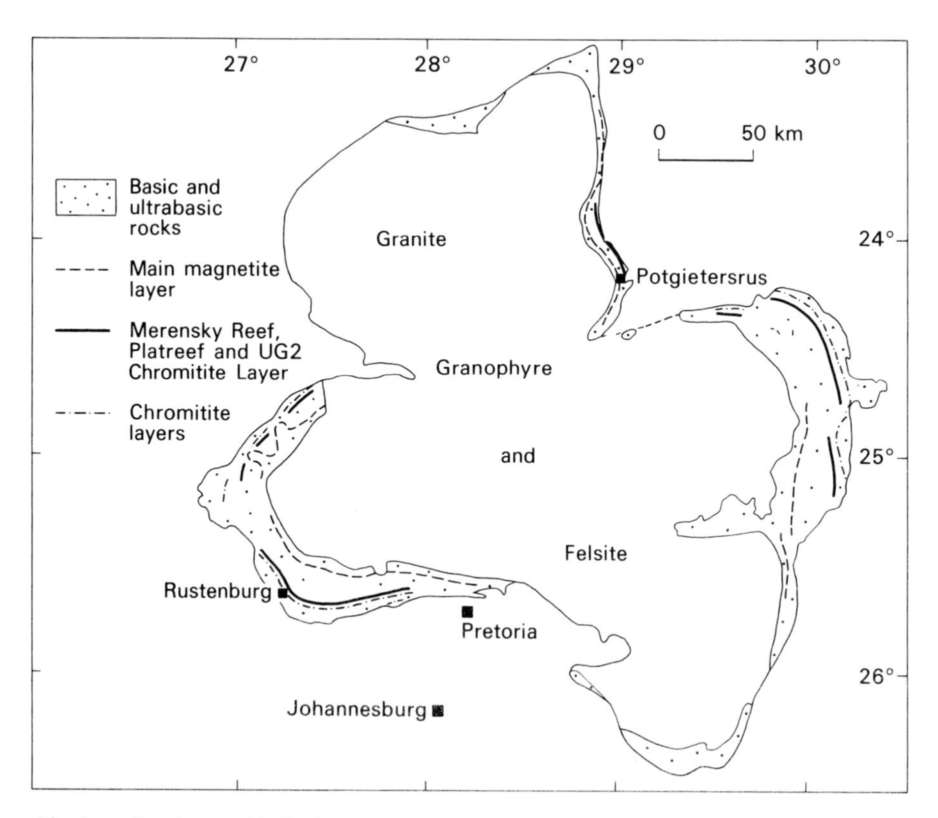

Fig. 8.1. Sketch map of the Bushveld Complex. (After van Gruenewaldt 1977.)

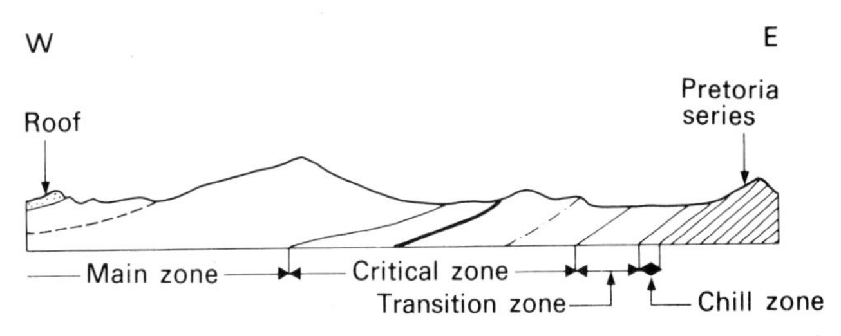

Section showing major zones in the Bushveld Complex, north of Steelport. Length of section, 30·5 km. (After Hall, 1932)

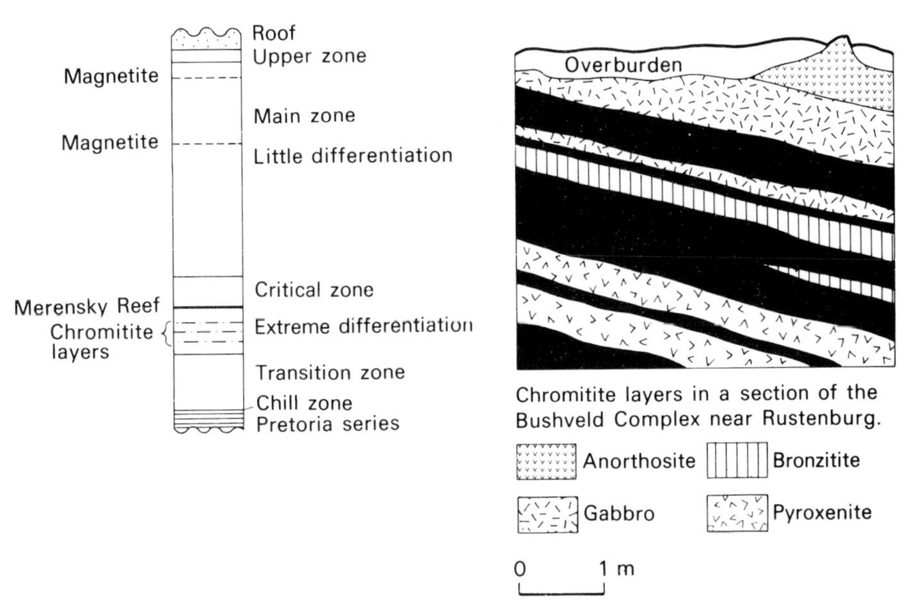

Chromitite layers in a section of the Bushveld Complex near Rustenburg.

Fig. 8.2. Sections showing the occurrence of economic minerals in the Bushveld Complex.

GREAT DYKE OF ZIMBABWE-RHODESIA

This is another layered complex 532 km long and 5-9.5 km broad. In cross-section the layers are synclinal (Fig. 8.3). Chromite layers occur along the entire length and individual layers extend across the entire width. The layers are in the range 5-45 cm and nearly all the chromite is the high chromium variety. Only layers 15 cm, or more, thick are mined.

PODIFORM DEPOSITS

These are lenticular or rudely tabular pods ranging from a few kilograms to several million tonnes. Most production comes from bodies containing 100 000 t or more. Throughout the world, less than a dozen bodies that contained 1 000 000 t or more of chromite are known. Most pod deposits are of high chromium type but these deposits are also the only source of high aluminium chromite. The chromite layers are in the range 1-40 cm, or above.

Podiform deposits occur in irregular peridotite masses or peridotite-gabbro complexes of the Alpine-type which are mainly restricted to orogenic zones such as the Urals and the Philippine island arc. Within many large complexes the chromite

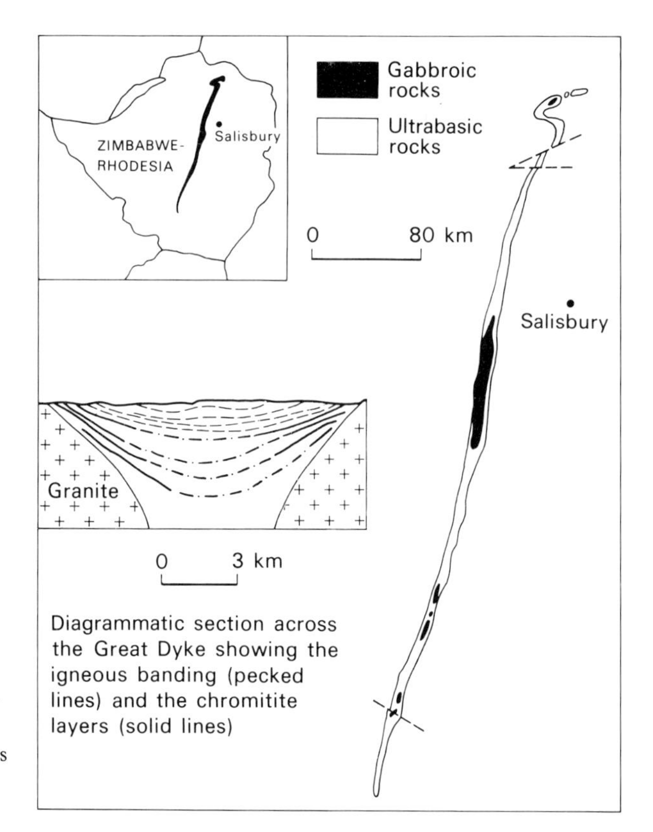

Fig. 8.3. Sketch diagrams illustrating the Great Dyke of Zimbabwe-Rhodesia and the occurrence of chromitite layers in it. (In part after Bichan 1969.)

deposits are generally near contacts between peridotite and gabbro, but where there is no gabbro the deposits appear to be distributed at random through the peridotite. The host intrusions are usually only a few tens of square kilometres, or less, in area. They are generally strongly elongated and lenticular in shape. Large numbers of these small intrusions occur in narrow zones (serpentinite belts) running parallel to regional thrust zones and the general trend of the orogen in which they occur. The intrusions are usually layered, but the layering does not often show the perfection nor the continuity of the stratiform deposits. They are short lenses rather than extensive sheets. Compositions range from dunite to gabbro and, whilst the average composition of stratiform hosts is close to gabbro, that of Alpine intrusions is near peridotite. There are two subtypes, harzbergite and lherzolite. It is very important for exploration purposes to note that it is the harzbergite subtype which carries chromite deposits (Jackson & Thayer 1972). Most podiform deposits are Palaeozoic or younger, many of them are Mesozoic or Tertiary. The largest known deposits are those of the Urals, the Philippines and Turkey. Significant deposits occur in Cuba, New Caledonia, Yugoslavia and Greece.

Whilst podiform deposits belong to mobile belts, stratiform deposits (apart perhaps from the Fiskenæsset Complex of West Greenland) were intruded into stable cratons. The Fiskenæsset Complex, being associated with what were oceanic basalts, may have been emplaced in stable oceanic crust. Thus, the two major types of chromite deposit belong to very different geological environments. Other differences are their general age—stratiform deposits with abundant chromite are

Precambrian—and number of deposits. Whilst there are numerous podiform deposits, only three stratiform deposits have produced chromite.

DISRUPTED STRATIFORM DEPOSITS

Tectonic dismemberment of stratiform deposits has led to their individual parts being identified as podiform deposits. Recognition of parts of a once continuous stratiform complex can be very important in mineral exploration as it will lead to the search for missing segments, whereas a similar programme in a podiform chromite district could be financially ruinous (Thayer 1973). The chromite deposits of Campo Formoso in Brazil were formerly thought to be isolated blocks. They have recently been shown to be parts of a highly-faulted layered complex about 18 km long.

GENESIS OF PRIMARY CHROMITE DEPOSITS

All chromite of economic interest is believed to have crystallized from basic or ultrabasic magmas, essentially simultaneously with the enclosing silicates. Chromite is denser than the magma from which it forms and therefore like the other early formed minerals it sinks to collect in layers on the floor of the magma chamber. In most stratiform complexes emplacement of the magma and crystallization took place in a stable cratonic environment and many delicate primary igneous and sedimentary features are preserved. They were, then, originally emplaced in the upper crust.

The host rocks of podiform deposits, on the other hand, probably originally crystallized in the mantle and were then incorporated into highly unstable tectonic environments in the crust by movement up thrusts and reverse faults. They are part of the ophiolite suite and were probably originally developed at mid-oceanic or back-arc spreading ridges. Despite their tectonic deformation they often preserve textures due to crystal settling, some identical with those in stratiform deposits, indicating a common origin. The transport of the peridotite and chromite from the upper mantle into the upper crust, where they now are, probably occurred by plastic flowage at high temperatures possibly over many kilometres and this fragmented the original layering and has produced many metamorphic features in the chromite and their host rocks. These peridotites have usually suffered a high degree of serpentinization in contrast to the stratiform deposits, in which it is relatively negligible.

CONSTITUTION OF CHROMITE CONCENTRATIONS

As will have been gathered from the foregoing, there are some chemical differences between the chromites from stratiform and podiform deposits. These, together with the relationship of chromites from the Fiskenæsset Complex of western Greenland are shown in Fig. 8.4. Apart from the differences brought out by the figure, it should be noted that the Fiskenæsset chromites are vanadiferous (about 1.5% V_2O_5) like the magnetites of the upper basic levels of the Bushveld Complex; and, like the Bushveld magnetite layers, the Fiskenæsset chromitites occur in the upper part of the intrusion. These differences are perhaps to be attributed to crystallization from a water-rich magma in an oceanic environment, compared with the dry magmas of the Bushveld and similar lopoliths.

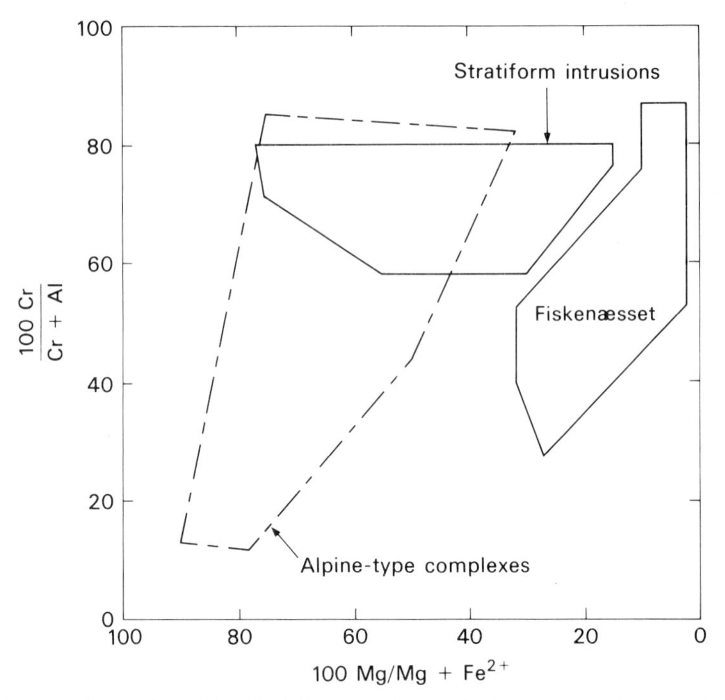

Fig. 8.4. Plot of mg against cr for spinels from stratiform and alpine-type complexes together with the Fiskenæsset Complex of western Greenland. (After Steel *et al.* 1977.)

Platinum group metals

These metals used to be produced entirely from placer deposits, but now primary deposits are more important. From 1778-1823 Columbia was the only producer, then in 1822 placers were also discovered in the Urals. In 1919, the recovery of by-product platinum metals from the Sudbury copper-nickel ores of Canada started. In 1924, the South African deposits in the Bushveld Complex were found and in 1956 South African production outstripped Canadian. The principal producers are now: (1) USSR, (2) South Africa, (3) Canada, (4) Columbia, (5) Alaska.

PRIMARY DEPOSITS

There are two principal types of deposit exemplified by those associated with the basic lopolith of Sudbury, Ontario, and the basic-ultrabasic lopolith of the Bushveld Complex. In the former, platinum minerals occur in liquation ores at the base of the intrusion. In the latter, platinum occurs in the Merensky Reef and similar layers some 2000 m above the base of the intrusion. Platinum-rich layers have recently been discovered in the Stillwater Complex of Montana. The platinum in these deposits occurs mainly as arsenides, sulphides and antimonides (Mertie 1969).

A third type of primary deposit is sometimes important. It consists of deposits of platinum metals in peridotites and perknites, commonly in dunite and serpentinite, less commonly in pyroxenic rocks containing no olivine. In these rocks the platinum occurs mainly as platinum alloys associated with lenticular masses of chromite or disseminated through the host rock in association with chromite. Such deposits have occasionally yielded small orebodies of phenomenally high grade, but are more important as source rocks for the formation of platinum placers.

The Sudbury copper-nickel ores will be covered in the next chapter. Attention will therefore be focused here on the Bushveld Complex.

BUSHVELD COMPLEX

In this lopolith there are three very extensive deposits: the Merensky Reef, the UG2 Chromitite Layer (both in the western and eastern Bushveld) and the Platreef in the Potgietersrus area. The first-named occurs in the Merensky Zone—an exceedingly persistent igneous zone traced for over 220 km in the western Bushveld where it has been proved to contain workable ore for over 110 km (Fig. 8.1). This zone is also well developed in two other districts. It is 0.6-11 m thick and generally consists of a dark-coloured norite.

The Merensky Reef is a thin sheet of coarsely crystalline pyroxenite with a pegmatitic habit which lies near the base of the Merensky Zone. It has a thickness of 0.3-0.6 m. Chromite bands approximately 1 cm thick mark the top and bottom of the reef. They are enriched with platinum metals relative to the reef. The platinum minerals are sperrylite, $PtAs_2$; braggite (Pt,Pd,Ni)S; stibiopalladinite, Pd_3Sb and laurite, RuS_2, together with some native platinum and gold. Grades at Rustenberg are 7.5-11 g t^{-1}. Grades elsewhere are lower. The average stoping width is 0.8 m. The exploitable ore contains 3.3×10^9 t. Sulphides (chalcopyrite, pyrrhotite, pentlandite, nickeliferous pyrite, cubanite, millerite and violarite) are also present, and copper (0.11%) and nickel (0.18%) are recovered.

The UG2 Chromitite Layer lies 150-300 m below the Merensky Reef. Little information about it is available. It is 90-150 cm thick, carries 3.5-19 g t^{-1} platinum together with copper and nickel values. Chromium will also be produced. The reserves are greater than for the Merensky Reef and total about 5.42×10^9 t. The Platreef is similar to the Merensky Reef, with which it has been correlated, but it is much thicker, being up to 200 m with rich mineralization over thicknesses of 6-45 m. Grades are very irregular. Ore reserves are about 4.08×10^9 t.

The South African ores are essentially platinum ores with by-product nickel and copper. At Sudbury, Ontario, the platinum metals are by-products of copper-nickel ores and the grades of the platinum metals are much lower, being about 0.6-0.8 g t^{-1}. Similar ores to those at Sudbury occur at Noril'sk in Siberia and the grade is said to average 2.2.g t^{-1}.

GENESIS OF PLATINUM DEPOSITS

The genesis of the platinum-rich layers in the Bushveld Complex is still very much an open question. The mystery centres on why the platinum and associated sulphides are concentrated into just a few thin layers. Was it that particular pulses of magma introduced into the chamber were especially rich in these materials? Or was a process such as that envisaged by Vermaak (1976) responsible?

Vermaak has suggested that anorthosites of the Critical Zone, which includes the Merensky Reef, were formed when plagioclase crystals floating upwards reached a temperature-density-compositional inversion in the magma where they formed stationary layers which grew by an underplating of rising plagioclase crystals. Platinoids, copper and nickel are not accommodated in silicate structures and these, rising through the cumulus fluids, would be trapped and concentrated together with volatiles beneath the plagioclase mat. At the same time, removal of plagioclase caused enrichment of basic magmatic and economic constituents in the trapped magma, whilst deposition of chromites depleted the magma in iron thus lowering the oxygen fugacity and increasing the activity of sulphur in the closed system. Thus, sulphide droplets might have formed and collected platinoids. The entrapped volatiles would have promoted crystal growth and this would account for the pegmatitic aspect of the Merensky Reef.

Titanium

The principal source of titanium is placer rutile. A few primary ilmenite deposits are, however, exploited. These are always associated with anorthosite or anorthosite-gabbro complexes and have generally been interpreted as magmatic segregation deposits.

At Allard Lake, Quebec, the ores grade 32-35% TiO_2 and occur in anorthosites. The Lac Tio deposit contains about 125×10^6 t ore. The ore is coarse-grained, with ilmenite grains containing exsolved hematite blebs reaching up to 10 mm in diameter. Gangue minerals are chiefly plagioclase, pyroxene, biotite, pyrite,

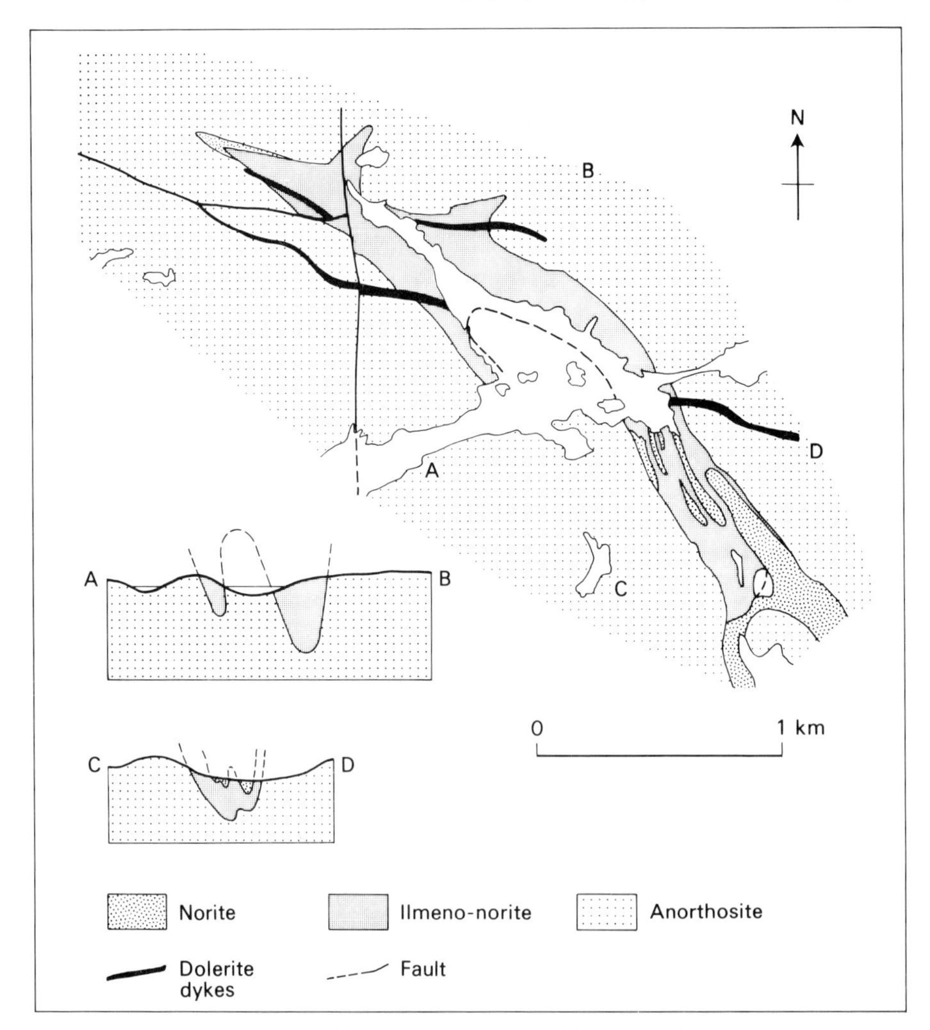

Fig. 8.5. Map and sections of the Tellnes titanium orebody. (After Dybdahl 1960.)

pyrrhotite and chalcopyrite. The exsolution hematite is too fine-grained to be separated by grinding from the ilmenite, and dilutes the ilmenite concentrate. The orebodies form irregular lenses, narrow dykes, large sill-like masses and various combinations of these forms. Some of these clearly cut the anorthosite and appear to be later in age. It has been suggested that the titanium ores and the anorthosite are differentiates of the same parent magma.

The world's largest ilmenite orebody is at Tellnes in the anorthosite belt of southern Norway about 120 km south of Stavanger. The deposit is boat-shaped, elongated north-west, 2.3 km long, 400 m wide and about 350 m deep (Fig. 8.5). It occurs in the base of a noritic anorthosite. Proven reserves are 300 × 10^6 t of 18% TiO_2, 2% magnetite and 0.25% sulphides (pyrite and Cu-Ni sulphides). The ilmenite carries up to 12% hematite as exsolution lamellae. The annual production is 2.76 × 10^6 t of ore, and extraction of this by opencast methods means that 1.36 × 10^6 t of waste have to be removed.

Iron

Many small to medium sized magnetite deposits occur in gabbroic intrusions but the really big tonnages occur in the stratiform lopoliths. Some of these deposits are likely to be worked for vanadium as well as for iron. This is the case with the vanadiferous magnetite in the Upper Zone of the Bushveld Complex (Fig. 8.1). This carries 0.3-2% V_2O_5. Values above 1.6% V_2O_5 are usually found in the Main Magnetite Layer and several thinner layers below it, but only the main layer (some 1.8 m thick) can be considered as ore. Based on a vertical limit of 30 m for opencast mining, the reserves are about 1030 × 10^6 t. Much more iron ore than this is present, but it is spoilt by a titanium content of up to 19%.

9

Orthomagmatic Copper-Nickel-Iron (-Platinoid) Deposits Associated with Basic and Ultrabasic Rocks

Introduction

Nickel metal production in the non-Communist world is currently running at about 536 000 t yr^{-1}. It is a metal which commands a high price, £3700 t^{-1} compared with about £900 t^{-1} for copper in May 1979. Nickel is produced from two principal ore-types: nickeliferous laterites and nickel sulphide ores. We are concerned with the latter deposit type in this chapter. These deposits usually carry copper, often in economic amounts, and sometimes recoverable platinoids. Iron is produced in some cases from the pyrrhotite concentrates which are a by-product of the dressing of these ores. Nickel sulphide deposits are not common and so just a few countries are important for production of nickel sulphide ores. Canada is pre-eminent, the USSR and Australia are the only other important producers. A small production comes from Zimbabwe-Rhodesia, South Africa, Botswana and Finland.

The mineralogy of these deposits is usually simple consisting of pyrrhotite, pentlandite $(Fe,Ni)_9S_8$, chalcopyrite and magnetite. Ore grades are somewhat variable. The lowest grade of working deposit in western countries appears to be an Outukumpu (Finnish) mine working 0.2% Ni ore. This low grade can be compared with the very high grade sections of some Western Australian deposits that run about 12% Ni. Of course, the overall grade for Australian deposits is less than this because lower grade ore is mined with these high grades. Therefore, although such high grades occur at Kambalda in the Western Mining Corporation mines, the ore reserves at June 1977 were 22 223 000 t running 3.19% Ni. In this mining camp the ore treated in 1976-7 totalled 1 479 000 t having a mill grade of 2.69% and a nickel recovery of 89.38%.

All nickel sulphide deposits are associated with basic or ultrabasic igneous rocks. There is both a spatial and a geochemical relationship in that deposits associated with gabbroic igneous rocks have a high Cu/Ni ratio (e.g. Sudbury, Ontario; Noril'sk, USSR), and those associated with ultrabasic rocks a low Cu/Ni ratio (e.g. Thompson Belt, Manitoba; Western Australian deposits). There are a number of different types of basic and ultrabasic rocks, not all of these have nickel sulphide deposits associated with them. Therefore, in exploring for these deposits it is important to know which classes of basic and ultrabasic rocks are likely to have associated nickel sulphide ores. Naldrett (1973) and Naldrett & Cabri (1976) have classified basic and ultrabasic igneous rocks from this point of view.

Classification of ultrabasic and basic bodies with special reference to nickel sulphide mineralization

The Naldrett & Cabri classification of these bodies is given in Table 9.1. The following remarks are largely based on their paper.

Table 9.1. Classification of ultrabasic and related basic bodies. (After Naldrett & Cabri 1976.)

Class	Examples	Remarks
A. Bodies emplaced in active orogenic areas	1. Bodies contemporaneous with eugeosynclinal volcanism	
	(i) Tholeiitic suite	
	(a) Picritic subtype Munro-Dundonald area, Ontario Pechenga, USSR	Gravity-differentiated flows, capped by hyaloclastite with 13-15 wt % MgO. Gravity differentiated sills
	(b) Anorthositic subtype Doré Lake Complex, Quebec Kamiskotia Complex, Timmins, Ontario	Some examples of this class are conformable, others appear to be discordant
	(ii) Komatiitic suite Munro-Dundonald area, Ontario Eastern Goldfields, Australia	Simple flows, spinifex-capped flows, differentiated flows, and sills, pipes. Composition of flow ranges from peridotite to basalt and of sills from dunite to anorthositic gabbro
	2. Alpine-type bodies	
	(i) Large obducted sheets New Caledonia Papua New Guinea	
	(ii) Ophiolite complexes Vourinos, Greece Troodos, Cyprus Bay of Islands, Newfoundland	
	3. Alaskan-type complexes Intrusions of Alaska and British Columbia, including Duke Island, Union Bay, Tulameen. Intrusions of Urals	
B. Bodies emplaced in non-orogenic areas	4. Large stratiformly layered complexes Bushveld, S.Africa Stillwater, Montana Duluth, Minnesota Sudbury, Ontario	
	5. Sills and sheets equivalent to flood basalts Palisades sill, New Jersey Insizwa-Ingeli intrusion, S. Africa Dufek intrusion, Antarctica	Generally occur in areas in which extrusion of flood basalts has occurred. The sills are chemically similar to the extruded basalts
	6. Medium- and small-sized intrusions Skaergaard, Greenland Rhum, Scotland Noril'sk-Talnakh area, USSR	
	7. Alkalic ultramafic rocks in ring complexes and kimberlite pipes	

(A) BODIES IN AN OROGENIC SETTING

Three broad groups of ultrabasic and basic bodies can be seen in this setting: bodies coeval with eugeosynclinal volcanism; syntectonic alpine-type bodies; and late tectonic, uplift stage, Alaskan-type concentric complexes.

The first group occurs in the Archaean and Proterozoic greenstone belts and can be divided into the tholeiitic and komatiitic suites. The tholeiitic suite contains the picritic and anorthositic classes. The anorthositic class is important for titanium mineralization but so far no substantial nickel mineralization has been found in rocks of this class. The picritic class is an important nickel ore carrier. Ultrabasic rocks in this class occur as basal accumulations in differentiated sills and lava-flows, and some have basal sulphide segregations. The Dundonald Sill of the Abitibi Greenstone Belt, Canada, is a good example. The tholeiitic activity in this and other areas was often contemporaneous with komatiitic volcanicity.

The komatiitic suite is a more important carrier of nickel mineralization. This suite was named by Viljoen & Viljoen (1969) and since little about it appears in textbooks on igneous petrology and recognition of its presence during nickel exploration is vital, a short discussion is included here. Komatiites range from dunite (> 40 wt % MgO) through peridotite (30-40%), pyroxene-peridotite (20-30%), pyroxenite (12-20%) and magnesian basalt to basalt, thus forming a magma series with the status of the tholeiitic or calc-alkaline series. Komatiites are both extrusive and intrusive. Ultrabasic members are believed to have crystallized from liquid with up to 35 wt % MgO and carrying 20-30% of olivine phenocrysts in suspension. In some flows and near surface sills, quench textures (probably due to contact with sea water) are present in the upper part consisting of platy and skeletal olivine and pyroxene growths. This is called spinifex texture. These flows clearly crystallized from highly magnesian undifferentiated magma extruded (in the case of 35 wt % MgO) at up to 1650°C.

It has been suggested by Naldrett & Cabri that before a rock sequence is assigned to this suite certain criteria must be met. These are (a) that lavas must be present, (b) olivine-rich rocks must be present in the sequence, (c) spinifex textures must be present in some bodies. Geochemical criteria can also be used (see below). Spinifex textures resemble those of silica-poor slags, having a low viscosity and a high rate of internal diffusion—ideal conditions for the sinking of sulphide droplets to form accumulations at the flow bottom. It is very unlikely that the sulphides were in solution when rapid crystallization started or they would have been trapped before they could sink to the flow bottom. Therefore they were probably already in droplet form when the magma was erupted.

The syntectonic alpine-type intrusions which form the second group of orogenic bodies do not carry commercial nickel deposits. Alaskan-type complexes form concentrically zoned intrusions which are well developed along the Alaskan panhandle. As a group they are distinguished from alpine-type ultrabasics or stratiform intrusions by highly calcic clinopyroxene, no orthopyroxene or plagioclase, much hornblende, more iron-rich chromite, and magnetite. The latter occurs in concentrations that are occasionally of economic interest. Similar bodies occur in the Urals, south central British Columbia and Venezuela.

(B) BODIES EMPLACED IN A CRATONIC SETTING

There are three main groups to be noted. The first is an important metal producer because it is that of the large stratiform complexes. In the last chapter, we noted the importance of the Bushveld Complex and its by-product nickel-copper won from the platinum-rich horizons. The Sudbury intrusion also belongs to this group and it hosts the world's greatest known concentration of nickel ores. The Duluth and Stillwater Complexes may also become producing areas in the future. The overall composition of this group is basic rather than ultrabasic. A lower ultrabasic zone is, however, usually present. Sudbury is a notable exception to this rule but it possesses a sublayer rich in ultrabasic xenoliths probably derived from a hidden layered sequence.

The second group includes sills and sheets equivalent to flood basalts such as the Palisades Sill, the large Dufek intrusion, Antarctica and the small Insizwa intrusion of South Africa. This group is not yet known to host economic deposits but sub-economic mineralization is known at Insizwa.

Thirdly, there is a group of medium- and small-sized intrusions such as Skaergaard, Rhum and Noril'sk. The last named is a very important host of nickel-copper and platinoid mineralization.

Magma-types in bodies having associated nickel sulphide and platinoid mineralization

Naldrett & Cabri (1976) applied field and geochemical criteria to the host rocks of ores from the principal nickel fields to determine their magmatic affinities. The field criteria are shown in Table 9.2. From this table it is clear that e, f and h are

Table 9.2. Field criteria for recognition of ultrabasic and related rocks as komatiites. (Modified from Naldrett & Cabri 1976.)

Orefield	Criteria (see key) 1	2	3
a. Sudbury	–	–	–
b. Bushveld	–	+	–
c. Noril'sk	+	+	–
d. Pechenga	+	+	–
e. Eastern Goldfields, W.A.	+	+	+
f. Abitibi, Ontario	+	+	+
g. Ungava nickel belt	+	+	–
h. Manitoban nickel belt	+ (?)	+	+
i. Shangani (Zimbabwe-Rhodesia)	+	+	–

1 Should be part of a sequence of lavas and shallow intrusions.
2 Some representatives should be olivine-rich.
3 Spinifex texture present.
\+ Criteria fulfilled.
– Criteria not fulfilled.

komatiitic; c, d, g and i fulfil two criteria but lack spinifex texture. This is not an absolute requirement and it is therefore worth looking at some geochemical data. In Fig. 9.1 komatiitic rocks tend to plot to the left of the solid line and in Fig. 9.2 they lie in or close to the area outlined with a pecked line. The two plots confirm that e, f and h are komatiitic and show that g and i also belong to this magma-type.

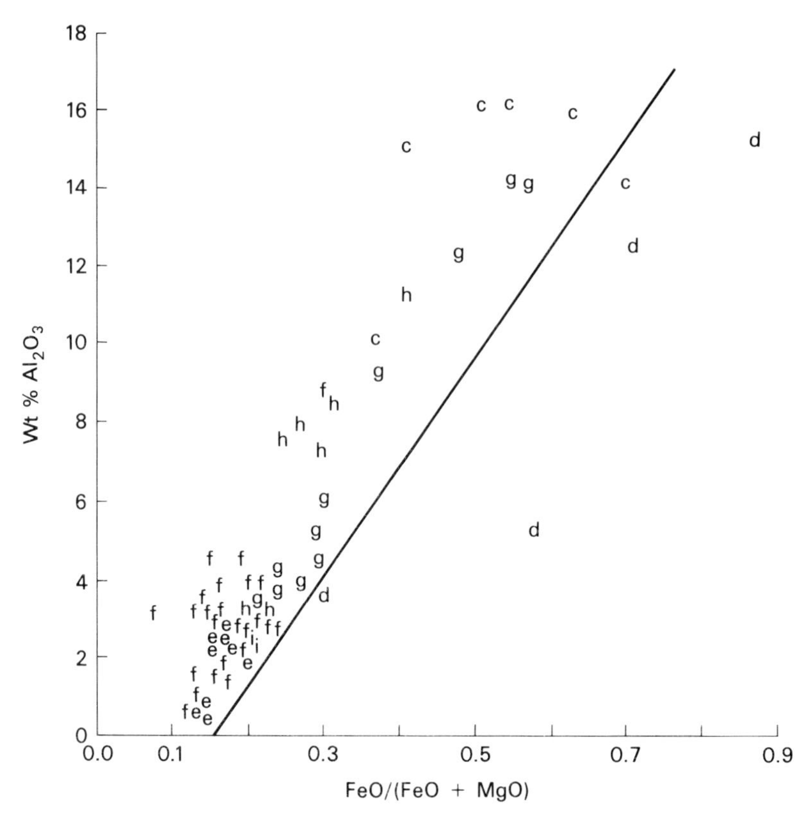

Fig. 9.1. Variation of Al_2O_3 with FeO/FeO + MgO in the host rocks of certain nickel sulphide orebodies. The key to the letters is given in Table 9.2. (After Naldrett & Cabri 1976.)

Pechenga, however, is out on both counts and Noril'sk is too high in TiO_2. These, with Sudbury, are tholeiitic.

Relationship of nickel sulphide mineralization to classes of ultrabasic and basic rocks

Fig. 9.3 indicates the relationship between known reserves plus past production of nickel in the main deposits of the world and the rock classes given in Table 9.1. Apart from the unique position of Sudbury, production from there being responsible for the entire Canadian section of classes 4 and 5, komatiitic magmatism [1(ii)] is clearly the most important. Tholeiitic volcanism is much less important. Deposits near the basal contacts of the Stillwater and Duluth complexes (class 4) are low grade disseminated deposits which are unlikely to be producers in the near future. Class 6 is represented by Noril'sk which, like Sudbury, has many unique features. Intrusions of classes 2, 3 and 7 are unlikely hosts for nickel sulphide orebodies on the basis of past experience.

Genesis of sulphur-rich magmas

For a rich concentration of magmatic sulphides it is necessary that (1) the host magma is saturated in sulphur, and (2) a reasonably high proportion of sulphide droplets can settle rapidly to form an orebody. Slow settling may give rise to a

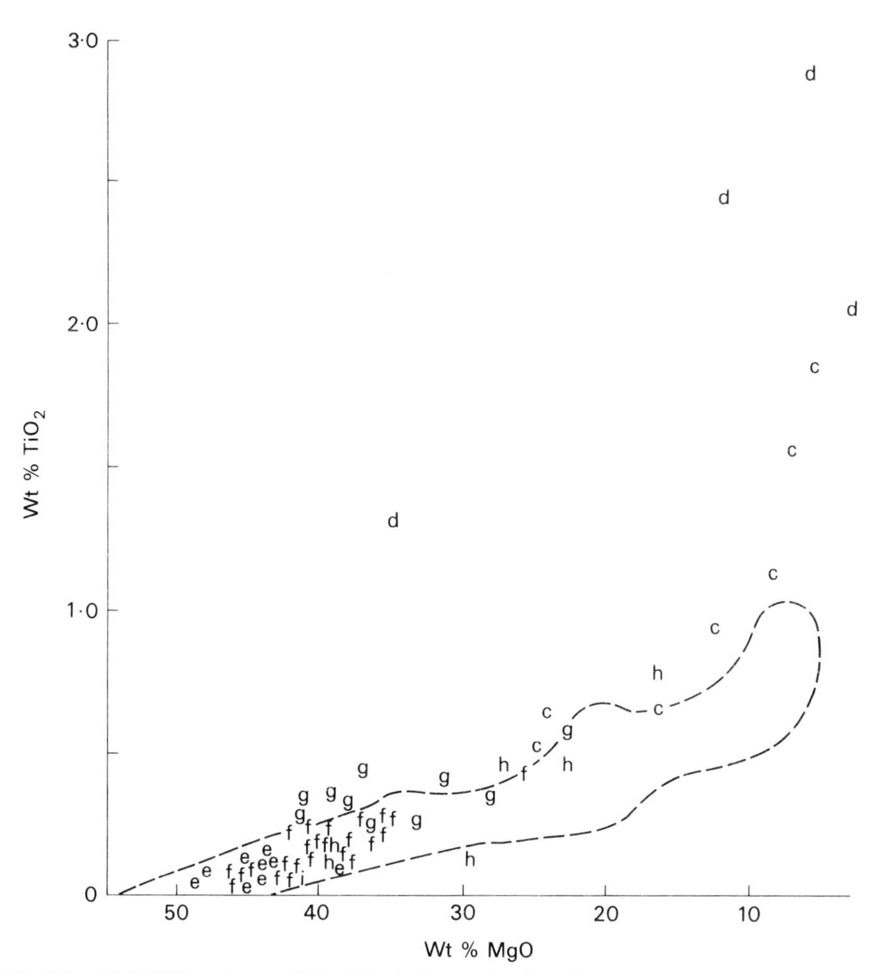

Fig. 9.2. Wt % TiO_2 against wt % MgO for the host rocks of certain nickel sulphide orebodies. The key to the letters is given in Table 9.2. (After Naldrett & Cabri 1976.)

disseminated, uneconomic ore. The production of a high proportion of immiscible sulphides is possible if the magma assimilates much sulphur from its country rocks, e.g. Duluth, Noril'sk, or if the magma carries excess amounts of mantle-derived sulphides, e.g. some komatiites. We can often differentiate between these two sources (crustal and mantle) by examining variations in $^{34}S/^{32}S$. These are reported in the delta notation (page 55). For this work the standard is troilite of the Cañon Diablo meteorite which is taken to be equivalent to the earth's mantle and for which $\delta^{34}S = 0$. Originally, all the earth's sulphur was in the mantle. Much of that transferred to the crust has undergone biogenic fractionation producing an enrichment in ^{34}S and giving $\delta^{34}S$ values as high as $+30$. It is now held to be an oversimplification to claim that for all natural sulphur a narrow range of $\delta^{34}S$ close to zero indicates a mantle source, and a wide range of $\delta^{34}S$ a crustal source. This statement is still probably true, however, for the majority of magmatic systems.

Let us look first at three examples where there is no evidence to suggest that the sulphur came from anywhere but the mantle. At Sudbury, $\delta^{34}S$ varies from -0.2 to $+5.9$ with a mode of 1.7. This is a narrow spread of values close to $\delta^{34}S = 0$ and

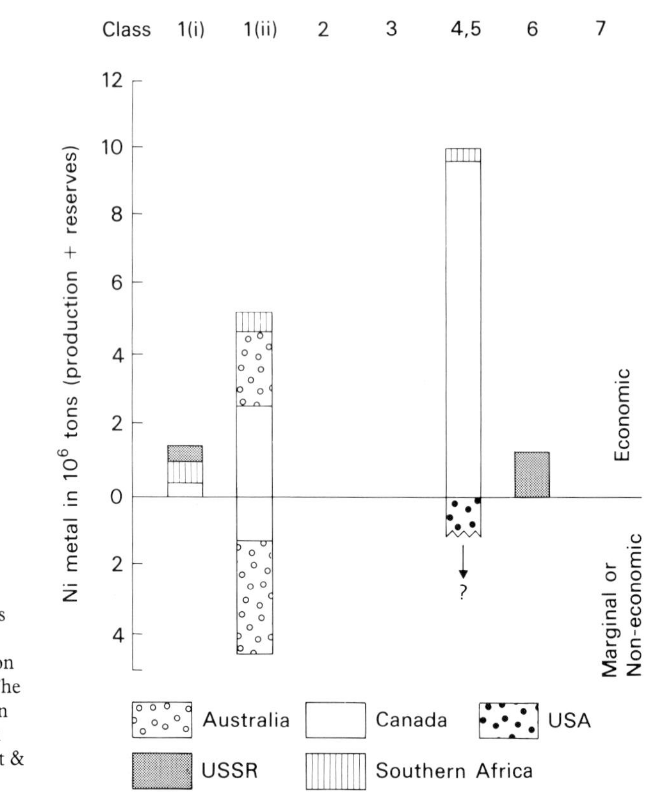

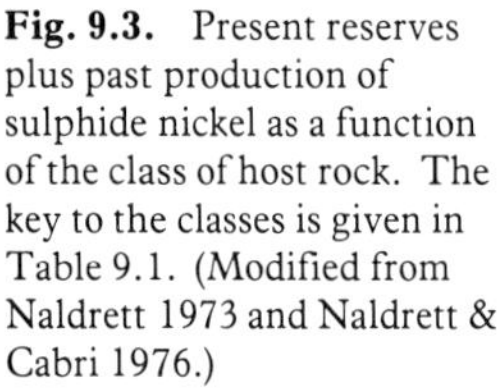

Fig. 9.3. Present reserves plus past production of sulphide nickel as a function of the class of host rock. The key to the classes is given in Table 9.1. (Modified from Naldrett 1973 and Naldrett & Cabri 1976.)

suggestive of a mantle origin. Similarly, ore associated with komatiites at the Alexo Mine, Ontario gives $\delta^{34}S$ = 4.4-6.2; again close to a zero value but perhaps showing a little temperature fractionation. At Noril'sk, the values are +7 to +17. The Noril'sk Gabbro is Triassic (an unusual and perhaps unique situation as practically all economic nickel mineralization is early to middle Precambrian). Its magma intruded through gypsum beds where it is thought to have picked up ^{34}S-rich sulphur. A comparable example is the Water Hen intrusion of the Duluth Complex with values of +11 to +16. This is believed to have gained its sulphur by assimilating sulphur-rich sediments of the Virginia Formation in which $\delta^{34}S$ ranges from +17 to +19 (Mainwaring & Naldrett 1977). Clearly, basic intrusions in sulphur-rich country rocks should be carefully prospected for nickel mineralization!

We must now consider the problem of how some mantle-derived magmas acquire a high sulphur content. This is discussed by Naldrett & Cabri (1976) with the aid of the relationships shown in Fig. 9.4 which shows the degree of melting of the mantle and the oceanic geothermal gradients for the present day and the Archaean. Consider mantle material at A, it is sufficiently above the solidus to yield a 5% partial melt. Any slight perturbation (such as a tectonic effect, accession of H_2O or magma from the descending slab of a Benioff Zone, normal convective overturn) would cause a diapir tens of kilometres across to rise adiabatically to B. Partial melting would increase to 30% and if this melt separated from the diapir it might rise to the surface along the non-adiabatic curve BF. The continued rise of the diapir would produce further partial melting but now the melt would form from

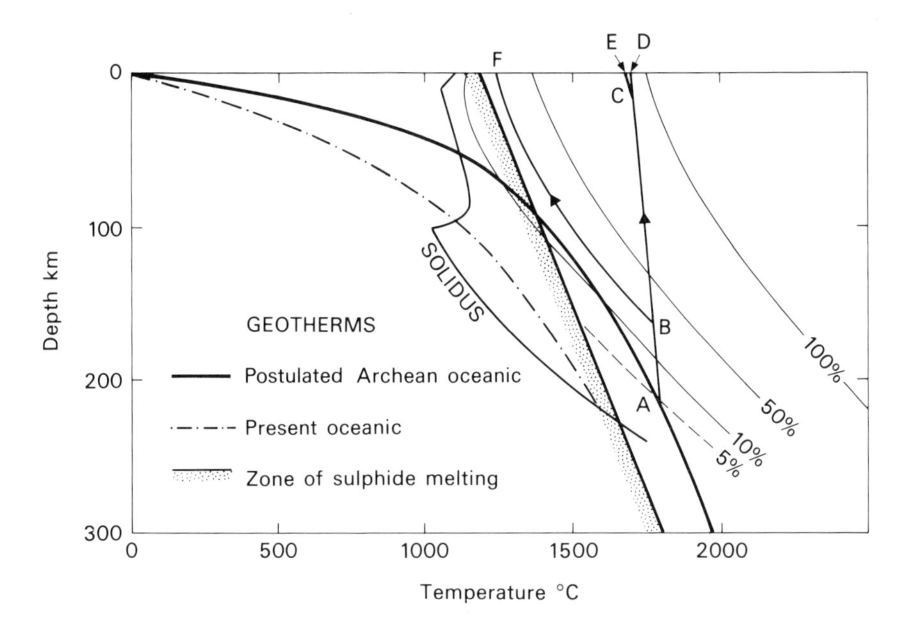

Fig. 9.4. A depth-temperature diagram showing the relationship between estimates of the modern and Archaean oceanic geotherms and melting relations of possible mantle material (pyrolite II + 0.2 wt % H_2O); drawn to illustrate the generation of komatiitic magma. (After Naldrett & Cabri 1976.)

depleted mantle and it would be much more magnesian since basaltic material has been removed. Partial melting might reach 30% at C (the contours hold for undepleted mantle only) to produce a much more magnesian (komatiitic) magma which might be intruded or extruded at E.

Fig. 9.4 shows that the zone of general sulphide melting intersects the Archaean geothermal gradient at about 100 km depth. Below this depth sulphides were molten and probably percolated downwards leaving a zone depleted in sulphides and producing a deeper enriched zone. This sulphur-enriched zone will lie at the level from which komatiitic magmas are ultimately derived. In this way, we can explain the relationship between komatiites and sulphide ores and the fact that they may be associated with tholeiitic picrites which themselves may carry sulphide ores as in the Abitibi region of Ontario.

Nickel mineralization in time and depletion of sulphur in the mantle

The only large nickel sulphide deposits which are younger than 1700 Ma are Duluth (1115 Ma) and Noril'sk (Triassic). As we have seen, both host intrusions probably acquired most of their sulphur by assimilation of crustal rocks. This restriction of orthomagmatic nickel deposits to the Archaean and early Proterozoic may reflect sulphur depletion of the upper mantle caused by many cycles of plate tectonic or similar processes.

Origin of the metals

There is no difficulty here. Ultrabasic and basic magmas are rich in iron and in trace amounts of copper, nickel and platinoid elements. These would be scavenged by the sulphur to form sulphide droplets. The reason why other chalcophile elements such as lead and zinc are not present in significant amounts in nickel sulphide ores has been explained by Shimazaki & MacLean (1976).

Examples of nickel sulphide orefields

THE VOLCANIC ASSOCIATION

Here we are mainly concerned with class 1(ii) (Table 9.1), the komatiitic suite. In some of these, sulphides occur at or near the base of the flow or sill suggesting gravitational settling of a sulphide liquid. Typical sections through two deposits are given in Fig. 9.5. These have certain features in common:

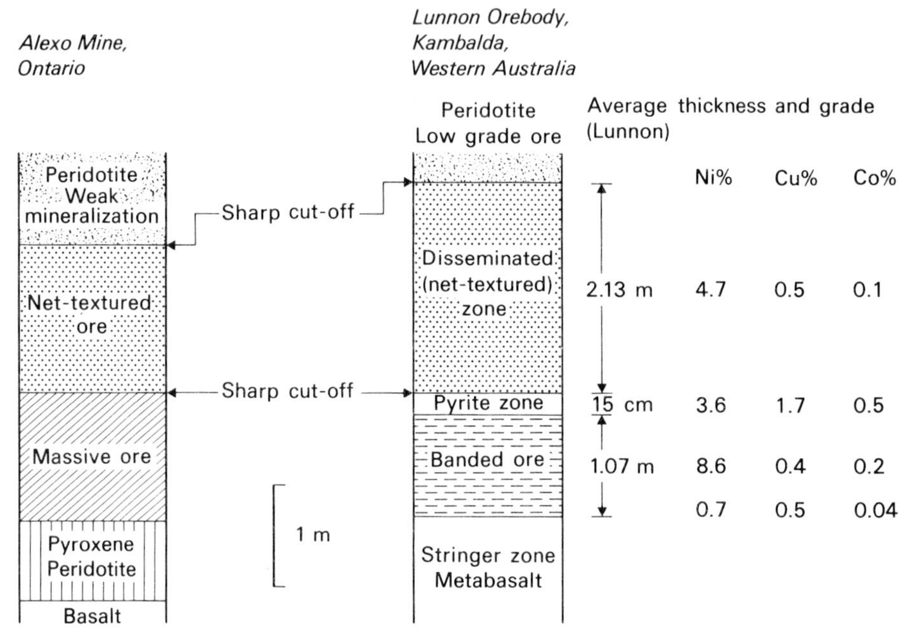

Fig. 9.5. Typical sections through two nickel sulphide orebodies associated with Archaean class 1(ii) ultrabasic bodies. The Alexo Mine is 40 km east-north-east of Timmins, Ontario. (After Naldrett 1973.)

(1) massive ore at the base (the banding in the Lunnon orebody is probably the result of metamorphism);

(2) a sharp contact between the massive ore and the overlying disseminated ore which consists of net-textured sulphides in peridotite;

(3) another sharp contact between the net-textured ore and the weak mineralization above it which grades up into peridotite with a very low sulphur content.

This situation is strongly reminiscent of the billiard ball model of sulphide segregation described on page 34.

The Eastern Goldfields Province of Western Australia is a typical Archaean region having a considerable development of greenstone belts (Fig. 9.6). The nickel sulphide deposits are of two main types. The first consists of segregations of massive and disseminated ores at the base of small lenslike peridotitic to dunitic flows or subvolcanic sills at the bottom of thick sequences of komatiitic flows, e.g. Kambalda, Windarra, Nepean, Scotia (Fig. 9.6). These are termed volcanic-type deposits (Barrett *et al.* 1977). The second type, dykelike deposits, occur in largely

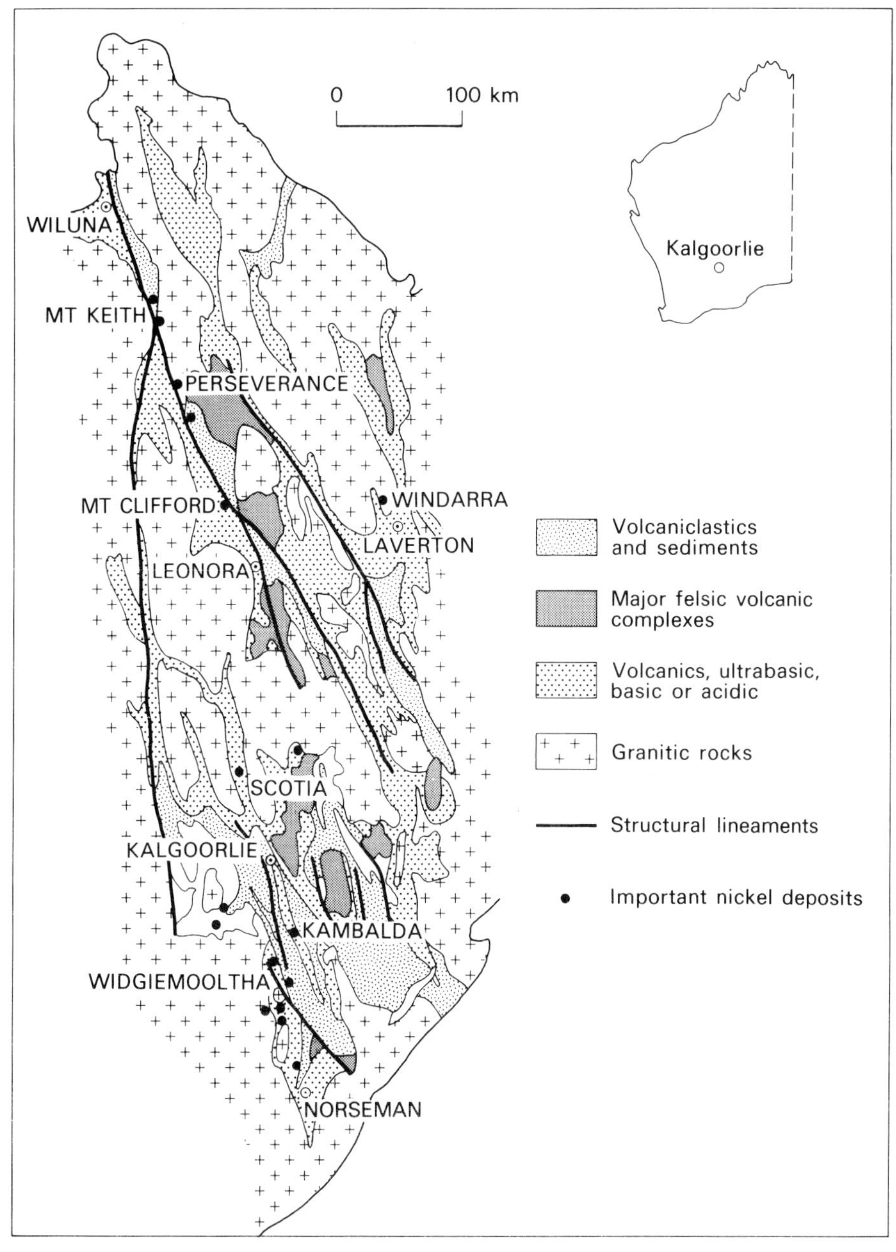

Fig. 9.6. Generalized geological map of the Eastern Goldfields Province of the Yilgarn Block showing some of the important nickel deposits. (Modified from Gee 1975.)

concordant but partially discordant dunitic intrusions emplaced in narrow zones up to several hundred kilometres in length, e.g. Perseverance, Mount Keith.

The volcanic-type deposits are commonly clustered around the periphery of granitoid periclines as at Kambalda (Fig. 9.7). These periclines trend north-north-westerly. Most ores are at, or close to, the basal contact of the mineralized ultrabasic sequence (Fig. 9.8) and they are commonly associated with, or even confined to

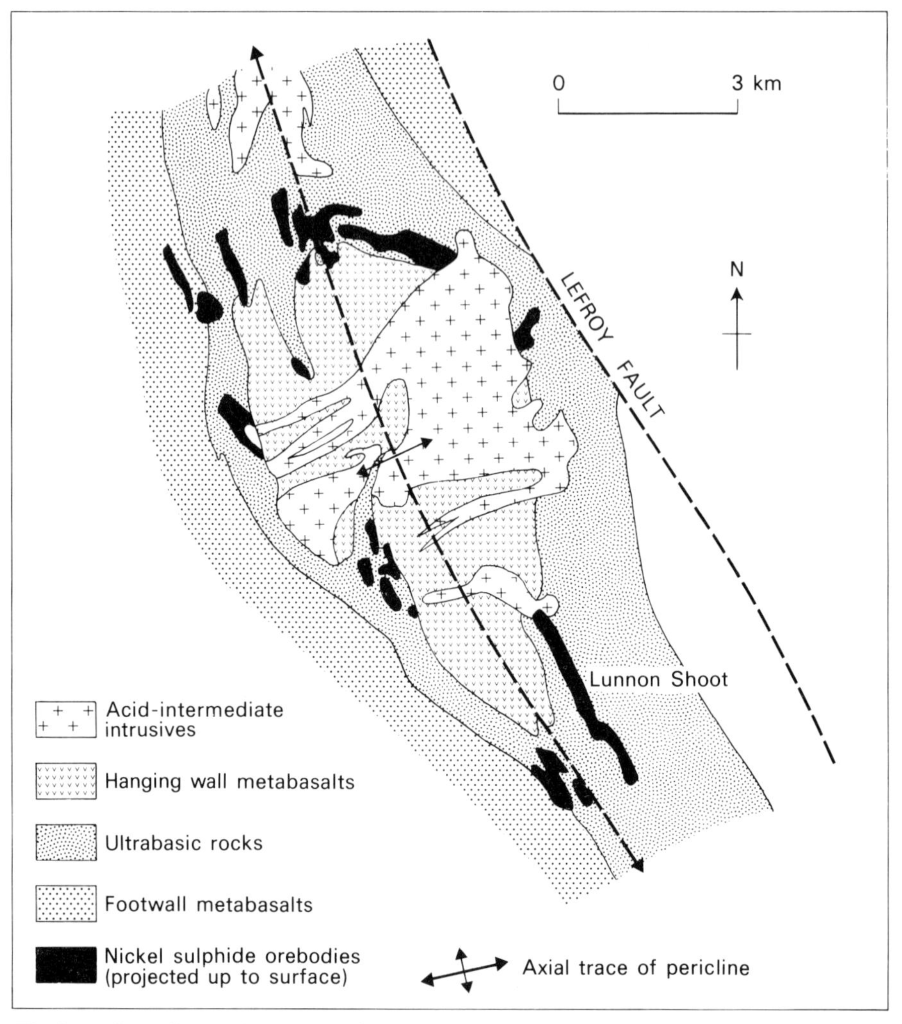

Fig. 9.7. Generalized geological map of the Kambalda Dome showing the positions of the sulphide orebodies.

embayments in the footwall. The orebodies are essentially tabular with their greatest elongation subparallel to the penetrative linear fabrics in the enclosing rocks and/or the trend and plunge of both regional and parasitic folds. The ores have been metamorphosed and complexly deformed and are only developed in amphibolite facies metamorphic domains. Barrett *et al.* (1977), whilst conceding that the initial concentration must have been by liquation, suggested that considerable further concentration occurred during the metamorphic episode. Their objections to Naldrett's billiard ball model are worthy of note.

The dykelike deposits are much larger than the volcanic-type, for example, at Perseverance 33 Mt averaging 2.2% nickel with a cut-off grade of 1% have been outlined and at Mount Keith 290 Mt grading 0.6%. They occur in long dunite dykes especially where these bulge out to thicknesses of several hundred metres, e.g. at Perseverance the host dyke thickens from a few metres to 700 (Fig. 9.9). The orebodies generally appear to be associated with areas of considerable serpentiniza-

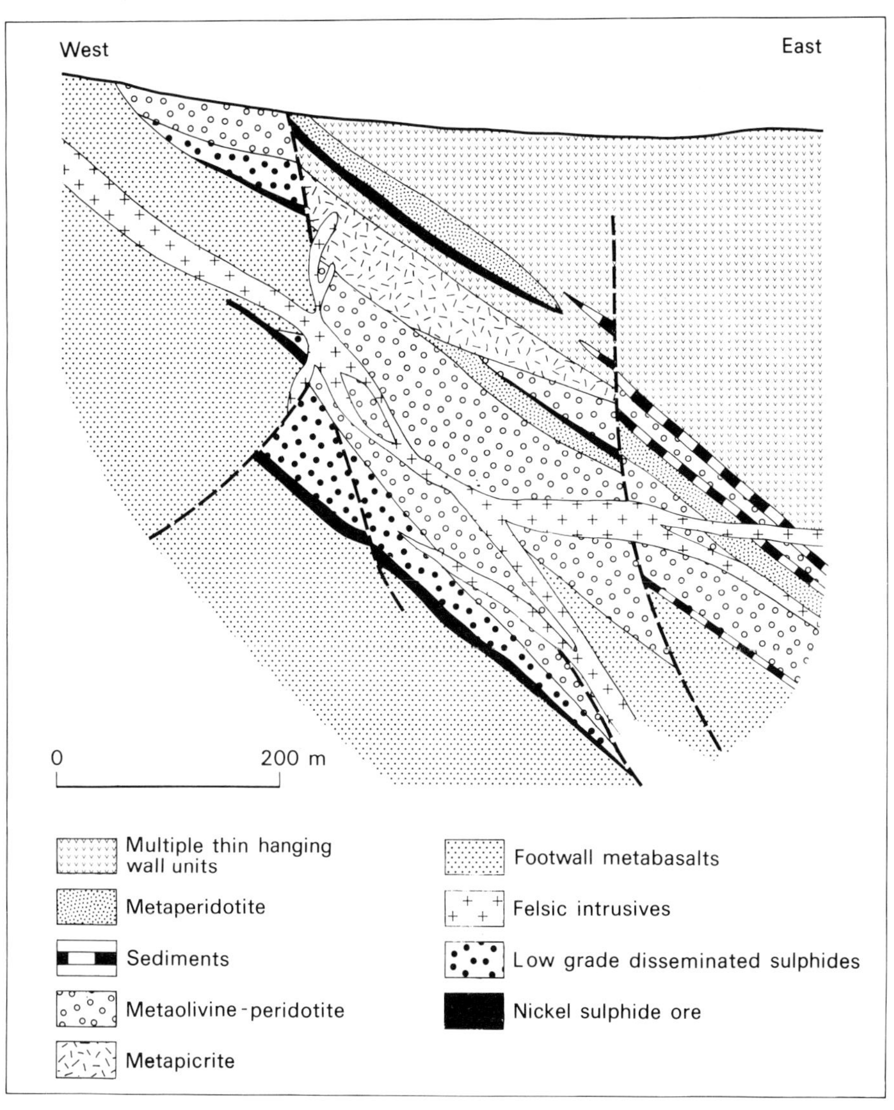

Fig. 9.8. Section through the Lunnon and neighbouring ore shoots, Kambalda, Western Australia. (Modified from Ross & Hopkins 1975.)

tion during which enrichment of the ores appears to have occurred. The ores are dominantly of disseminated type though some massive sulphides occur as at Perseverance.

PLUTONIC ASSOCIATION

(a) *Sudbury, Ontario.* The Sudbury Basin lies just north of Lake Huron near the boundary of the Superior and Grenville Provinces of the Canadian Shield. The basin is 60 × 27 km (Fig. 9.10). Its most obvious feature is the so-called Nickel Irruptive, which consists of a lower (outer) layer of augite-norite and an upper layer of granophyre. There is a transition zone of quartz-gabbro between the two (Naldrett *et al.* 1970). The Irruptive is believed to have the shape of an irregular

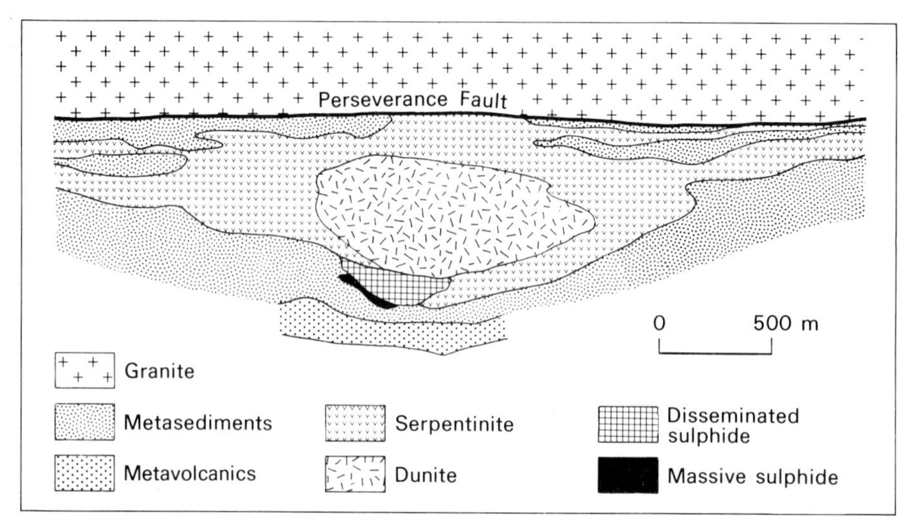

Fig. 9.9. Generalized geology of the Perserverance deposit. Many faults have been omitted. (Modified from Martin & Allchurch 1975.)

lopolith (Popelar 1972). At the base of the norite there is a discontinuous zone of inclusion- and sulphide-rich norite intrusions known as the sublayer. In the so-called offsets (steep to vertical dykes which appear to penetrate downwards into the footwall from the base of the main norite), the inclusion-rich, sulphide-bearing rock is a quartz-diorite. The sublayer and offsets are at present the world's largest source of nickel as well as an important source of copper, cobalt, iron, platinum and eleven other elements.

Inside the Irruptive is the Basin *sensu stricto* consisting of a volcaniclastic-like sequence (Onaping Formation), a manganese-rich slate sequence (Onwatin Formation) and the Chelmsford Formation—a carbonaceous and arenaceous proximal turbidite. These formations form the Whitewater Group which has only suffered slate grade regional metamorphism. No rocks which can be correlated with the Whitewater Group have been positively identified outside the basin.

To the south-east of the Irruptive there are metasedimentary and metavolcanic rocks of the Huronian Supergroup that were deposited unconformably upon the migmatitic Archaean basement which is exposed along the north-western side of the Irruptive. The Huronian has suffered a much higher grade of metamorphism than the Whitewater Group and it contains a number of basic and acid plutonic intrusions. It forms a southward facing homocline and carries a penetrative foliation caused by flattening. The rocks of the Basin are also considerably deformed (Brocoum & Dalziel 1974) and finite strain analysis suggests that the Chelmsford Formation was shortened by 30% in a north-westerly direction; this means that the Chelmsford Formation and probably the Sudbury Basin were almost circular prior to deformation.

The origin of the Sudbury Basin and the structure along which the Irruptive was intruded have been debated since the area was first mapped about the turn of the century. There is now an important school of thought initiated by Dietz (1964) that regards the Basin as having resulted from a meteoritic impact. Shatter cones are present in the surrounding rocks, shock metamorphic effects can be seen in the

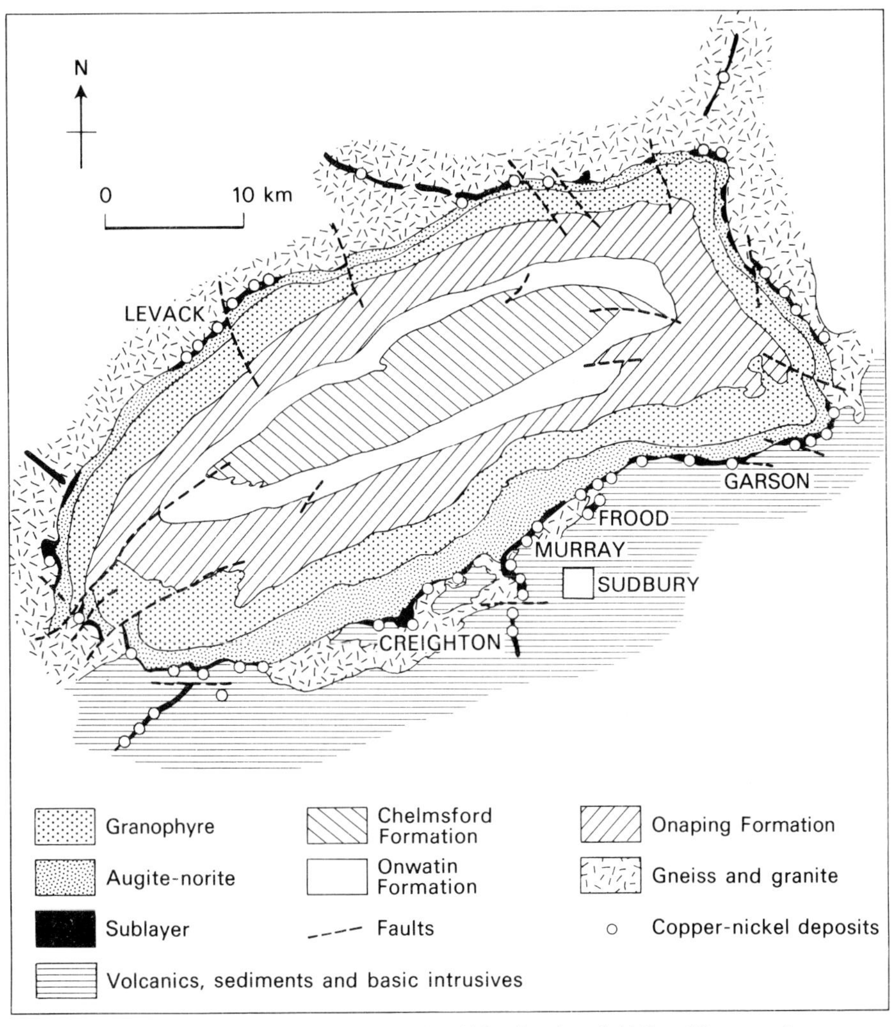

Fig. 9.10. Geological map of the Sudbury district. (After Souch *et al.* 1969 and Brocoum & Dalziel 1974.)

base of the Onaping Formation and in the Archaean rocks north of the Basin. The Onaping Formation is regarded by this school as being a fall-back breccia resulting from the impact. Geologists arguing against the meteorite impact hypothesis interpret the Onaping Formation as an ignimbrite deposit (Stevenson 1972), and Stevenson (1961) mapped a quartzite unit believed to be a basal quartzite lying beneath the Onaping Formation. This school generally regards the Irruptive as having been intruded along an unconformity at the base of the Whitewater Group. Another controversial rock-type is the 'Sudbury Breccia'. This consists of zones, a few centimetres to several kilometres across, of brecciated country rocks which are sometimes hosts for the orebodies.

The discovery of mineralization during the construction of the Canadian Pacific Railway in 1883 has led to the development of over forty mines and the total declared ore reserves of the district from the time of the original discovery to the

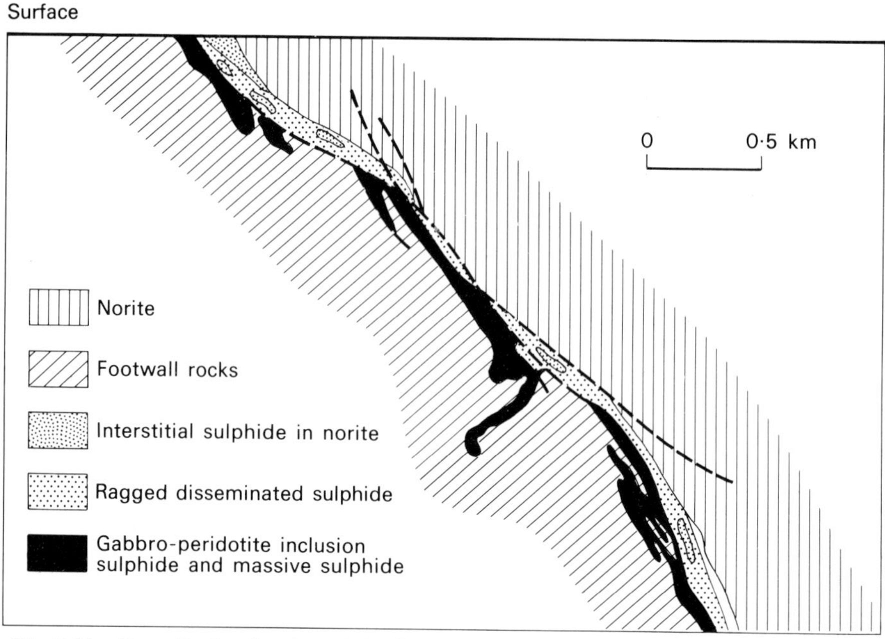

Fig. 9.11. Generalized section through the Creighton ore zone looking west. (After Souch *et al.* 1969.)

present are of the order of 930 Mt. Of these, about 500 Mt have been exploited but new reserves are constantly being blocked out, just about keeping pace with production. According to Lang (1970), the grade of ores worked in the past was about 3.5% nickel and 2% copper. Today, with large-scale mining methods, the grade worked is around 1% for both metals.

The orebodies occur in the sublayer which is clearly a later intrusion than the overlying norite with which it has a sharp contact. The sublayer magma was rich in sulphides and inclusions of peridotite, pyroxenite and gabbro. The sulphides tended to sink into synclinal embayments in the footwall giving a structural control of the mineralization. The Creighton ore zone has the greatest number of ore varieties (Souch *et al.* 1969). It plunges north-westward down a trough at the base of the Irruptive for at least 3 km (Fig. 9.11) and consists of a series of ore-types. The hanging wall quartz-norite above the sublayer occasionally contains enough interstitial sulphide to form low grade ore. In the upper part of the sublayer, ragged disseminated sulphide ore occurs. It consists of closely packed inclusions (several millimetres to ten centimetres in size) in a matrix of sulphides and subordinate norite. The sulphide content increases downwards as does the ratio of matrix to inclusions, with a concomitant increase in inclusion size up to one metre, resulting in an ore called gabbro-peridotite inclusion sulphide. This ore changes towards the footwall into massive sulphide containing fragments of footwall rocks. It is called inclusion massive sulphide and it is discontinuous and also forms stringers and pods in the footwall.

The Frood-Stobie orebody is an example of an orebody in an offset dyke. This parallels the footwall of the Irruptive and it has been suggested that it may once have been continuous with the Irruptive at a higher level. It is a huge orebody,

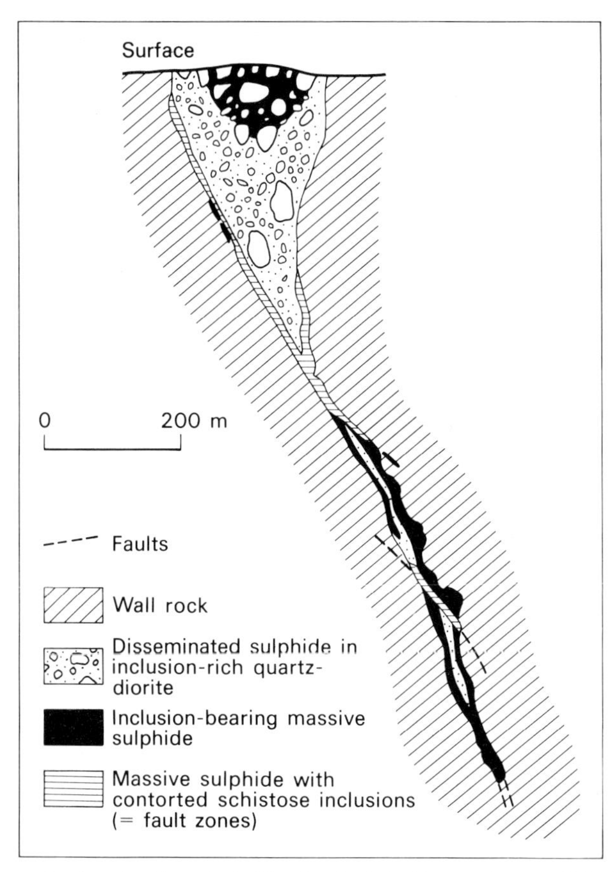

Fig. 9.12. Generalized section through the Frood orebody looking south-west. (After Souch *et al.* 1969.)

1.3 km long, 1 km deep and nearly 300 m across at its widest point. It consists of a wedge-shaped body of inclusion-bearing quartz-diorite (Fig. 9.12) with disseminated sulphide partially sheathed by inclusion massive sulphide. In the lower half, the ore described by Hawley (1962) as immiscible-silicate-sulphide ore occurs (Fig. 4.1). This ore-type grades into massive ore outwards and downwards. Inclusions in the quartz-diorite vary from a few centimetres to many metres in length. The largest found was one of peridotite 45 metres long!

(b) *The Noril'sk USSR deposits.* This region has the largest reserves of copper-nickel ore in the Soviet Union. It lies deep in northern Siberia near the mouth of the Yenisei River. The country rocks are carbonates and argillaceous sediments of the early and middle Palaeozoic overlain by Carboniferous rocks with coals, Permian and a Triassic basic volcanic sequence. The associated gabbroic intrusions form sheets, irregular masses and trough-shaped intrusions depending on their location in the gentle folds of the sediments lying beneath the volcanic sequence (Glazkovsky *et al.* 1977).

The Noril'sk I deposit occurs in a differentiated layered dominantly gabbroic intrusion which extends northwards for 12 km and is 30-350 m thick. In cross section it is lensoid with steep sides (Fig. 9.13). The copper-nickel sulphides form

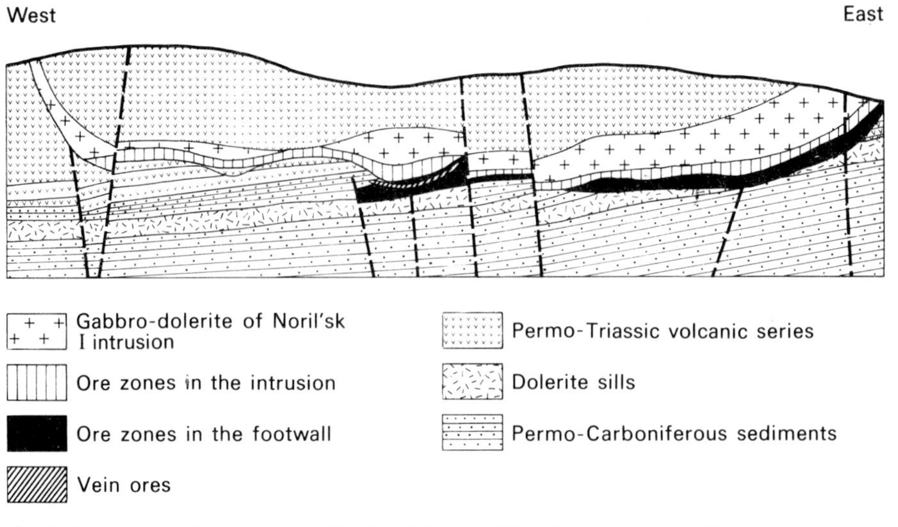

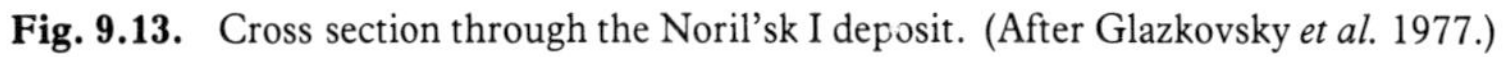

Fig. 9.13. Cross section through the Noril'sk I deposit. (After Glazkovsky *et al.* 1977.)

breccia, disseminated and massive ores at the base of the intrusion and vein orebodies developed in the footwall rocks and the basal portion of the intrusion. Like Sudbury there is a high Cu/Ni ratio and (Pt + Pd):Ni = 1:500.

10

The Carbonatite Environment

Carbonatites

Carbonatite complexes consist of intrusive magmatic carbonates and associated alkaline igneous rocks. They belong to alkaline igneous provinces and are generally found in stable cratonic regions with major rift faulting such as the East African Rift Valley and the St. Lawrence River Graben. They range in age from Proterozoic to Recent. There are exceptions where carbonatite complexes are not directly associated with any alkalic rocks, e.g. Sangu Complex, Tanzania; Kaluwe, Zambia, and not all alkalic rock provinces and complexes have associated carbonatites. Verwoerd (1964) pointed out that about 90% of carbonatites occur in areas of Precambrian granite or gneiss. However, Barker (1969) has shown that in North America 66% of known alkaline rock occurrences are in Phanerozoic orogenic belts. There is no doubt that, taking a world-wide view, the majority of carbonatites occur in the marginal parts of stable cratonic regions or are closely related to large-scale rift structures. An apparent exception of economic importance is the Mountain Pass carbonatite of California. This occurs within the Rocky Mountain area, but it is a Precambrian intrusion into Precambrian gneisses so that its present tectonic environment is a function of considerably later deformation (Moore 1973), and this may be the reason why it is not now in an obviously alkaline igneous province.

Carbonatites together with ijolites and other alkalic rocks commonly form the plutonic complex underlying volcanoes which erupted nephelinitic lavas and pyroclastics. Surrounding such complexes there is a zone of fenitization (mainly of a potassium nature and producing orthoclasites) which has a variable width. The emplacement of carbonatites was effected in stages (Le Bas 1977). Often the dominant rock is made up of early C_1 (sövite) intrusions. This is usually emplaced in an envelope of explosively brecciated rocks. C_1 carbonatite is principally calcite with apatite, pyrochlore, magnetite, biotite and aegirine-augite. C_2 carbonatite (alvikite) usually shows marked flow banding and is medium- to fine-grained compared with the coarse-grained sövite. C_3 carbonatite (ferrocarbonatite) contains essential iron-bearing carbonate minerals and commonly some rare-earth and radioactive minerals. C_4 carbonatites (late stage alvikites) are usually barren. Sövites form penetrative stock-like intrusions and C_2 and C_3 carbonatites form cone-sheets and dykes. The intrusion of C_1 and most C_2 carbonatites is preceded by intense fenitization; other carbonatites show little or no fenitization.

Economic aspects

The most important products of carbonatites are phosphorus (from apatite), niobium (pyrochlore) and rare-earth elements (monazite, bastnäsite). To date, only one carbonatite complex is a major producer of copper (Palabora, South Africa), but others are known to contain traces of copper mineralization, e.g. Glenover, South

Africa; Callandar Bay, Canada and Sulphide Queen, Mountain Pass, USA (Jacobsen 1975). Palabora also produces uranium, thorium, zirconia and magnetite, whilst the ultramafic rocks of the complex contain economic deposits of apatite and vermiculite. Other economic minerals in carbonatites include fluorite, baryte and strontianite and the carbonatites themselves are useful as a source of lime in areas devoid of good limestones.

THE MOUNTAIN PASS OCCURRENCES, CALIFORNIA

These lie in a belt about 10 km long and 2.5 km wide. One of the deposits, the Sulphide Queen carbonate body, carries the world's greatest concentration of rare-earth minerals. The metamorphic Precambrian country rocks have been intruded by potash-rich igneous rocks and the rare-earth-bearing carbonate rocks are spatially and probably genetically related to these granites, syenites and shonkinites (Olsen *et al.* 1954). The rare-earth elements are carried by bastnäsite and parisite. These minerals are in veins that are most abundant in and near the largest shonkinite-syenite body. Most of the 200 veins that have been mapped are less than 2 m thick. One mass of carbonate rock, however, is about 200 m in maximum width and about 730 m long. This is claimed to be the largest known orebody of rare-earth minerals in the world. It is called the Sulphide Queen Mine, not because of a high sulphide content, but because it is situated in Sulphide Queen Hill.

Carbonate minerals make up about 60% of the veins and the large carbonate body, they are chiefly calcite, dolomite, ankerite and siderite. The other constituents are baryte, bastnäsite, parisite, quartz and variable small quantities of 23 other minerals. The rare-earth content of much of the orebody is 5-15%.

THE PALABORA IGNEOUS COMPLEX

This lies in the Archaean of the north-eastern Transvaal. It resulted from an alkaline intrusive activity in which there were emplaced in successive stages pyroxenite, syenite and ultrabasic pegmatoids (Palabora Mining Co. Staff 1976). The first intrusion was that of a micaceous pyroxenite, kidney-shaped in outcrop (but forming a pipe in depth) and about 6 × 2.5 km. Ultrabasic pegmatoids were then developed at three centres within the pyroxenite pipe. In the central one, foskorite (magnetite-olivine-apatite rock) and banded carbonatite were emplaced to form the Loolekop carbonate-foskorite pipe which is about 1.4 × 0.8 km (Fig. 10.1). Fracturing of this pipe led to the intrusion of a dykelike body of transgressive carbonatite and the development of a stockwork of carbonatite veinlets. The zone along which the main body of transgressive carbonate was emplaced suffered repeated fracturing and mineralizing fluids migrated along it depositing copper sulphides which healed the fine discontinuous fractures. These near-vertical veinlets occur in parallel-trending zones up to 10 m wide, although individually the veinlets are usually less than 1 cm wide and do not continue for more than 1 m. Diamond drilling has shown that the orebody continues beyond 1000 m below the surface. Initial ore reserve estimates were of several million tons grading about 0.7% copper.

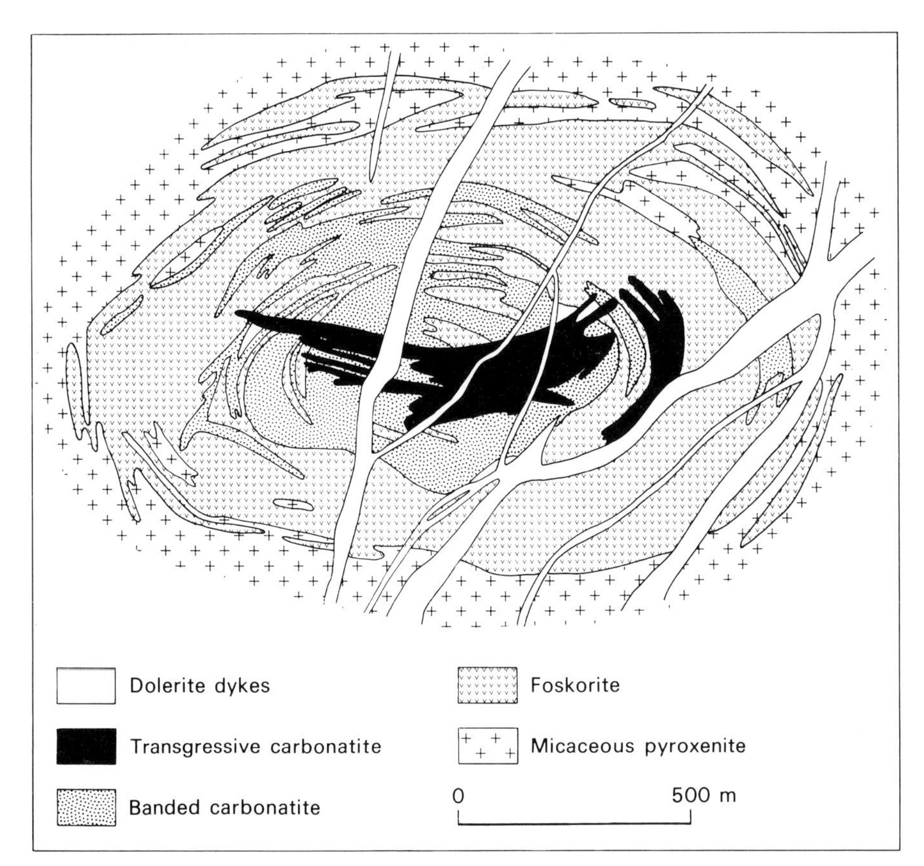

Fig. 10.1. Geology of the 122 m level, Loolekop Carbonatite Complex, Palabora, South Africa. (Modified from Jacobsen 1975.)

11

Pyrometasomatic Deposits

Some of the important features of these deposits have already been mentioned, these features, and others, are summarized in Table 11.1. The irregular morphology of their orebodies and the chief elements won from them have been discussed on page 10. On pages 42 to 43 the reader will find a short description of their mineralogy followed by a discussion of their genesis, where it was pointed out that experimental work indicated that dissolution of the host marble and precipitation of sulphide could be effected by the oxidation of bisulphide solutions at 400-450°C. Such high or higher temperatures are indicated by the nature of the gangue minerals and the work of Milovskiy *et al.* (1978). Working on the scheelite-molybdenite-chalcopyrite-bearing skarns of Chorukh-Dayron, they found from a study of fluid inclusions that ore deposition began with the crystallization of scheelite from highly

Table 11.1. General characteristics of pyrometasomatic deposits.

Depth of formation	Variable, one to several km
Temperature of formation	From 350-800°C
Occurrence	Adjacent to or partially inside deep-seated intrusive rocks. Often emplaced in carbonate rocks, occasionally in hornfelses, schists or gneisses
Nature of ore zones	Extremely irregular, tongues of ore may project along any available planar structure—bedding, joints, faults, etc. Distribution within the contact aureole is often apparently capricious. Structural changes may cause abrupt termination of the orebodies
Ores of	Fe, Cu, W, C, Zn, Pb, Mo, Sn, etc.
Ore minerals	*Magnetite*, specularite, *graphite*, gold, chalcopyrite, *pyrrhotite*, *scheelite*, *wolframite*, galena, sphalerite, pyrite, molybdenite, cassiterite
Gangue minerals	High temperature skarn minerals, e.g. grossularite, hedenburgite, idocrase, epidote, actinolite, wollastonite, diopside, forsterite, anorthite, etc. Quartz if present crystallized as high form. Carbonates
Wall rock alteration	Widespread development of skarn, tactite and/or marble
Textures and structures	Usually coarse-grained, may be banded due to replacement of bedded limestones
Zoning	Paragenetic sequence is usually: silicates; scheelite, magnetite, cassiterite; base-metal sulphides. These groups often show a zonal arrangement—sulphides outermost. Pb and Zn usually persist to greater distances from the contact than does Cu. Barren silicate zones may be present next to the intrusive
Examples	Fe of Marmora, Ontario; Cornwall, Penn.; Cu of Mackay District, Idaho; Concepciòn del Oro district, Mexico; Zn of Oslo area, Norway; W of Gold Hill, Utah; King Is. Scheelite, Tasmania; Sn of Kramat Pulai, Malaya

concentrated solutions (50-60 wt % of salts) at temperatures of 475-400°C and pressures of the order of 161 600 kPa. Metamorphic studies indicate that the previously crystallized calc-silicate gangue in pyrometasomatic deposits formed at temperatures approaching those of the associated magmatic intrusion.

Pyrometasomatic deposits and their host rocks usually show a zoning of both the silicate and ore minerals. This has been well documented for the magnetite deposits of Cornwall, Pennsylvania, by Lapham (1968), and Theodore (1977) has discussed the zoning of skarn deposits associated with porphyry copper deposits. One of his examples is reproduced in Fig. 11.1. The mineral zones in the Ely, Nevada, cupriferous skarns generally parallel the igneous contact. The bulk of the copper in the skarn was deposited in veinlets which cut the andradite-diopside rocks. The alteration envelopes along these veins contain actinolite-calcite-quartz-nontronite assemblages. These relationships indicate that a clay-sulphide stage was superposed on the earlier calc-silicate rocks.

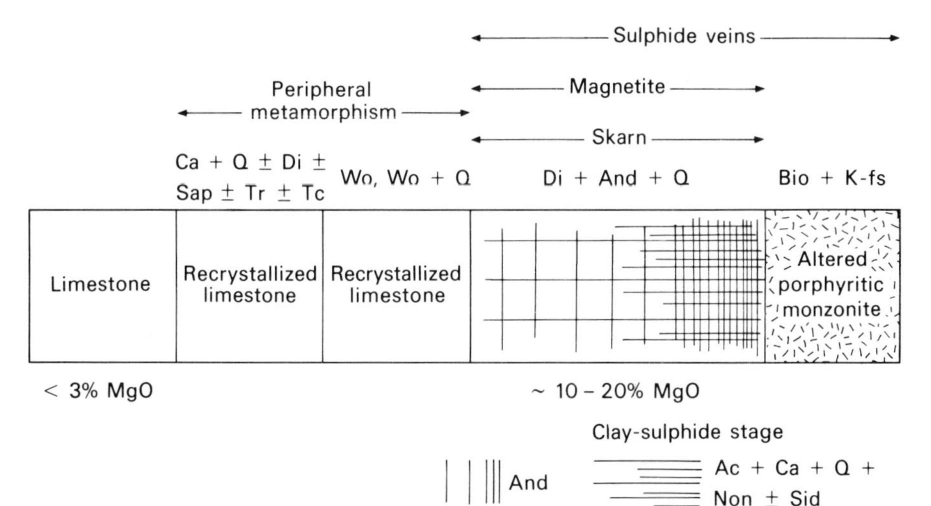

Fig. 11.1. Zonation of mineral assemblages at the Ely, Nevada, deposit. The width of the zones is schematic and the density of the patterns indicates the relative amounts of andradite (And) and the clay-sulphide stage. Abbreviations: Ac, actinolite; And, andradite; Bio, secondary biotite; Ca, calcite; Di, diopside; K-fs, secondary potash feldspar; Non, nontronite; Q, quartz; Sap, saponite; Sid, siderite; Tc, talc; Tr, tremolite, and Wo, wollastonite. (Modified from Theodore 1977.)

Some examples of pyrometasomatic deposits

COPPER CANYON, NEVADA

At this deposit, skarn has replaced calcareous shale or argillite beds just above the Golconda Thrust (Fig. 11.2) producing a flat-lying tabular zone of andradite-rich rock in which most of the mineralization occurred. Although the andradite rock stretches at least 400 m from the granite porphyry stock, only that part within 180 m of the contact contains ore. Silicate zones in this deposit are symmetrical about the andradite rock thus forming zones at right angles to the present igneous contact. For this and other reasons, Theodore (1977) suggests that the porphyry or the present site of the porphyry was not the locus from which the skarn-forming fluids emanated.

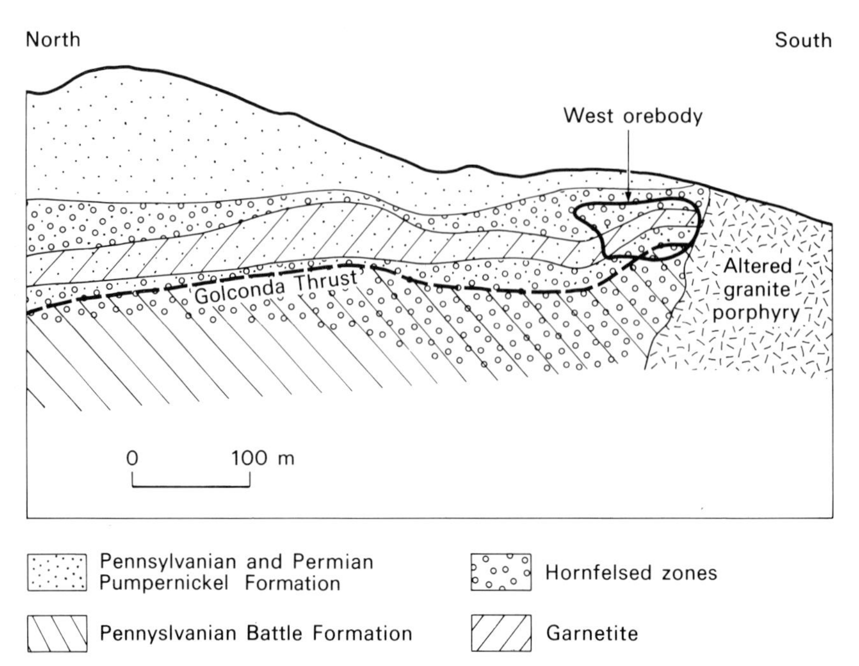

Fig. 11.2. Cross section through the copper-gold-silver pyrometasomatic deposit at Copper Canyon, Nevada. Rocks peripheral to the hornfelsed zones carry secondary mica.
(Modified from Theodore 1977.)

THE MEMÉ MINE, NORTHERN HAITI

Frequently at the contacts of skarns and intrusions there is a completely gradational contact and this is the case at the Memé copper mine where a large block of Cretaceous limestone has been surrounded by quartz-monzonite. Mineralization was preceded by extensive magmatic assimilation that formed zones of syenodiorite and granodiorite around the limestone. Following the crystallization of the magma, the limestone and neighbouring parts of the intrusion were replaced by skarn. The intrusion-derived skarn contains large quantities of diopside which distinguishes it from the marble-derived skarn. Both skarn types have a completely gradational contact (Kesler 1968).

Mineralization followed skarn formation and consisted of the introduction of hematite, magnetite, pyrite, molybdenite, chalcopyrite, bornite, chalcocite and digenite, in that paragenetic order. These occur as replacement zones. The main skarn and ore development is along the lower contact with the limestone block (Fig. 11.3). Skarn formation took place at between 480 and 640°C. Exsolution textures suggest that the minimum temperature of copper-iron sulphide deposition exceeded 350°C and the youngest ore minerals crystallized above 250°C. The grade is about 2.5% Cu.

THE KING ISLAND SCHEELITE DEPOSITS

King Island, which lies between Australia and Tasmania at the western approach to Bass Strait contains a number of important tungsten deposits. These are scheelite-bearing skarns developed in Upper Proterozoic to Lower Cambrian sediments which have been intruded by a granodiorite and an adamellite of lower Carboniferous age. These granites are thought to be related to the tin- and tungsten-bearing

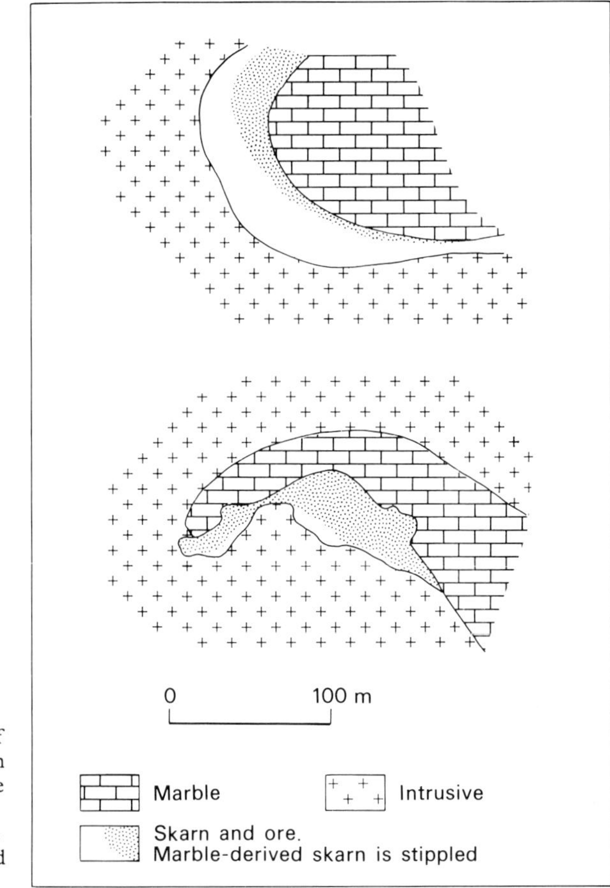

Fig. 11.3. Geological map of the 1500 ft level (above) and an east-west section (below) of the Memé Mine, Haiti. Note the concentration of skarn and ore beneath the marble. (Modified from Kesler 1968.)

granites of the Aberfoyle and Storey's Creek district of Tasmania (Danielson 1975). Up to 1972 5.67 Mt averaging 0.53% WO_3 had been mined and reserves at 1975 were put at 7 Mt averaging 0.75%.

Andradite skarns were formed by selective replacement of limestone beds. In places, these carry scheelite grains in sufficient quantity to form orebodies. In other places, the skarn is mineralized but below ore grade. Irregular relicts of marble occur in the skarn demonstrating its metasomatic origin. A section through the Bold Head orebodies is given in Fig. 11.4.

THE MARY KATHLEEN URANIUM DEPOSIT, NORTHERN QUEENSLAND

This occurs in Lower Proterozoic rocks which form a north-south trending synclinal block about 3 km wide bordered on both sides and probably underlain by granite. The syncline consists of metamorphosed lime-rich sediments and quartzites. The western granite is gneissic, the eastern granite is a massive, clearly intrusive granodiorite showing metasomatism along its contact. It is thought to be the origin of the mineralizing solutions (Hughes & Munro 1965). The main host rock of the orebodies is a metamorphosed conglomerate consisting of clasts of quartzite and feldspathic rocks in a fine-grained matrix of diopside and feldspar; orebodies also occur in scapolite-diopside granulites above and below the metaconglomerate.

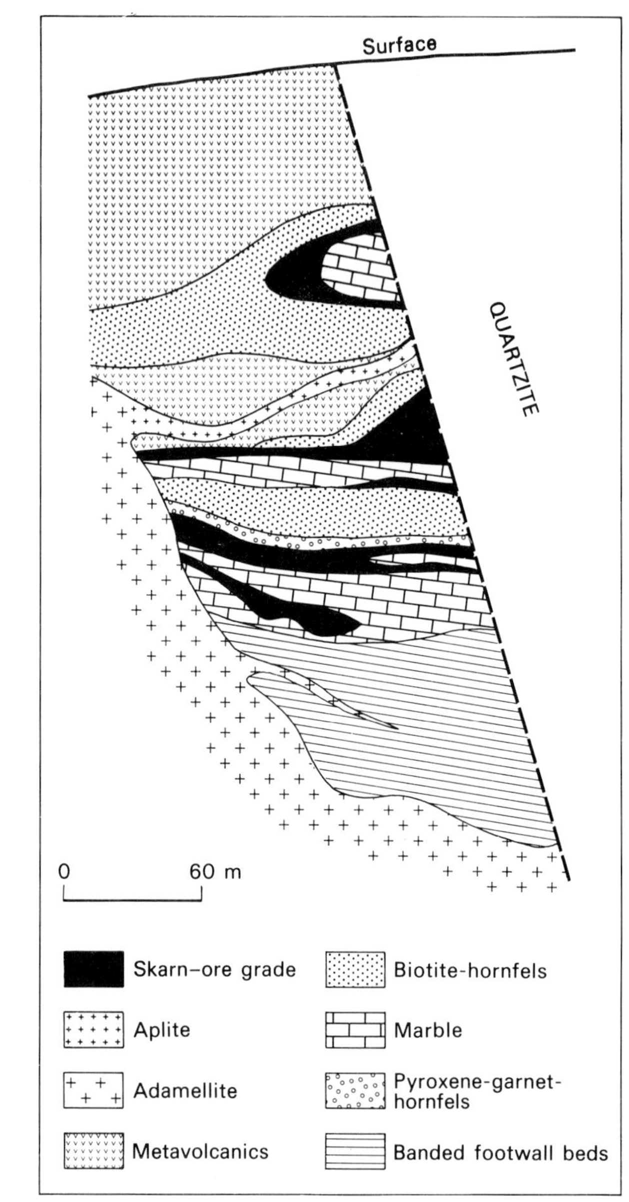

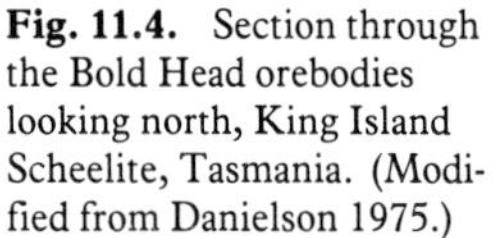

Fig. 11.4. Section through the Bold Head orebodies looking north, King Island Scheelite, Tasmania. (Modified from Danielson 1975.)

Much garnet is developed in the neighbourhood of the orebodies and in the orebodies themselves which also carry much allanite, apatite and stillwellite. The orebodies are lenticular (Fig. 11.5) and at depth appear to be broadly conformable with the stratigraphy. Uraninite is the only primary uranium mineral. It is disseminated through the rocks as ovoid grains which have their greatest concentration in allanite and stillwellite. Ore reserves plus mined ore amounted in 1975 to 9.483 Mt grading 0.131% U_3O_8 (Hawkins 1975).

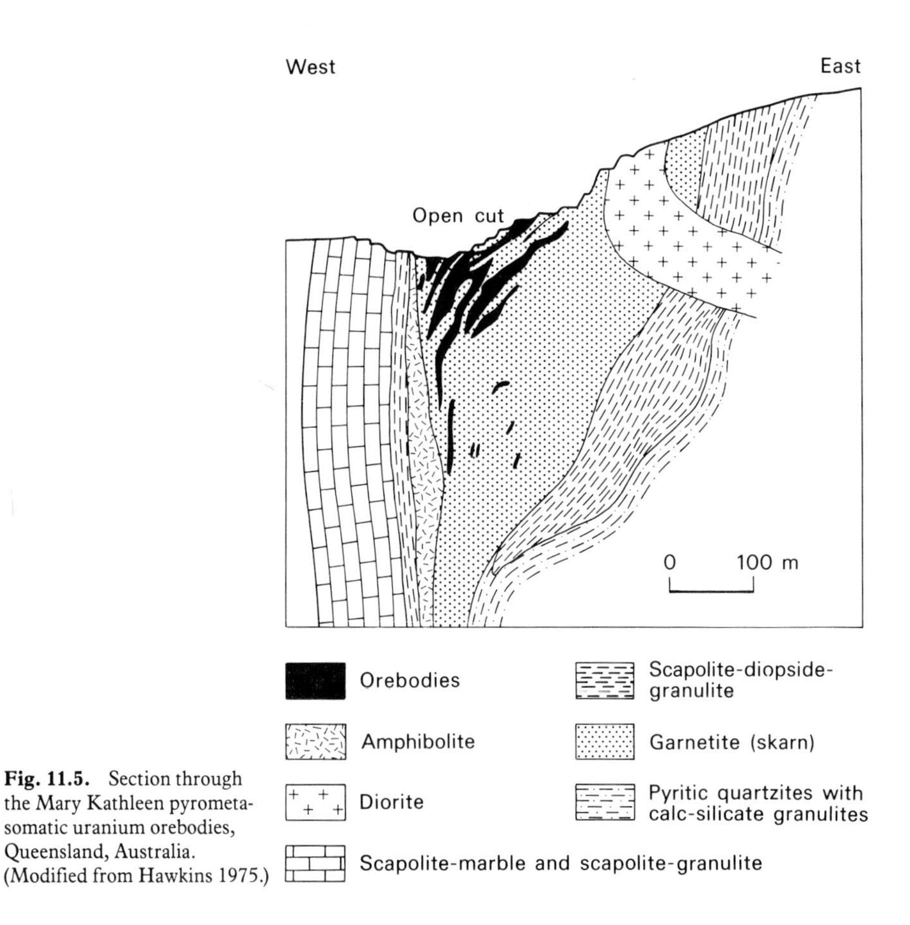

Fig. 11.5. Section through the Mary Kathleen pyrometasomatic uranium orebodies, Queensland, Australia. (Modified from Hawkins 1975.)

12

Disseminated and Stockwork Deposits Associated with Plutonic Intrusives

We are concerned in this chapter with low grade, large tonnage deposits which are principally mined for copper, molybdenum and tin. These deposits are normally intimately associated with intermediate to acid plutonic intrusives. All are characterized by intense and extensive hydrothermal alteration of the host rocks. The ore minerals in these deposits are scattered through the host rock either as what is called disseminated mineralization, which can be likened to the distribution of seeds through raspberry jam, or they are largely or wholly restricted to quartz veinlets which form a ramifying complex called a stockwork (Fig. 12.1). In many deposits or parts of deposits both forms of mineralization occur (Fig. 12.2).

The first copper deposits of this type to be mined on any scale are in some of the south-western states of the USA. These are associated with porphyritic intrusives often mapped as porphyries. The deposits soon came to be called copper porphyries, the name by which they are still generally known. On the other hand, more or less identical molybdenum deposits have been known as disseminated molybdenums although they are also called molybdenum stockworks or porphyry molybdenums. Similar tin deposits are more usually called tin stockworks, though the term porphyry tins has been used. The student will find all these names in present use. Whereas economic porphyry coppers and molybdenums are usually extremely large

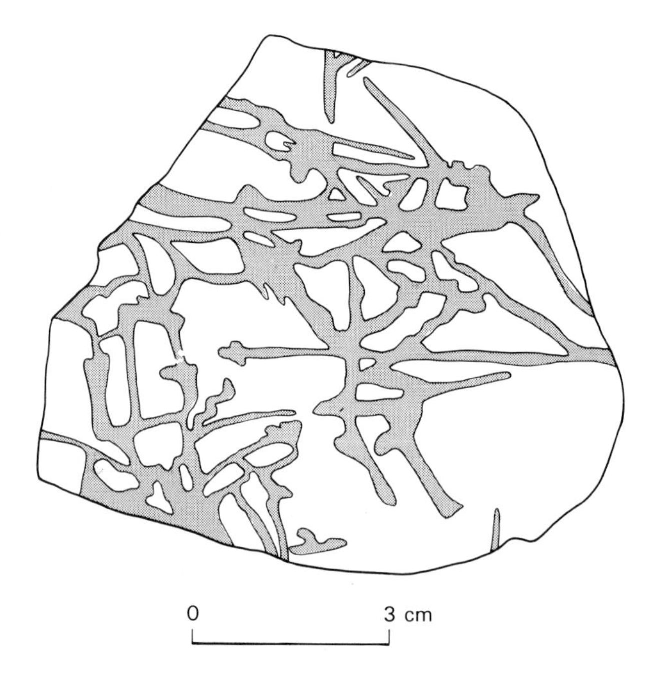

Fig. 12.1. Stockwork of molybdenite-bearing quartz veinlets in granite which has undergone phyllic alteration. Run of the mill ore, Climax, Colorado.

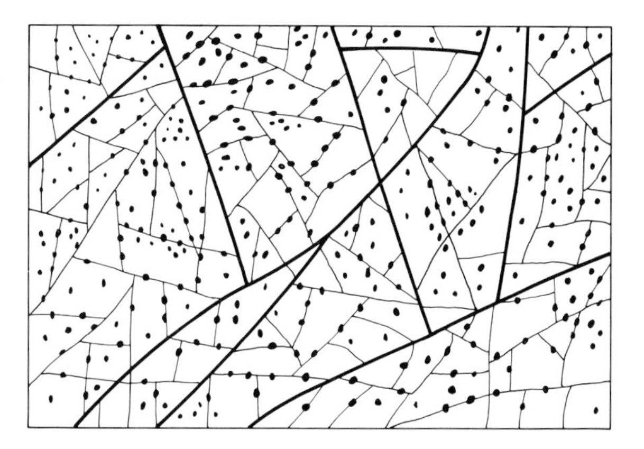

Fig. 12.2. Schematic drawing of a stockwork in a porphyry copper deposit. Sulphides occur in veinlets and disseminated through the highly altered host rock.

orebodies (50-500 million tonnes is the common size range), tin stockworks and greisens are much smaller—2-20 million tonnes being the common size range. All three types of metal deposit may yield important by-products. Amongst these are molybdenum and gold from porphyry coppers; tin, tungsten and pyrite from the Climax molybdenum deposit (other porphyry molybdenums tend to be without useful by-products); and tungsten, molybdenum, bismuth and fluorite from tin stockworks.

Porphyry coppers annually provide over 50% of the world's copper and over 100 deposits are in production. They are situated in orogenic belts in many parts of the world. Porphyry molybdenums in production are far fewer, about ten, and account for over 70% of world production. Tin stockworks and greisen are much less important, most tin production coming from placer and vein deposits.

Porphyry copper and molybdenum deposits are closely related and, although the next section is mainly devoted to porphyry coppers, mention will be made of some salient points concerning porphyry molybdenums.

Porphyry copper deposits

GENERAL DESCRIPTION

As has been indicated above, these are large low grade stockwork to disseminated deposits of copper which may also carry minor recoverable amounts of molybdenum and gold. Usually they are copper-molybdenum or copper-gold deposits. They must be amenable to bulk mining methods, that is open pit or, if underground, block caving. Most deposits have grades of 0.4-1% copper and total tonnages range up to 1000 million with a few giants being even larger than this (Fig. 12.3). Selective mining is of course impossible and host rock, stockwork and disseminated mineralization have to be extracted *in toto*. In this way, some of the largest man-made holes in the crust have come into being.

The typical porphyry copper deposit is a cylindrical stock-like, composite mass having an elongate or irregular outcrop about 1.5 × 2 km, often with an outer shell of equigranular medium-grained rock. The central part is porphyritic—implying a period of rapid cooling to produce the finer grained groundmass—the porphyry part of the intrusion. This raises the problem of how a late phase of rapid cooling could occur in the hot centre of the intrusion, insulated as it would be by the only just

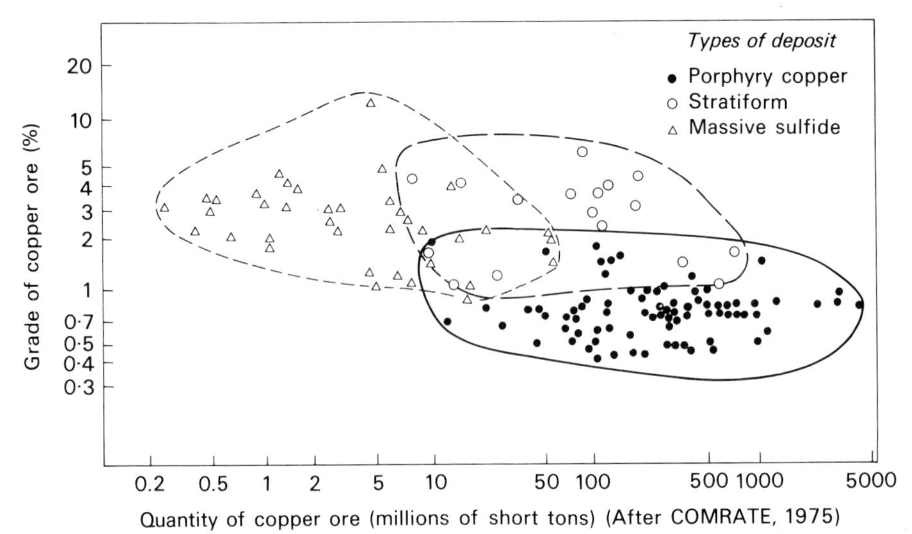

Fig. 12.3. Tonnage-grade relationships in porphyry copper, stratiform and massive sulphide deposits.

solidified and still hot outer portion. This problem will be considered in a later section.

PETROGRAPHY OF THE HOST INTRUSIONS

The most common hosts are acid plutonic rocks of the granite clan ranging from adamellite through granodiorite to tonalite. However, diorite through monzonite (especially quartz-monzonite) to syenite (sometimes alkalic) are also important host rock-types. Suggestions made in the past that diorite hosts only occur in island arcs have proved to be incorrect. Silica-poor hosts occur in both British Columbia and the central Andes.

INTRUSION GEOMETRY

The host intrusions usually appear to be passively rather than forcefully emplaced, stoping and assimilation being the principal mechanisms. They can be divided into three classes.

(i) A class in which the ore-related intrusive is simply an isolated stock. A variation on this theme could be a sill or a series of dykes or irregular bodies.

(ii) In this class we no longer have a discrete isolated stock. The host is now a late stage unit of a composite, co-magmatic intrusion often batholithic in dimensions. Examples belonging to this class occur in both continental and island arc settings.

(iii) This class is not yet known to carry economic mineralization but it is clearly related to porphyry copper deposits. Occurrences belonging to this class take the form of extensive alteration zones carrying weak mineralization and occurring in the upper parts of equigranular intrusions.

HYDROTHERMAL ALTERATION

In 1970, Lowell & Guilbert described the San Manuel-Kalamazoo orebody (Arizona) and compared their findings with 27 other porphyry copper deposits. From this study they drew up what is now known as the Lowell-Guilbert model.

In this invaluable and fundamental work they demonstrated that the best reference framework to which we can relate all the other features of these deposits is the nature and distribution of the zones of hydrothermal wall rock alteration. They claimed that generally four alteration zones are present as shown in Fig. 12.4.

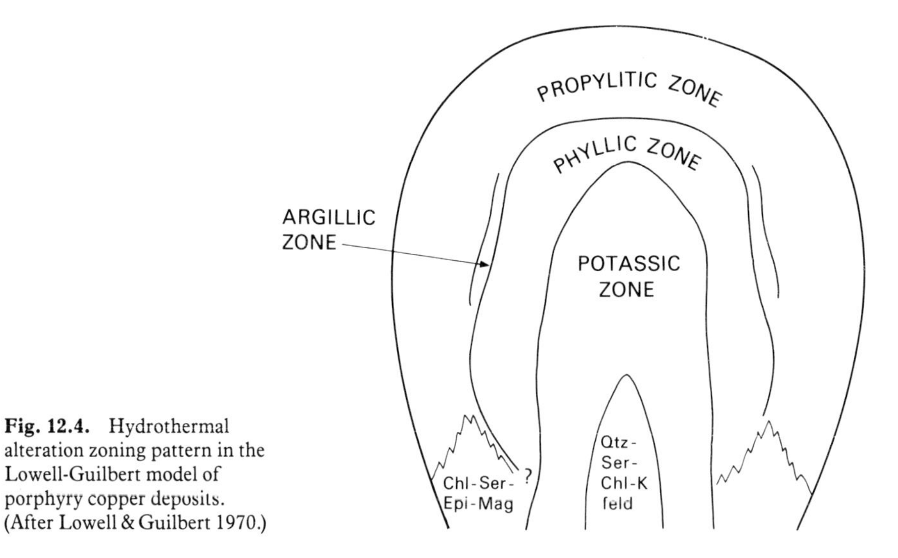

Fig. 12.4. Hydrothermal alteration zoning pattern in the Lowell-Guilbert model of porphyry copper deposits. (After Lowell & Guilbert 1970.)

These are normally centred on the porphyry stock in coaxial zones which form concentric but often incomplete shells. These zones are frequently used as a guide to ore in exploring porphyry copper deposits. In the Lowell-Guilbert model they are as follows.

(a) *The potassic zone.* This zone is not always present. When present it is characterized by the development of secondary orthoclase and biotite or by orthoclase-chlorite and sometimes orthoclase-biotite-chlorite. Sericite may also be present. These secondary minerals replace the primary orthoclase, plagioclase and mafic minerals of the intrusion. Anhydrite may be prominent in this zone. The secondary potash feldspar is generally more sodic than the primary potash feldspar. It may also be present in the quartz veinlets forming the stockwork. There is often a low grade core to this zone in which chlorite and sericite are prominent.

(b) *The phyllic zone.* This is alteration of the type known in other deposits as sericitization and advanced argillic alteration. It is characterized by the assemblage quartz-sericite-pyrite and usually carries minor chlorite, illite and rutile. Pyrophyllite may also be present. Carbonates and anhydrite are rare. The inner part of the zone is dominated by sericite—further out clay minerals become more important. The sericitization affects the feldspars and primary biotite. Alteration of the latter mineral produces the minor rutile. These are silica-generating reactions, so much secondary quartz is produced (silicification). The contact with the potassic zone is gradational over tens of metres. When the phyllic zone is present it possesses the greatest development of disseminated and veinlet pyrite.

(c) *Argillic zone.* This zone is not always present. It is the equivalent of what is called intermediate argillic alteration in other deposits. Clay minerals are prominent with kaolin being dominant nearer the orebody, and montmorillonite further away. Pyrite is common, but less abundant than in the phyllic zone. It usually occurs in veinlets rather than as disseminations. Primary biotite may be unaffected or converted to chlorite. Potash feldspar is generally not extensively affected.

(d) *Propylitic zone.* This outermost zone is never absent. Chlorite is the most common mineral. Pyrite, calcite and epidote are associated with it. Primary mafic minerals (biotite and hornblende) are altered partially or wholly to chlorite and carbonate. Plagioclase may be unaffected. This zone fades into the surrounding rocks over several hundreds of metres.

Obviously, in many deposits the behaviour of these zones in depth is poorly known and for some deposits there are no data at all. What evidence there is suggests that the zones narrow in depth and quartz-potash feldspar-sericite assemblages become more frequent, with chlorite replacing biotite.

HYPOGENE MINERALIZATION

The ore may be found in three different situations. It may be (i) totally within the host stock, (ii) partially in the stock and partially within the country rocks, or (iii) in the country rocks only. The most common shape for the orebody in the examples analysed by Lowell & Guilbert (1970) is that of a steep walled cylinder (Fig. 12.5). Stubby cylindrical to flat conical forms and gently dipping tabular shapes are also known. The orebodies are usually surrounded by a pyrite-rich shell.

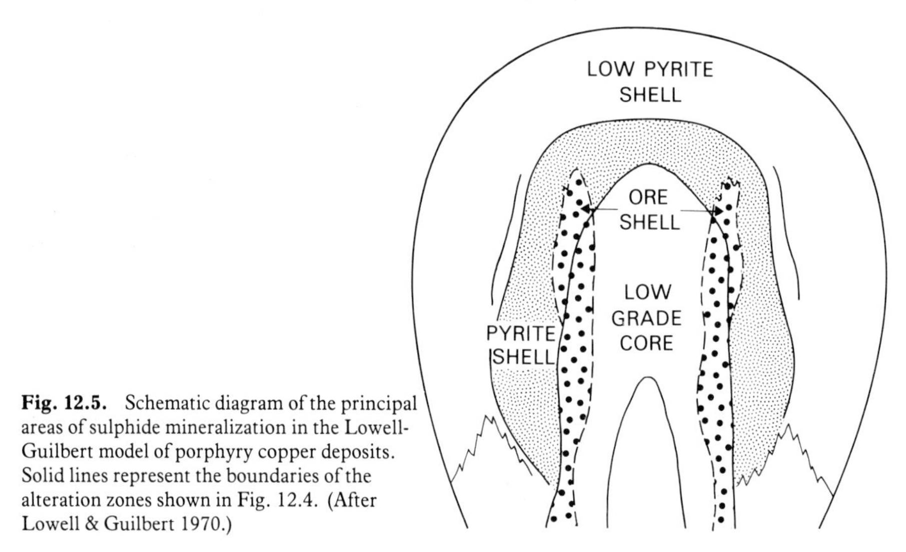

Fig. 12.5. Schematic diagram of the principal areas of sulphide mineralization in the Lowell-Guilbert model of porphyry copper deposits. Solid lines represent the boundaries of the alteration zones shown in Fig. 12.4. (After Lowell & Guilbert 1970.)

Like the alteration, the mineralization also occurs in concentric zones, this time of sulphide-oxide assemblages. Hypogene copper and molybdenum occur in both the potassic and phyllic zones (though not always together). Copper is best developed at and near the boundary between these two zones. Mineralization is not commonly found in the argillic zone. Supergene mineralization may occur in all three zones. Neither hypogene nor supergene mineralization are normally found in the propylitic zone. According to Lowell & Guilbert the zones of mineralization may be outlined as follows.

(a) *Innermost alteration zone.* This is generally coincident in space with the potassic alteration zone. It is variable in width but is generally several hundred metres across. There is a relatively low sulphide content with pyrite running about

10% and pyrite/chalcopyrite about 3/1. This zone may reach ore grade and in wholly hypogene deposits is usually the main ore carrier. It forms the low-grade centre in some deposits. The mineralization is in a disseminated form rather than in veinlets. The principal sulphides are chalcopyrite, pyrite and molybdenite.

(b) *Inner alteration zone.* This frequently corresponds with the phyllic zone and forms the main ore shell. Pyrite may run quite high, 10-15%, with pyrite/chalcopyrite about 12.5/1. Copper may reach ore grade and occurs mainly in veinlets. Disseminations are subordinate. The opaque mineral assemblage is pyrite-chalcopyrite-molybdenite-bornite-chalcocite-sphalerite-magnetite-enargite.

(c) *Intermediate alteration zone.* This corresponds approximately with the argillic zone, and mineralization may overlap from the previous zone, but generally this zone is outside the mineable ground. Pyrite is the main sulphide with pyrite/chalcopyrite about 23/1. Most of the sulphide occurs in veinlets. The opaque mineral assemblage is pyrite-chalcopyrite-bornite with traces of chalcocite, galena, enargite, sphalerite, molybdenite, tennantite and wolframite.

(d) *Outer alteration zone.* This generally corresponds with the propylitic alteration zone and the sulphide mineralization consists principally of pyrite. Sparse chalcopyrite may be present together with variable amounts of bornite, molybdenite, magnetite, specularite, rhodochrosite, sphalerite, galena and rhodonite.

In these last three zones of mineralization, veinlet usually predominates over disseminated mineralization.

BRECCIA ZONES AND PIPES

Breccia zones and pipes are common in a number of deposits and are often mineralized. Some deposits consist mainly of mineralized breccia pipes. These breccias may occur within the porphyry body or in its wall rocks. Some appear to be the result of hydrothermal activity—fluidized breccias with rounded clasts and rock flour cements—whilst others appear to be angular collapse breccias. The former are frequently referred to as pebble dykes.

VERTICAL EXTENT OF PORPHYRY BODIES

Sillitoe (1973) suggested that porphyry copper deposits occur in a subvolcanic environment associated with small high-level stocks and he emphasizes their close association with subaerial calc-alkaline volcanism. He envisaged the host pluton as being overlain by a stratovolcano (Fig. 12.6). Sillitoe contended that evidence from Chilean and Argentinian deposits shows that the propylitic alteration extends upwards into the stratovolcano which is itself likely to carry native sulphur deposits. The zones of alteration close upwards. The potassic zone dies out, sericitic and argillic alteration become important and the upper limit of economic mineralization is reached. At the same time, the porphyry stock becomes smaller in size and hydrothermal breccias appear over large areas.

In the lower parts of porphyry copper deposits the available evidence suggests that there is often a downward transition from porphyry into an equigranular plutonic rock of similar composition which forms part of a pluton of much larger dimensions. With this textural change the mineralization dies out in depth.

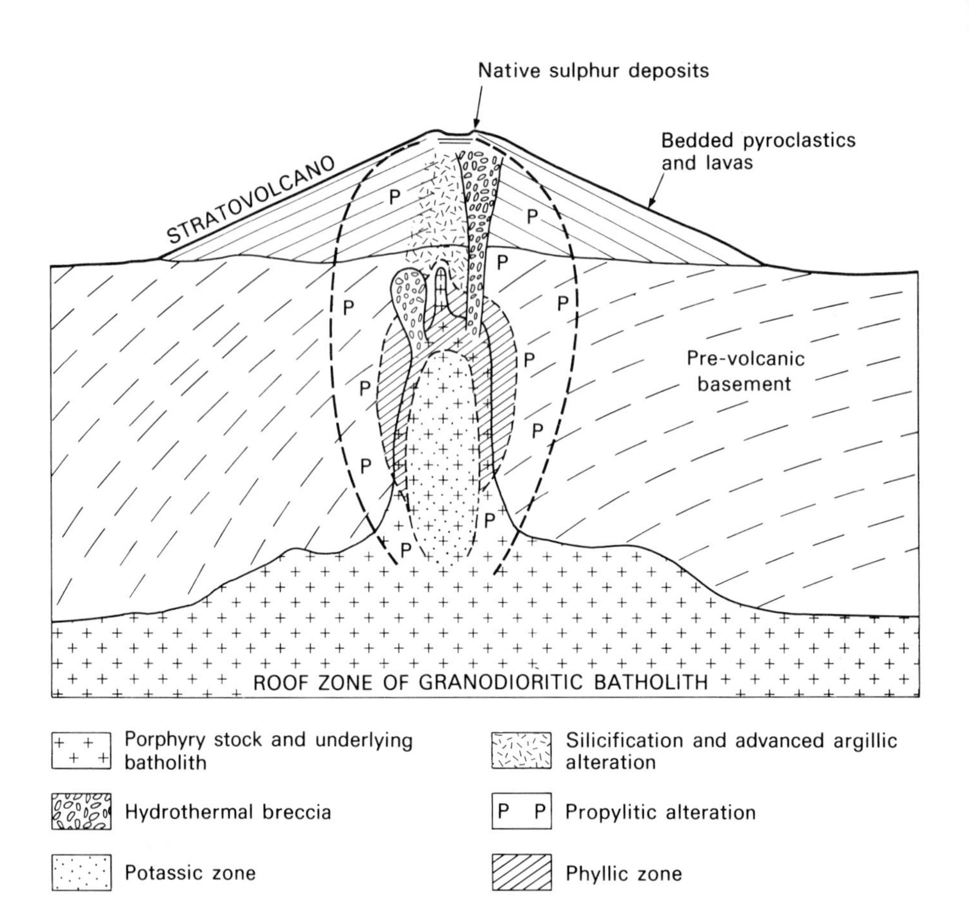

Fig. 12.6. Diagrammatic representation of a simple porphyry copper system on the boundary between the volcanic and plutonic environments. (After Sillitoe 1973.)

THE DIORITE MODEL

Subsequent to Lowell & Guilbert's classic work it has been recognized that some porphyry copper deposits are associated with intrusives having low silica/alkali ratios. Various names have been suggested for this type. The one which has won general recognition is 'diorite model', although the host pluton may be a syenite, monzonite, diorite or alkalic intrusion (Hollister 1975). Diorite model deposits differ in a number of ways from the Lowell-Guilbert model; one of the main reasons appears to be that sulphur concentrations were relatively low in the mineralizing fluids. As a result, not all the iron oxides in the host rocks are converted to pyrite and much iron remains in the chlorites and biotites. Excess iron tends to occur as magnetite which may be present in all alteration zones.

(a) *Alteration zoning.* The phyllic and argillic alteration zones are usually absent so that the potassic zone is surrounded by the propylitic zone. This zonal pattern is present in both island arc and continental porphyry copper deposits. In the potassic zone, biotite may be the most prominent potassium mineral and when orthoclase is not well developed, plagioclase may be the principal feldspar.

(b) *Mineralization.* The main difference from the Lowell-Guilbert model is that significant amounts of gold may now occur and molybdenum/copper is usually low. The fractures containing gangue silicate minerals and copper sulphides may be

Table 12.1. Comparison of Lowell-Guilbert and diorite models of porphyry copper deposits.

Feature	Lowell-Guilbert model	Diorite model
Host pluton		
Common rock-types	Adamellite, granodiorite, tonalite	Syenite, monzonite
Rarer rock-types	Quartz-diorite	Diorite
Alteration		
Central core area	Potassic	Potassic
Peripheral to core	Phyllic Argillic Propylitic	Propylitic
Mineralization		
Quartz in fractures	Common	Erratic
Orthoclase in fractures	Common	Erratic
Albite in fractures	Trace	Common
Magnetite	Minor	Common
Pyrite in fractures	Common	Common
Molybdenite	Common	Rare
Chalcopyrite/bornite	3 or greater	3 or less
Dissemination of chalcopyrite	Present	Important
Gold	Rare	Important
Structure		
Breccia	May occur	Rare
Stockwork	Important	Important

devoid of quartz. On the other hand, chlorite, epidote and albite are fairly common.

(c) *Comparison of the Lowell-Guilbert and diorite models.* Table 12.1 (based on Hollister 1975) contrasts the principal features of each model and lists further features of the diorite model.

REGIONAL CHARACTERISTICS OF PORPHYRY DEPOSITS

The distribution of porphyry copper and molybdenum deposits is shown in Fig. 12.7. From this map it can be seen that the majority of porphyry deposits are associated with Mesozoic and Cenozoic mountain belts and island arcs. The major exceptions are the majority of the USSR deposits and the Appalachian occurrences of the USA. These exceptions belong to the Palaeozoic. Only two or three porphyry deposits have so far been found in the Precambrian. These facts are of great importance from the exploration point of view.

(a) *South-western USA.* The deposits of this region form an oval cluster which contrasts with the linear belts of deposits in the Andes and elsewhere (Lowell 1974). There are about 88 deposits, with 30 in production. The ages range from 20 to 163 Ma with a peak at 58-72 Ma (Laramide orogeny). Some deposits lie along marked lineaments with a tendency to be developed at their intersections. This is not, however, true of all the deposits. They are spread over a considerable area and some are so far inland that if the genesis of the host intrusives is to be linked to a subduction zone then it is necessary to postulate very low dipping or multiple subduction zones.

(b) *Northern (Canadian) Cordillera of America.* In this complex region two main groups of deposits are present. (i) Triassic deposits in hosts intruded into volcanic

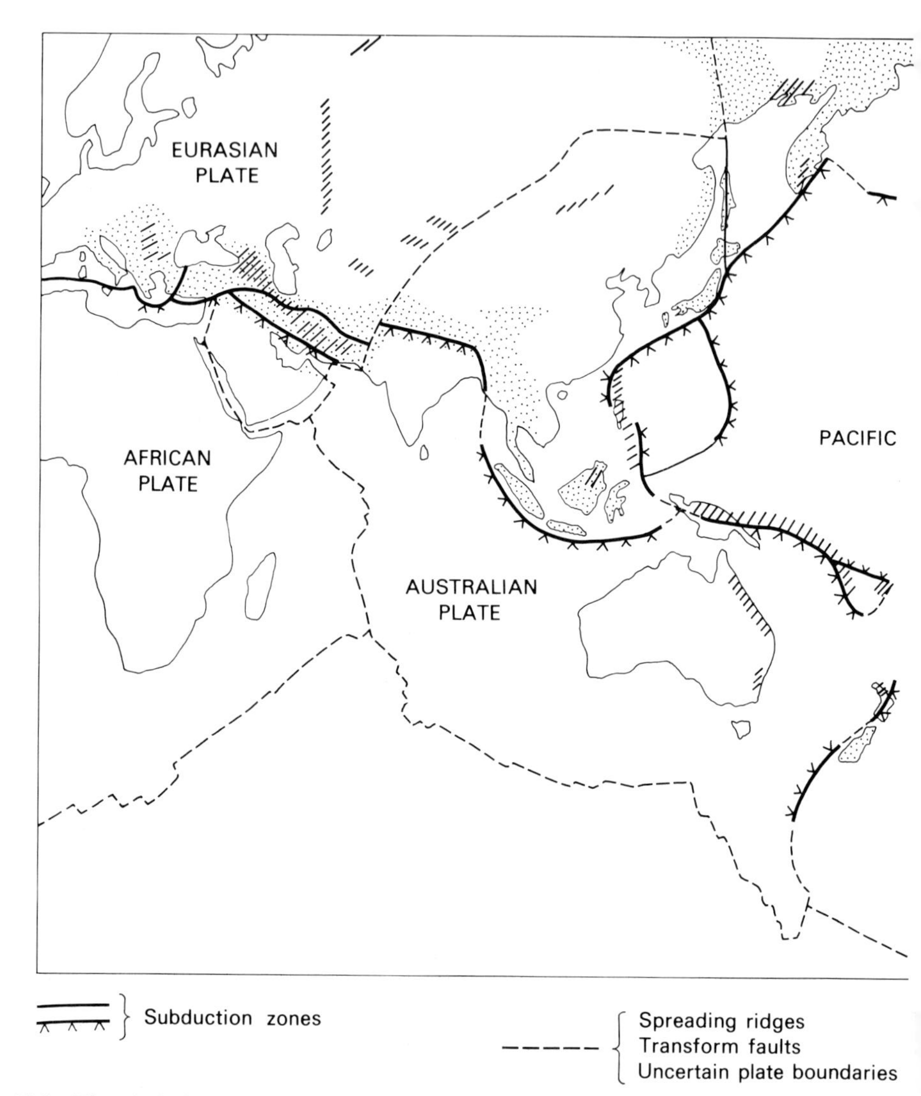

Fig. 12.7. The principal porphyry copper and molybdenum regions of the world. Also shown are present plate boundaries and Mesozoic-Cenozoic mountain belts.

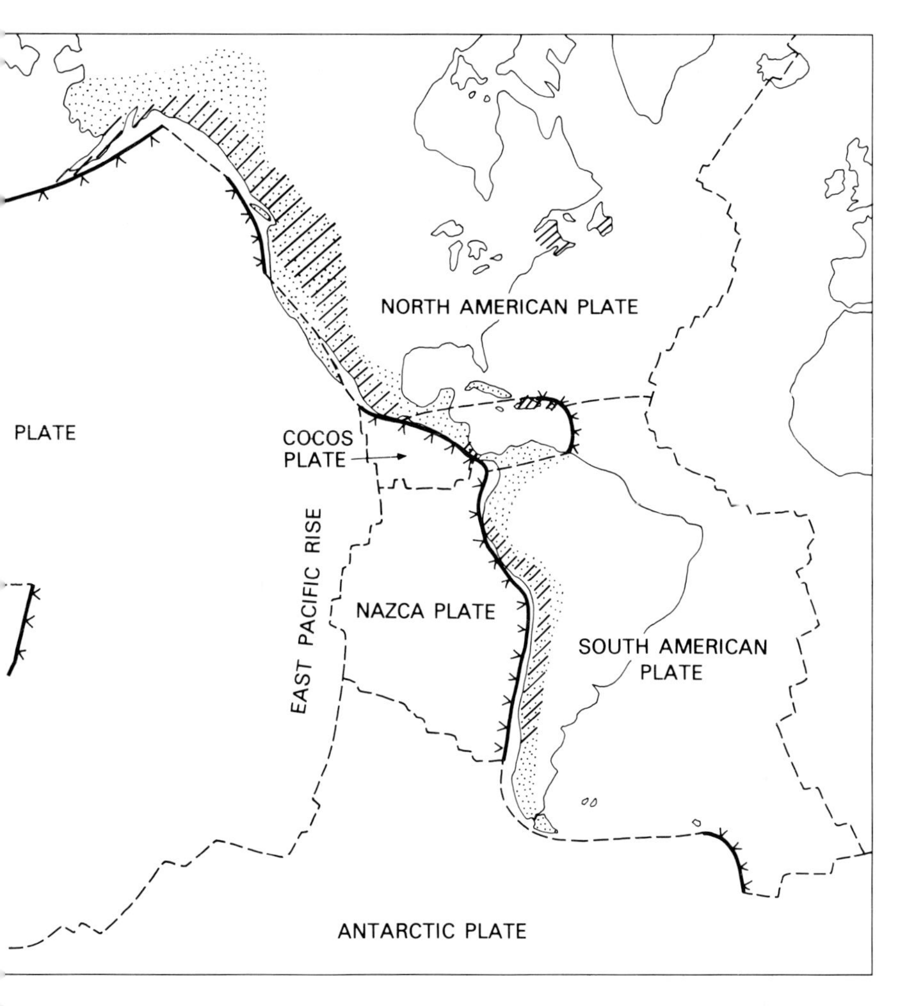

NORTH AMERICAN PLATE
PLATE
COCOS
PLATE
EAST PACIFIC RISE
NAZCA PLATE
SOUTH AMERICAN
PLATE
ANTARCTIC PLATE
Mesozoic and Cenozoic
mountain belts
Regions with porphyry copper
and molybdenum deposits

rocks that include oceanic crust. (ii) Tertiary deposits near areas of Tertiary volcanic rocks belonging to a later stage of crustal evolution. Generally speaking, the deposits with a higher molybdenum content are developed where batholithic plutonism or some other evidence of thickening of the continental crust occurs. Deposits with little or no molybdenum and significant gold generally tend to occur among the rocks belonging to the earlier geosynclinal stages.

(c) *Appalachian orogen.* According to Hollister *et al.* (1974), porphyry deposits were first developed in this region during the Cambrian and Ordovician and later in the Devonian and Carboniferous. Considering all the deposits together, copper porphyries formed first followed by the coeval development of separate copper and molybdenum porphyries with a final period when only molybdenum porphyries were formed. Accompanying this change was a variation in the composition of the magmas associated with the mineralization from quartz-monzonitic to granitic. The copper and molybdenum porphyries either lack or have a very small develop-

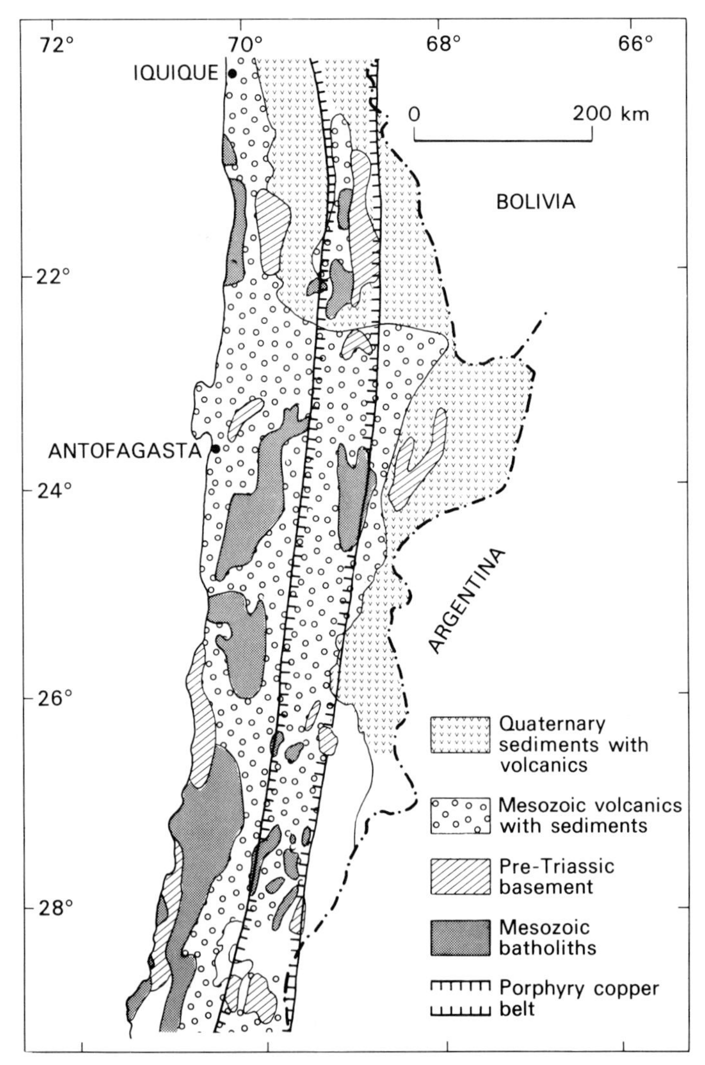

Fig. 12.8. Map of part of the Chilean porphyry copper province.

ment of the phyllic zone. This may be an effect of deep erosion. On the other hand, this fact and the low content of pyrite may indicate a deposit type transitional to the diorite model.

Prior to continental drift, this orogen was continuous with the Caledonian Province of the British Isles where porphyry copper mineralization has been found in North Wales (Cambro-Ordovician) and Scotland (Devonian).

(d) *The Andean province.* As Fig. 12.8 shows, this has a most marked linear distribution of deposits. These occur in a region where erosion has not cut down so deeply as it has in the coastal range batholiths to the west, but has cut down more deeply than in the volcanic belt to the east. There is a strong fault control, many deposits being associated with tear faults and what may be continental continuations of transform faults. The age range of the deposits is 59-4.3 Ma, nearly all have a significant molybdenum content and the underlying crust appears to be entirely continental.

(e) *The South-western Pacific island arcs.* The island arcs of this province generally lack continental crust and gold is usually an important by-product and molybdenum uncommon. Lowell-Guilbert and diorite model deposits occur together. Most of the deposits occur in arcs containing thick sequences of pre-ore rocks. The host porphyries have penetrated to high structural levels and are all very young (less than 16 Ma). The deposits formed at a late stage of arc evolution just before volcanic activity ceased.

(f) *Porphyry copper and molybdenum deposits in the USSR.* The majority of Soviet deposits are Palaeozoic with a peak in the Carboniferous. The distribution of the major fields is shown in Fig. 12.7. The most important are in Kazakstan, the Caucasus, Uzbekistan and the Batenevski Range in Siberia. Most, if not all, deposits are related to present or former subduction zones and about a quarter are molybdenum-poor copper porphyries which have formed in island arc settings. Copper-molybdenum deposits lie along continental or microcontinental margins like those of the south-western USA and the Andes. Porphyry molybdenum deposits in any given area are always younger than associated porphyry copper deposits (Laznicka 1976).

METAL ABUNDANCES IN PORPHYRY COPPER DEPOSITS

The vast majority of deposits can be divided into two classes according to whether the principal accessory metal is molybdenum or gold (Kesler 1973). Generally speaking, copper-gold deposits appear to be concentrated in island arcs and copper-molybdenum where continental crust is present. There are, however, some notable exceptions, e.g. Cerro Colorado, one of the largest island arc deposits, has a high molybdenum/gold ratio. Other exceptions have been noted by Titley (1978). The crustal setting clearly has some control on the deposit composition, but it is not the only factor.

GENESIS OF PORPHYRY COPPER DEPOSITS

The most striking characteristic of porphyry copper deposits when compared with other hydrothermal orebodies is their enormous dimensions. The size and shape of these deposits imply that the hydrothermal solutions permeated very large volumes of rock, including country rocks, as well as the parent pluton. That at least some of these solutions originated in the host pluton is suggested by the existence of crackle brecciation.

(a) *Crackle brecciation and its origin.* Crackle brecciation is the name given to the fractures which are usually healed with veinlets to form the stockwork mineralization. The zone of crackle brecciation is usually circular in outline and always larger than the orebodies. It fades out in the propylitic zone. It is often less well developed near the centre of the deposit, particularly if potassic alteration is present.

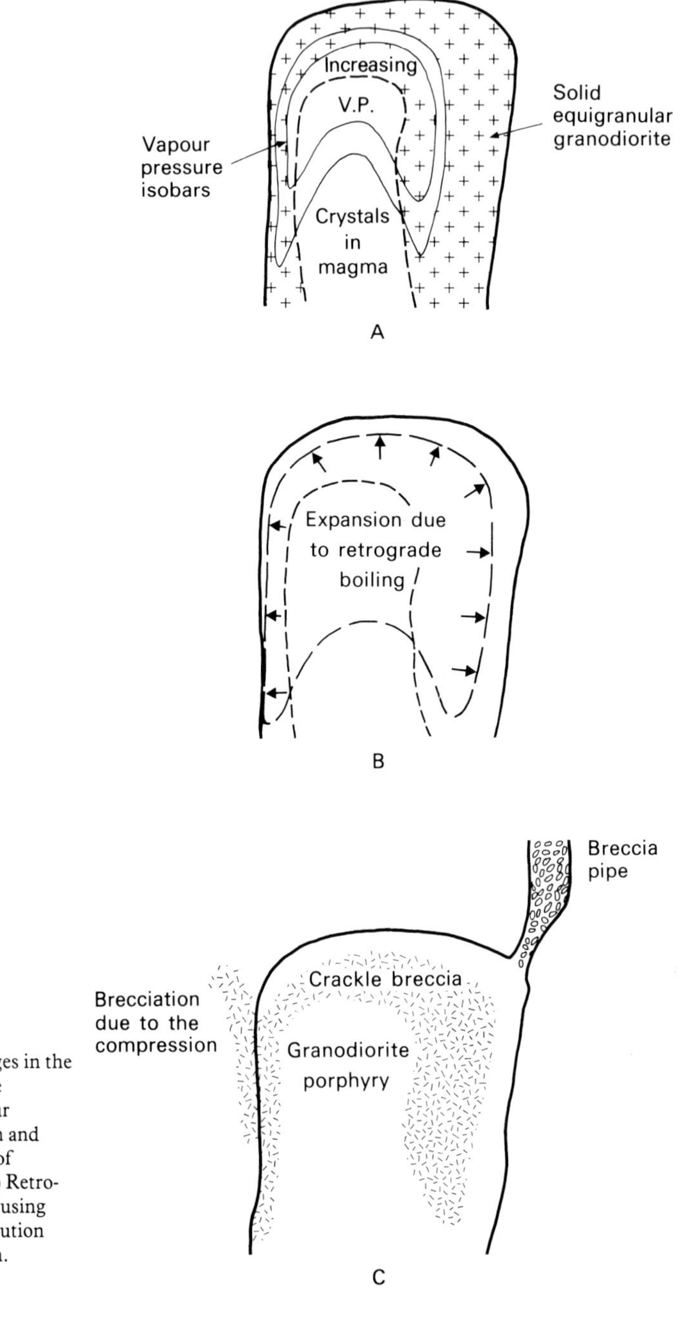

Fig. 12.9. Three stages in the development of crackle brecciation. (A) Vapour pressure building up in and around upper portion of magmatic fraction. (B) Retrograde boiling occurs causing expansion. (C) Distribution of resulting brecciation. (After Phillips 1973.)

This brecciation is thought to be due to the expansion resulting from the release of volatiles from the magma (Phillips 1973).

The host magmas of porphyry copper deposits appear to have reached to within 0.5-2 km of the surface before equigranular crystallization commenced in their outer portions. The intrusions would then be stationary and the confining pressure would not fluctuate. With the steady development of crystallization, however, anhydrous minerals are forming and the liquid magma becomes richer in volatiles leading to an increase in the vapour pressure. If the vapour pressure rises above the confining pressure, then what is called retrograde boiling will occur. A rapidly boiling liquid will separate. If retrograde boiling occurs in a largely consolidated rock, the vapour pressure has to overcome the tensile strength of the rock as well as rising above the confining pressure. This will result in expansion and extensive and rapid brecciation (Fig. 12.9). The reason for this is that water released at a depth of about 2 km at 500°C would have a specific volume of 4 and, if 1% by weight formed a separate phase, it would produce an increase in volume of about 10%. At shallower depths the increase would be even greater. Evidence for the development of retrograde boiling in porphyry copper deposits is common. It comes from the widespread occurrence of liquid-rich and gas-rich fluid inclusions in the same thin section (Chapter 3).

It is important to notice that whilst the crystallization of solid phases is an exothermic one, bubble formation is an endothermic process. Rapid nucleation and the adiabatic expansion of the vapour would absorb a great deal of heat taking up the latent heat of crystallization and significantly lowering the temperature of the system. This would result in a second phase of rapid cooling in the central part of the intrusion which would considerably increase the number of nucleation sites producing a period of rapid crystallization which, in turn, would be responsible for the fine-grained groundmass and hence the porphyritic nature of the intrusion.

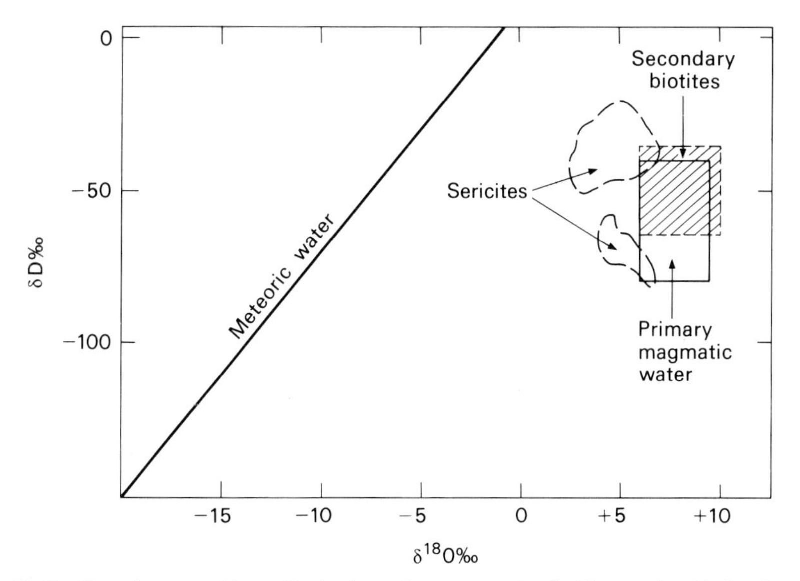

Fig. 12.10. Isotopic compositions of hydrothermal waters associated with secondary biotites from the potassic zones and sericites from the phyllic zones of five porphyry copper deposits. (Modified from Sheppard 1977.)

(b) *Evidence from isotopic investigations.* Further evidence of a magmatic derivation of at least some of the hydrothermal solutions comes from stable isotope investigations (Sheppard 1977). Waters in equilibrium with potassium silicate alteration assemblages and formed at 550-700°C are isotopically indistinguishable from primary magmatic waters (Fig. 12.10). On the other hand, waters associated with sericites from the phyllic zone of alteration are depleted in ^{18}O relative to the biotites of the potassic zone. Comparison with Fig. 4.10 suggests that connate waters from the country rocks were involved in the sericitization; in other words, meteoric water played a significant role in the hydrothermal fluids responsible for the phyllic alteration. The isotopic data for advanced and intermediate argillic alteration show an identical pattern to that for the phyllic alteration data. Field and microscopic evidence suggest that the phyllic and argillic alterations were later than the potassium silicate and propylitic alterations and were superimposed to varying degrees upon them. These two stages of development are depicted in Fig. 12.11.

It appears that after intrusion of the porphyry body solidification occurs and a magmatic-hydrothermal solution evolves. This solution reacts with the porphyry and to a varying extent with the surrounding country rocks giving rise to the development of a central zone of potassium silicate alteration. The introduction of much of the metals and sulphur probably accompany this stage. Further out from the intrusion, thermal gradients set up a convective circulation of water in the country rocks and this is responsible for the propylitic alteration (diorite model conditions).

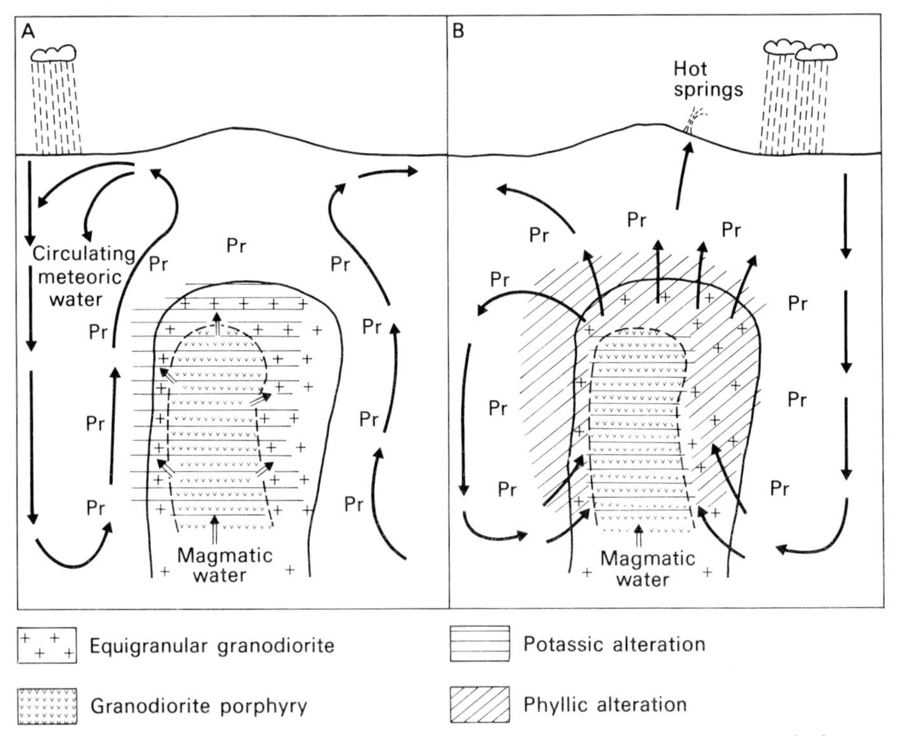

Fig. 12.11. Diagrammatic sections through a porphyry copper deposit showing two stages in the development of the hydrothermal fluids leading to the formation of a Lowell-Guilbert model deposit (B). Pr, propylitic alteration. (Modified from Sheppard 1977.)

When the intrusion cools, this meteoric-connate hydrothermal system may encroach upon and mix with the waning magmatic system leading to the development of lower temperature minerals; sericite, pyrophyllite and clay minerals. These would replace in particular the feldspar and biotite of the outer part of the original potassium silicate zone. The relatively rapid gradients in pH, temperature, salinity, etc., across the interface between these two hydrothermal systems probably accounts for the concentration of copper around the boundary zone between the potassium silicate and phyllic zones. With this second stage of alteration, a Lowell-Guilbert model deposit comes into being.

As can be seen in Fig. 12.10, the depletion of ^{18}O in the rock minerals is quite small in most investigated porphyry copper deposits, when compared with primary magmatic water values. This implies that meteoric water/rock ratios during sericitic and argillic alteration were not very high.

Porphyry molybdenum deposits

GENERAL DESCRIPTION

These have many features in common with porphyry copper deposits, some of these features have been touched on above. A useful summary account of the deposits in the Western Cordillera of North America has been given by Clark (1972). Average grades are 0.15-0.5 MoS_2. Host intrusions vary from quartz-diorite through quartz-monzonite and granodiorite to granite. Stockwork mineralization is more important than disseminated mineralization and the orebodies are associated with simple, multiple or composite intrusions or with dykes or breccia pipes. Cylindrical ore zones are known but the two biggest deposits, Climax and Henderson in Colorado (Figs 12.12 and 12.13), have been described as annular caps or umbrella-shaped

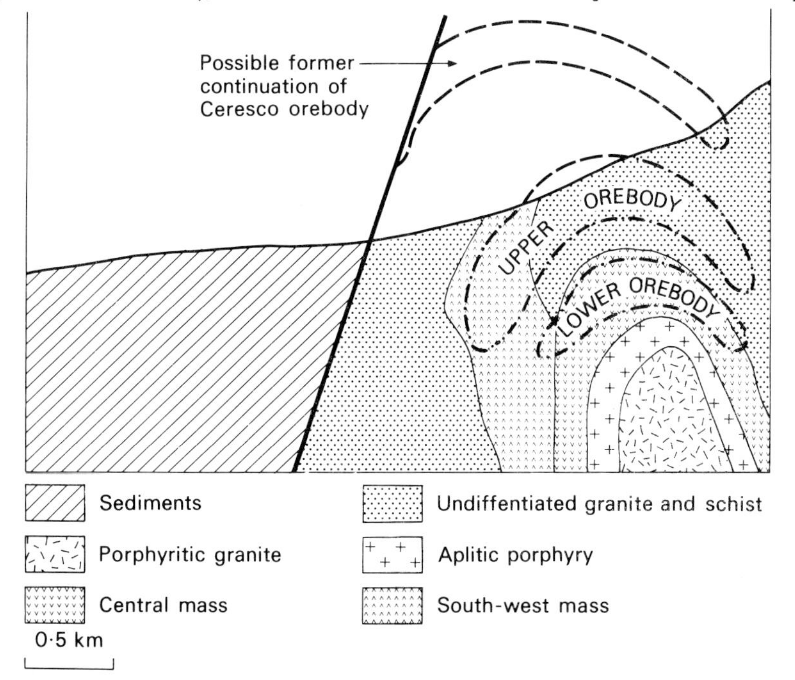

Fig. 12.12. Generalized geological section through the Climax molybdenum mine, Colorado. (Modified from Hall *et al.* 1974.)

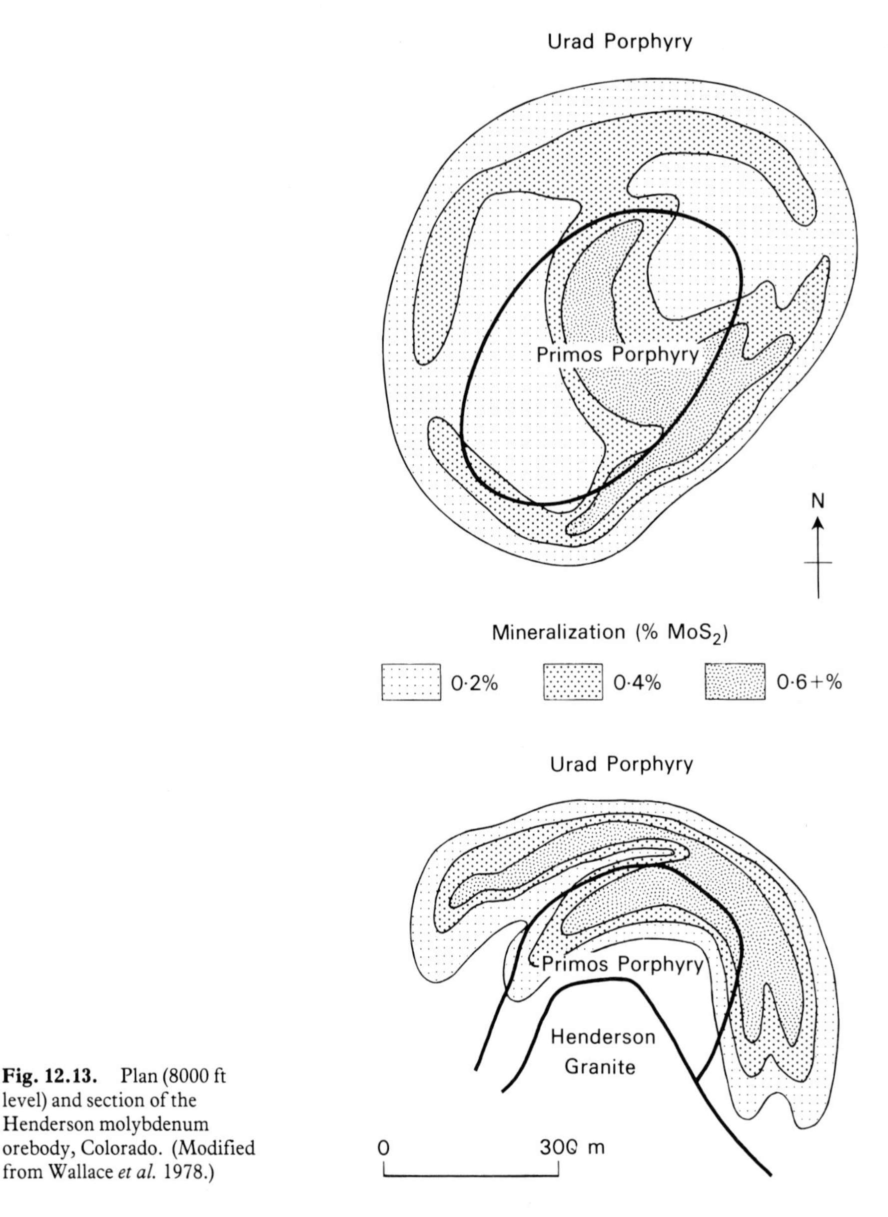

Fig. 12.13. Plan (8000 ft level) and section of the Henderson molybdenum orebody, Colorado. (Modified from Wallace *et al.* 1978.)

bodies (Wallace *et al.* 1978). The molybdenite occurs in (a) quartz veinlets carrying minor amounts of other sulphides, oxides and gangue, (b) fissure veins, (c) fine fractures containing molybdenite paint, (d) breccia matrices and, more rarely, (e) disseminated grains. Supergene enrichment, which can be very important in porphyry coppers, does not occur.

HYDROTHERMAL ALTERATION

The alteration patterns are very similar to those found in porphyry copper deposits with potassic alteration and silicification being predominant. The most detailed study is on the Urad and Henderson deposits (Wallace *et al.* 1978) where there is a central potassic zone carrying secondary potash feldspar and biotite. Succeeding this are quartz-topaz, phyllic, argillic and propylitic zones. There is a silicified zone which lies largely within the potassic zone. The Henderson orebody is roughly coincident with the potassic and silicified zones. A prominent pyrite zone, carrying 6-10% pyrite, is developed around the Henderson orebody and a similar less distinct zone is present at Climax. Peripheral pyrite zones have been reported from a number of other porphyry molybdenums. They appear to coincide mainly with the phyllic zone.

GENESIS OF PORPHYRY MOLYBDENUM DEPOSITS

The close spatial association of the orebodies of most deposits with the potassic zone of alteration suggests a magmatic source and Wallace *et al.* (1978) argue that, for the Henderson and Climax deposits, this must have been unexposed master reservoirs that fed the columns of exposed host intrusions. They consider that the volumes of the host intrusions are far too small to have supplied the large tonnage of molybdenum which is present in the deposits.

Porphyry tin deposits

Much primary tin has been won in the past from stockworks at Altenberg and Zinnwald in Germany, from many deposits in Cornwall, England, and from deposits in New South Wales, Tasmania and South Africa. Eroded stockworks in Indonesia and Malaysia have provided much of the alluvial cassiterite in those countries. Exploitation of such deposits in recent times has been mainly in Germany and New South Wales. At Ardlethan in New South Wales a quartz-tourmaline stockwork occurs in altered granodiorite carrying secondary biotite and sericite. With a grade of 0.45% tin this deposit is economic. The majority of tin stockworks in the world, however, only run about 0.1% tin and they are not at present mineable at a profit.

Tin stockworks in the countries mentioned above belong to the plutonic environment. Recently, Sillitoe *et al.* (1975) have described porphyry tin deposits from the subvolcanic section of the Bolivian tin province south of Oruro. Large volumes of rock grade 0.2-0.3% tin. They have shown that these deposits have much in common with porphyry copper deposits. There is pervasive sericitic alteration that grades outwards into propylitic alteration and pyrite halos are present in two deposits. The parent stocks are thought to have been overlain by stratovolcanoes when they were emplaced. Major differences include the absence of a potassic zone of alteration, the association with stocks having the form of inverted cones rather than upright cylinders and the presence of swarms of later vein deposits.

13

Stratiform Sulphide and Oxide Deposits of Sedimentary and Volcanic Environments

In this chapter we are concerned with a class or classes of deposit whose origin is at present hotly debated. As was pointed out in Chapter 2, there are many types of stratiform deposit. This chapter is mainly devoted to sulphide deposits and the related oxide deposits can only be mentioned *en passant*. The latter do not include bedded iron and manganese deposits, placer deposits and other ores of undoubted sedimentary origin which are dealt with in Chapter 16. The related oxide deposits are of tin and uranium (and possibly iron) with which may be grouped certain tungsten deposits.

Concordant deposits referred to in Chapter 2 which belong to this class include the Kupferschiefer of Germany and Poland, Sullivan, British Columbia, the Zambian Copperbelt and the large group of volcanic massive sulphides. There appears to be a possible gradation in type and environment from deposits such as the Kupferschiefer, which are dominantly composed of normal sedimentary material and which occur in a non-volcanic sedimentary environment, through deposits such as Sullivan, which are richer in sulphur and may have some minor volcanic formations in their host succession, to the volcanic massive sulphide deposits which are mainly composed of sulphides and occur in host rocks dominated by volcanics. Stanton (1972) considered that these and other deposits formed a 'spectrum of occurrence' and treated them as one class. Other workers, e.g. Barnes (1975) and Solomon (1976) have felt it better to divide these deposits into two classes, the first class being those developed in a sedimentary environment where sedimentary controls are important, and the second being the volcanic massive sulphide deposits in which exhalative processes were important during genesis. Such a division introduces difficulties when dealing with deposits showing only a weak link with volcanism but where exhalative processes may have been important, e.g. Sullivan. However, this division will be followed in this chapter as the author feels that deposits such as the Kupferschiefer and the Zambian Copperbelt are sufficiently different from the volcanic massive sulphide deposits to warrant some differentiation.

Stratiform sulphide deposits of sedimentary affiliation

GENERAL CHARACTERISTICS

The majority of these deposits occur in non-volcanic marine or deltaic environments. They are widely distributed in space and time, i.e. from the Proterozoic to the Tertiary, and can vary in tonnage from several hundred millions down to subeconomic sizes. In shape, they are broadly lensoid to stratiform with the length at least ten times the breadth. There is often more than one ore layer present. The degree of deformation and metamorphism varies with that of the host rocks, suggesting a pre-metamorphic formation. They are frequently organic-rich, particularly those in shales, and usually contain a less complex and variable suite of minerals and recoverable metals than volcanic massive sulphide deposits. The sulphides have a small grain size so that fine and often costly grinding is necessary to liberate them from the gangue. They may show a shore to basinward zoning of Cu + Ag → Pb → Zn (Barnes 1975).

SOME EXAMPLES

(a) *The European Kupferschiefer.* This is probably the world's best known copper-rich shale. It is of late Permian age and has been mined at Mansfeld (East Germany) for almost 1000 years. The Kupferschiefer underlies about 600 000 km^2 in Germany, Poland, Holland and England (Fig. 13.1). Copper concentrations greater

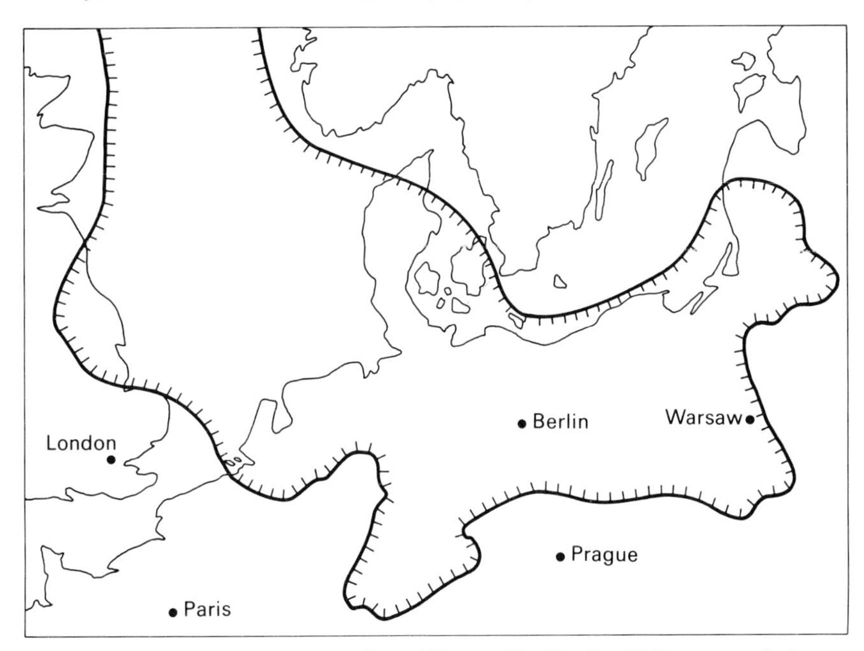

Fig. 13.1. Extent of the Zechstein Sea in Central Europe. The Kupferschiefer occurs at the base of the Upper Permian (Zechstein).

than 0.3% occur in about 1% and zinc concentrations greater than 0.3% in about 5% of this area. Thus, although all the Kupferschiefer is anomalously high in base metals, ore grades are only encountered in a few areas. The most notable recent discoveries have been in southern Poland where deposits lying at a depth of 600-1500 m have been found during the last two decades. Here, the Kupferschiefer varies from 0.4 to 5.5 m in thickness. Average copper content is around 1.5% and reserves at 1% Cu amount to some 1500 Mt making Poland the leading copper producer in Europe. The area underlain by these deposits is approximately 30 × 60 km.

The Kupferschiefer consists of thin alternating layers of carbonate, clay and organic matter with fish remains which give it a characteristic dark grey to black colour. The Kupferschiefer is the first marine transgressive unit overlying the non-marine Lower Permian Rotliegendes, a red sandstone sequence, and it is overlain by the Zechstein Limestone and this is overlain, in turn, by a thick sequence of evaporites. The Kupferschiefer and the Zechstein evaporates may represent a tidal marsh (sabkha) environment which developed as the sea transgressed desert sands.

The copper and other metals are disseminated throughout the matrix of the rock as fine-grained sulphides (principally bornite, chalcocite, chalcopyrite, galena, sphalerite). Typical features of the mineralization are shown in Fig. 13.2. A zone

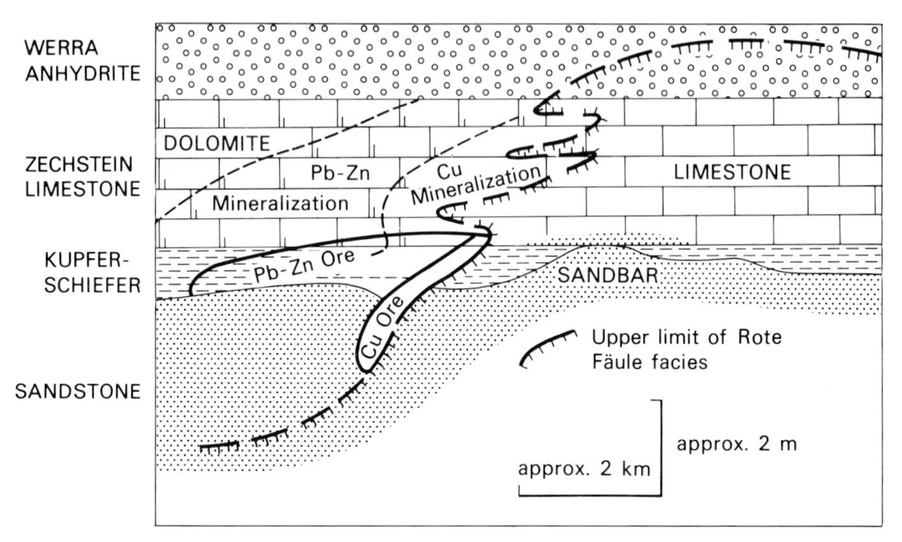

Fig. 13.2. Diagrammatic section through orebodies in the basal Zechstein with the Rote Fäule facies alteration gently transgressing the bedding above an area of sandbars formed by marine reworking of the Rotliegendes. Sulphide mineralization occurs in the unoxidized zone adjacent to the Rote Fäule with copper nearest to it and lead-zinc further away. (After Brown 1978.)

of superposed diagenetic reddening known as the Rote Fäule facies transgresses the stratigraphic horizons. Copper mineralization lies directly above the Rote Fäule and the copper zone is overlain, in turn, by lead-zinc mineralization. This relationship to the Rote Fäule has meant that the delineation of this facies is the most important feature of the search for new orebodies.

Most geologists regard the Kupferschiefer as the classic example of a syngenetic base-metal ore deposit. The most important evidence for this hypothesis being the continental extent of the mineralization at one particular horizon. Many sources for the base metals have been postulated:

(1) upwelling nutrient-rich sea water (Brongersma-Sanders 1969);

(2) hydrothermal solutions reaching the surface of the Harz Mountains and draining out to sea (Ekiert 1958);

(3) erosion of older mineral deposits (Richter 1941);

(4) submarine metal-rich hydrothermal springs (Dunham 1964), or

(5) leaching of the underlying redbeds (Deans 1950, Wedepohl 1971).

Recently, Brown (1978) has suggested a syngenetic or diagenetic enrichment in pyrite followed by a later influx of base metals. This would account for the common observation of the replacement of pyrite by copper sulphides. A comprehensive review of the Kupferschiefer in East Germany is given by Jung & Knitzschke (1976).

(b) *The Zambian Copperbelt.* This is part of the larger Central African Copperbelt of Zambia and Shaba (Zaïre) which produces 20% of the world's copper. In 1970, Zambia produced 686 000 t of copper, the average grade of ore mined was 3.8% and the estimated commercial reserves were 27.2 Mt of copper. Shaba produced 386 000 t copper in 1970 from ore averaging 4.2% Cu+Co and estimated copper reserves were 18.1 Mt. Almost all the copper mined in 1970 came from restricted horizons within the late Proterozoic Katangan sediments of the Lufilian Arc (Fig. 13.3). The Katangan rests unconformably on a granite-schist-quartzite basement and the lowermost Katangan sediments fill in the valleys of the pre-Katangan land surface. Most mineralization in Zambia and south-eastern Shaba occurs in the Ore Formation. This lies a few metres above the level at which the pre-Katangan topography became filled in. Shale or dolomitic shale forms the host rock for about 60% of the mineralized ground and the shale orebodies form a linear group to the south-west of the Kafue Anticline (Fig. 13.4). Arkose-arenite hosted ores occur mainly to the north-east of the anticline, e.g. Mufulira (see page 14). The footwall succession consists of quartzites, feldspathic sandstones and conglomerates of both aquatic and aeolian origin. The Ore Formation, generally 15-20 m thick, is succeeded by an alternating series of arenites and argillites. These and the rocks below them make up the Lower Roan Group. All the rocks and their contained copper minerals have suffered low to high grade greenschist facies metamorphism and many of the so-called shales are biotite-schists. In places they are tightly folded (Fig. 13.5).

Copper, together with minor amounts of iron and cobalt, occurs mainly in the lower part of the Ore Formation as disseminated bornite, chalcopyrite and chalcocite. In places, mineralization passes for short-distances into the underlying beds. Both the upper and lower limits of mineralization are usually sharply defined. The sulphide minerals show a consistent zonal pattern with respect to the strandline from barren near-shore sediments to chalcocite in shallow water, to bornite with carrollite and chalcopyrite, then chalcopyrite and finally pyrite (in places with sphalerite) in the deeper parts of marine lagoons and basins. A synsedimentary origin of the mineralization is suggested by various lines of evidence (Fleischer *et al.* 1976) but particularly by the evidence from slump breccias that mineralization was present before slumping occurred. These authors postulated that the supply of copper came from springs or streams draining a hinterland of aeolian sands and terrestrial red beds, and that it was selectively precipitated by hydrogen sulphide in bodies of standing water to form the zone of mineralization described above. Binda (1975) described detrital grains of bornite and bornite-bearing rock fragments from Mufulira and suggested that clastic sedimentation played a role (of yet unknown importance) as an ore-forming process in the Copperbelt. From an epigenetic viewpoint, Raybould (1978) has pointed out that the Zambian Copperbelt and other major Proterozoic stratiform copper and lead-zinc deposits such as McArthur River, Northern Territory; Mount Isa, Queensland; Sullivan, British Columbia; White Pine, Michigan, etc., appear to have developed in intracratonic and cratonic-margin rift systems of mid to late Proterozoic age. He suggested that the mineralization resulted directly from deep-seated processes coeval with the rifting and not from surface weathering. Annels (1979) has suggested that diagenetic metal chloride-enriched brines were responsible for the mineralization at Mufulira.

(c) *The White Pine Copper Deposit, Northern Michigan.* A different type of copper deposit is found in Precambrian strata at White Pine (Fig. 13.6). This district

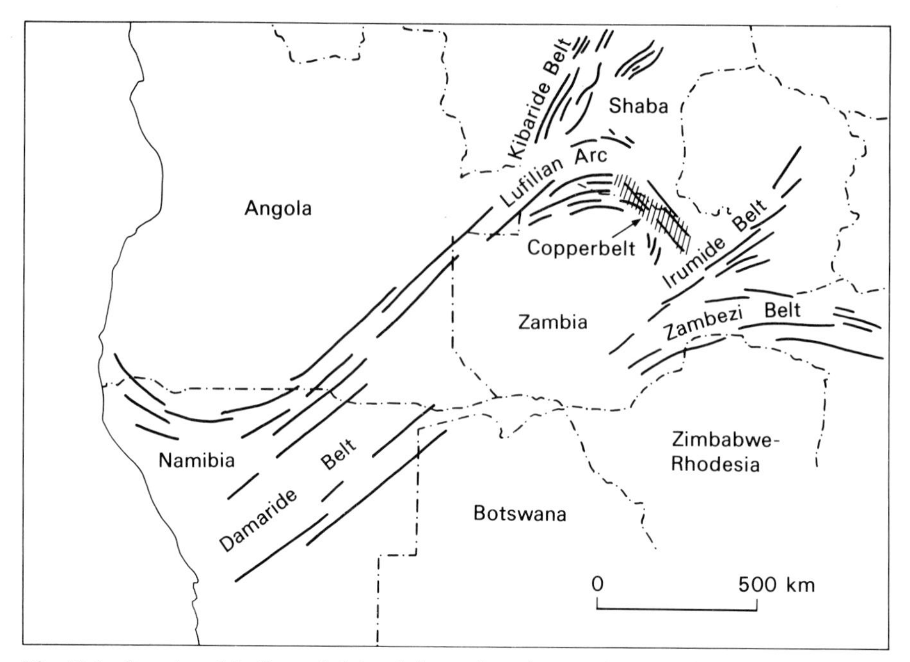

Fig. 13.3. Location of the Copperbelt in relation to the main tectonic trends of Central Africa. (After Raybould 1978.)

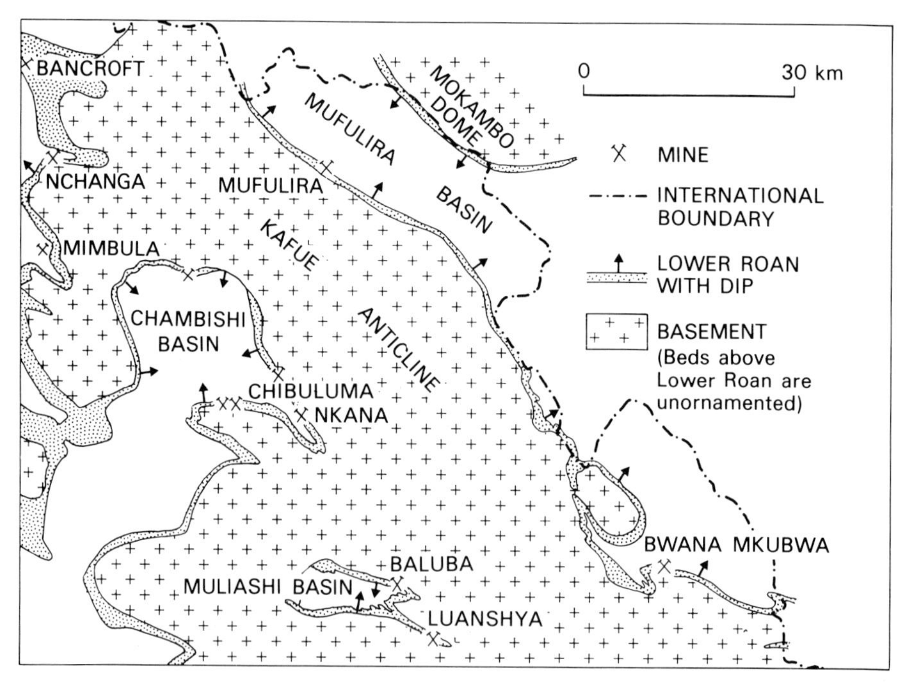

Fig. 13.4. Location map for the Zambian Copperbelt showing the regional geology. (Modified from Fleischer *et al.* 1976.)

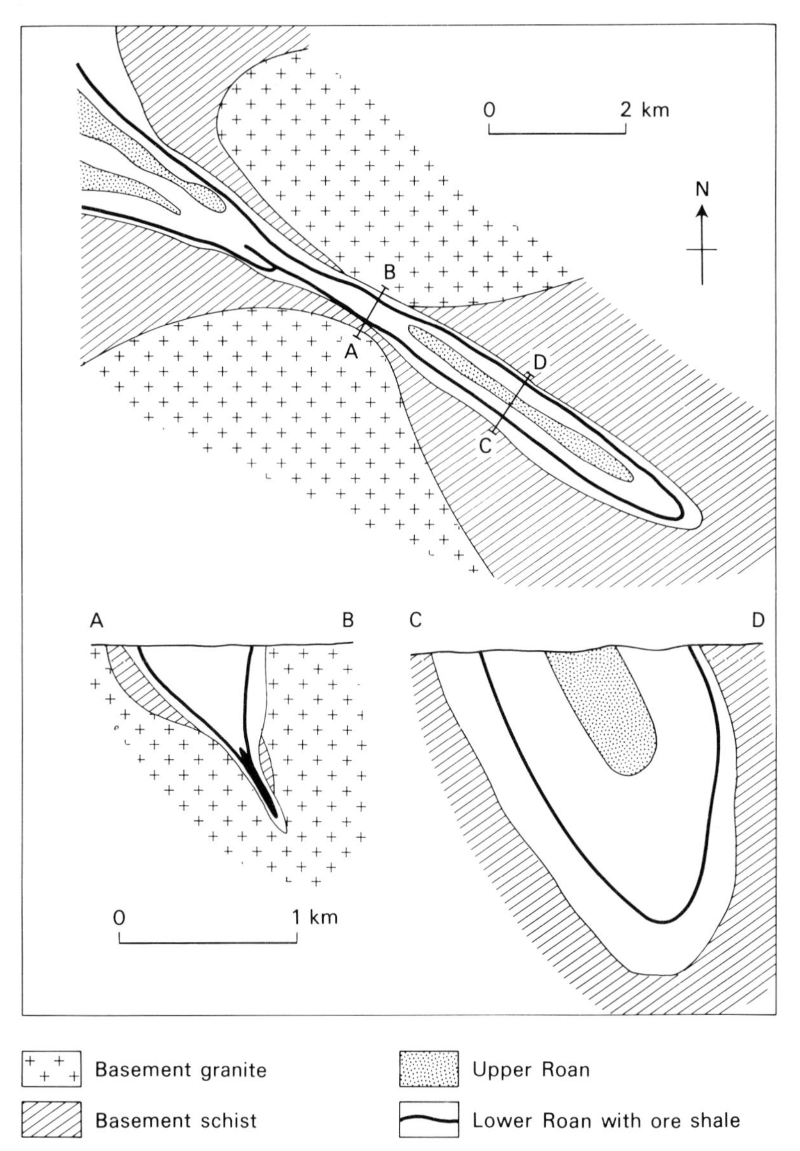

Fig. 13.5. Sketch map and sections of the Luanshya Deposit, Zambia. (Modified from Dixon 1979.)

supplies about 5% of the copper mined in the USA. Ore grade is about 1.2% copper as chalcocite and native copper. These minerals occur in the Nonesuch Shale which was deposited about 1000 Ma ago. This formation is about 150-200 m thick but only the lower 8-15 m contain significant copper mineralization (Burnie *et al.* 1972). The name 'Nonesuch Shale' is misleading because much of the formation is siltstone or sandstone. Copper mineralization is almost invariably confined to individual lithogical units and the content changes with sedimentary facies variations. The Nonesuch Shale conformably overlies the Copper Harbor Conglomerate, the upper beds of which are also locally cupriferous.

The Nonesuch Shale has been divided into three subzones. A basal Cupriferous Zone contains chalcocite and native copper, but only traces of bornite, chalcopyrite

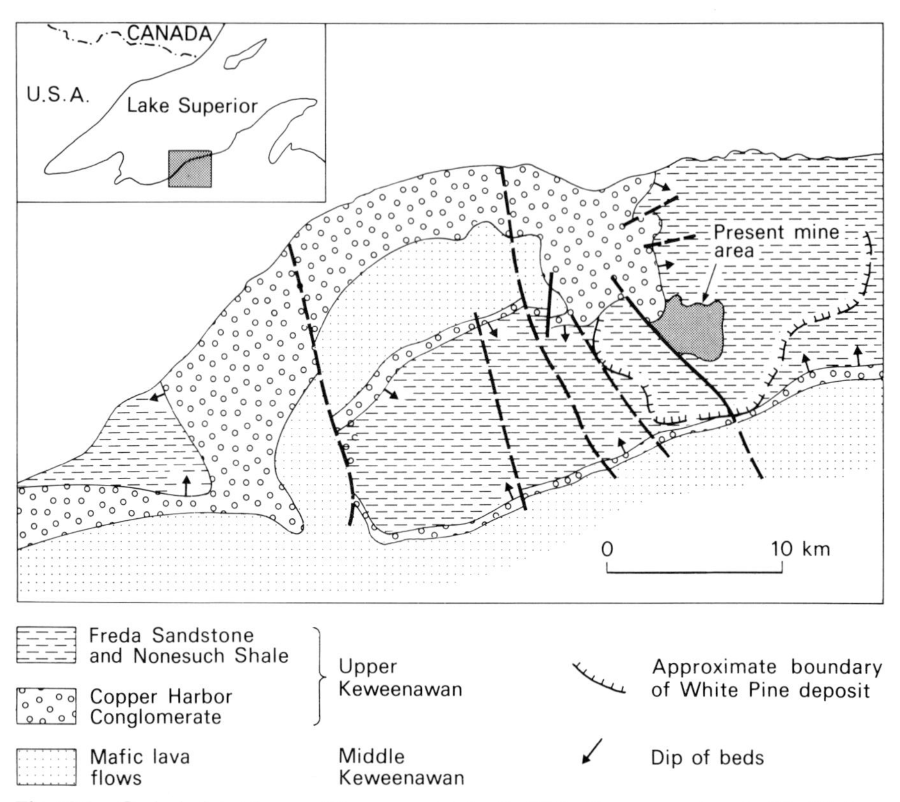

Fig. 13.6. Geological sketch map of the area containing the White Pine copper deposit. (Modified from Burnie *et al.* 1972 and White 1971.)

and pyrite. Next comes a transition zone with a gradation in the mineralogy from chalcocite through bornite and chalcopyrite to pyrite, lead-zinc minerals also occur. This is overlain by copper-poor pyritous shale and siltstone petrologically similar to the copper-bearing rocks. Within the cupriferous zone, maximum copper concentrations are found in siltstone and shale units immediately overlying sandstones. The sandstones are usually copper-poor.

Ideas concerning the genesis of this deposit show an interesting evolution. The original White Pine Mine worked ore from both the Nonesuch Shale and the Copper Harbor Conglomerate along the White Pine Fault and, because the mine was on a structural feature, it was assumed that the ore was epigenetic and deposited by hydrothermal solutions which rose up the fault. The discovery that the ore persisted along the base of the Nonesuch Shale over many tens of kilometres squared led to a concept of a syngenetic origin. Further study showed that locally ore cuts across bedding leading to a more sophisticated concept of genesis—that copper-bearing solutions circulated through beds that contained syngenetic pyrite and that the copper replaced the iron. A sulphur isotopic study by Burnie *et al.*, (1972) showed a wide range of $\delta^{34}S$ values characteristic of syngenetic sulphides formed by reduction of sea water sulphate.

Volcanic massive sulphide deposits

Some attention has already been paid to these deposits in Chapters 2 and 4. This section will therefore be used to amplify and add to what has already been written. The reader is consequently recommended to read pages 17 to 18 and 46 to 50 before reading this section.

GENERAL CHARACTERISTICS

Some of the more important features have been covered in the above sections of this book and only additional material is included here. The mineralogy of these deposits is fairly simple, the major minerals, in order of abundance, being: pyrite, pyrrhotite, sphalerite, galena, chalcopyrite (bornite and chalcocite are occasionally important); minor arsenopyrite, magnetite and tetrahedrite-tennantite may be present. The gangue is principally quartz. Occasionally, carbonate is developed: chlorite and sericite may be locally important.

The geochemical division into iron, iron-copper, iron-copper-zinc and iron-copper-zinc-lead deposits has already been touched on but it must be emphasized that while we may find pyrite deposits without any appreciable copper, copper is never found on its own. Similarly, if we find lead, we will have zinc and at least accessory copper too. With zinc will come copper and perhaps lead. Hutchinson (1973) suggested that these variations can be linked to crustal evolution. Thus, pyrite-sphalerite-chalcopyrite deposits are most numerous and best developed in Archaean greenstone belts, perhaps because they were generated under conditions of thin proto-crust, possibly by degassing of poorly differentiated mantle which was still sulphur-rich. Although they recur in younger volcanic successions, they become scarcer and smaller in later geological time. In subsequent epochs, their place is partially taken by the Cyprus and lead-bearing Kuroko types which are rare or absent in the Archaean. For instance, the lead-bearing type first appears in any volume in the Proterozoic, whilst in Phanerozoic orogens the Cyprus-type is common in the ophiolites that are developed in ridge-rift environments. It is probably also developed early in island arc evolution (Evans 1976b), during later parts of which Besshi and Kuroko deposits are formed. It is important to note that precious metals are also produced from some of these deposits, indeed in some Canadian examples they are the prime product. Thus both Besshi and Kuroko types may produce silver and gold whilst the Cyprus type may have by-product gold.

The vast majority of massive sulphide deposits are zoned. Galena and sphalerite are more abundant in the upper half of the orebodies whereas chalcopyrite increases towards the footwall and grades downward into chalcopyrite stockwork ore (Fig. 2.15). This zoning pattern is only well developed in the polymetallic deposits. As the number of mineral phases decreases, so the zonation tends to become obscure and may not be in evidence at all in Cyprus type deposits.

Textures vary with the degree of recrystallization. The dominant original textures appear to be colloform banding of the sulphides with much development of framboidal pyrite, perhaps reflecting colloidal deposition. Commonly however, recrystallization, often due to some degree of metamorphism, has destroyed the colloform banding and produced a granular ore. This may show banding in the zinc-rich section whereas the chalcopyrite-pyrite ores are rarely banded. Angular inclusions of volcanic host rocks are occasionally present and soft sediment structures (slumps, load casts) are sometimes seen. Graded bedding has also been reported from some deposits (Sangster & Scott 1976).

Wall rock alteration is usually confined to the footwall rocks. Chloritization and sericitization are the two commonest forms. The alteration zone is pipe-shaped and contains within it and towards the centre the chalcopyrite-bearing stockwork. The diameter of the alteration pipe increases upward until it is often coincident with that of the massive ore. Metamorphosed deposits commonly show alteration effects in the hanging wall. This is probably due to the introduction of sulphur released by the breakdown of pyrite in the orebody. The sulphur, by

Fig. 13.7. Distribution of dacite lava domes and Kuroko deposits, Kosaka District, Japan. (Modified from Horikoshi & Sato 1970.)

reacting with pore solutions, could give rise to extensive hydrogen ion production. Of the many massive sulphide ores in sedimentary environments few are underlain by alteration pipes. One example is Sullivan, British Columbia, (Fig. 2.11) which has a brecciated and tourmalinized zone beneath the massive ore. This, too, was presumably a feeder stockwork and its presence suggests that Sullivan must be grouped with the volcanic massive sulphide deposits.

The close association with volcanic domes has already been noted. The Kuroko deposits of the Kosaka district, Japan, are a good example (Fig. 13.7). All the Japanese Kuroko deposits are associated with Miocene volcanics and fossiliferous sediments developed along the eastern margin of a major geosyncline. Mineralization occurred during a limited period of the Middle Miocene over a strike length of 800 km in the Green Tuff volcanic region. Within this region more than a hundred Kuroko-type occurrences are known, but most are clustered into eight or nine districts. Their host rocks are acid pyroclastic flows which are also centred on domes, particularly those showing evidence of explosive activity (Fig. 13.7). Kuroko ores have a consistent stratigraphical succession of ore and rock types and an idealized deposit (Fig. 13.8) contains the following units:

hanging wall: upper volcanic and/or sedimentary formation;
ferruginous quartz zone: chiefly hematite and quartz;
baryte ore zone;
Kuroko or black ore zone: sphalerite-galena-baryte;
Oko or yellow ore zone: cupriferous pyrite ores; about this level, but often

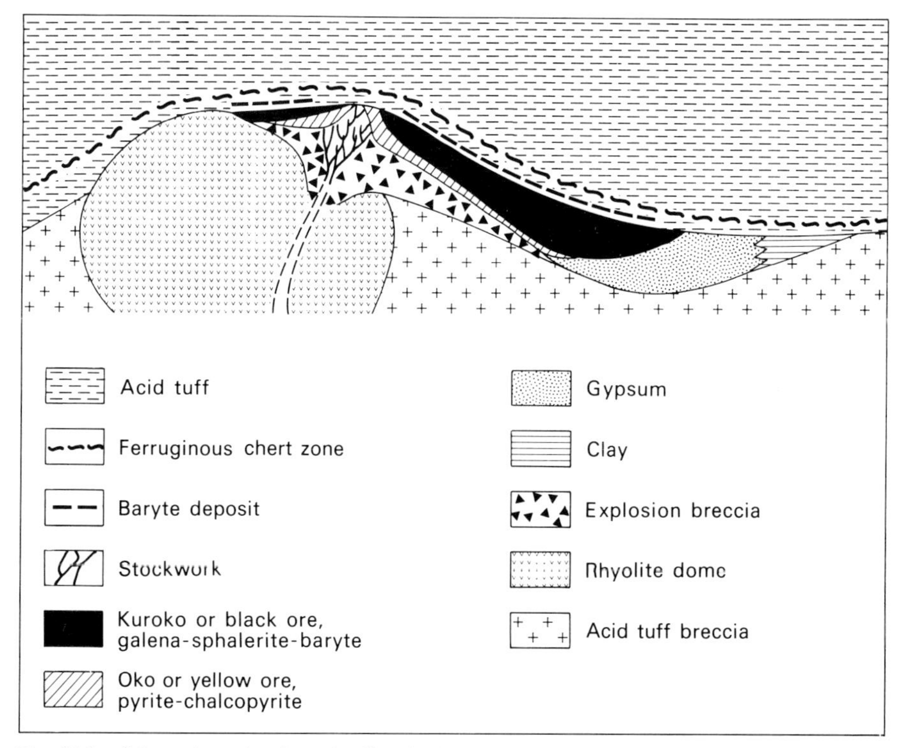

Fig. 13.8. Schematic section through a Kuroko deposit. (Modified from Sato 1977.)

towards the periphery of the deposit, there may be the Sekkoko zone of anhydrite-gypsum-pyrite;

Keiko or siliceous ore zone: copper-bearing, siliceous, disseminated and/or stockwork ore;

footwall: silicified rhyolite and pyroclastic rocks.

GENESIS

As was indicated in Chapter 4, there is by no means any unanimity regarding the origin of these deposits. Stanton (1978) objected to the division of this group into Cyprus, Besshi and Kuroko types as he considers the volcanic-exhalative ores to be one continuous spectrum, which show a progressive geochemical evolution that accompanies that of the calc-alkaline rocks in island arcs. Whilst agreeing that these two concomitant evolutionary trends are self-evident and probably complementary, the author of this text believes that it is useful to divide these deposits into different ore types just as we divide the accompanying volcanics into different rock types. That apart, as Stanton has pointed out, these two trends in island arcs take the form of pyritic copper deposits associated in time with both tholeiitic and ultrabasic rocks followed, with the change to calc-alkaline volcanism, by larger and more zinc-rich deposits which accompany andesitic to dacitic activity. Later lead-bearing deposits accompany the further development of the calc-alkaline trend in the form of dacites, rhyodacites and rhyolites. When analyses of the accompanying igneous rocks from an island arc carrying massive sulphide deposits, such as the Solomon Islands, are plotted on an AFM diagram, three magmatic

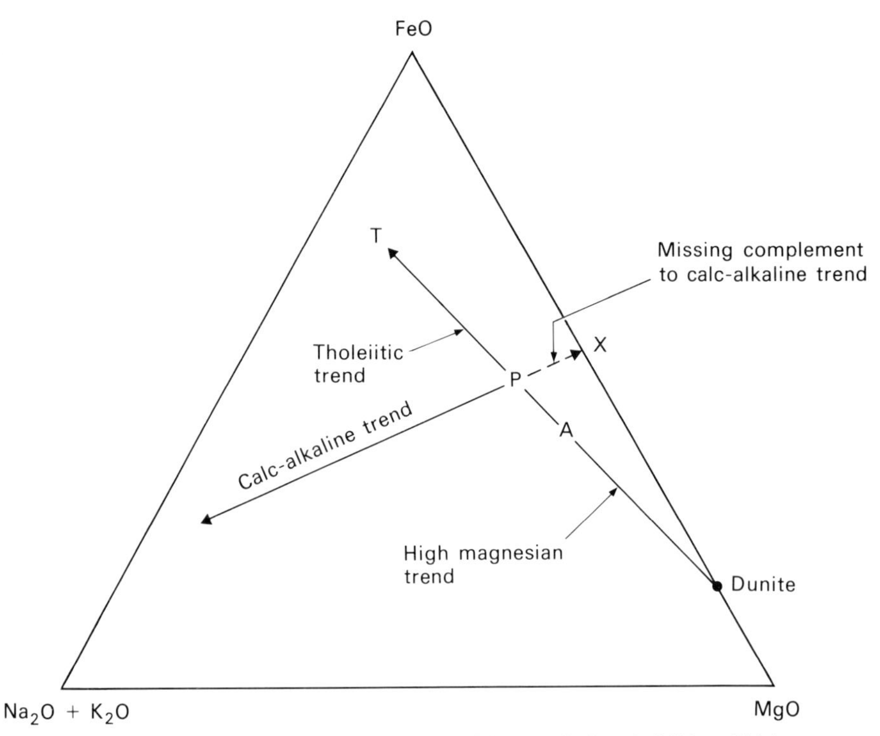

Fig. 13.9. AFM diagram showing the general nature of the calc-alkaline, tholeiitic and high magnesian suites together with the position of the missing complement to the calc-alkaline trend. (Modified from Stanton 1978.)

trends appear (Fig. 13.9). These are the tholeiitic PT, *with a complementary high magnesian trend towards dunite* PA and the calc-alkaline trend which Stanton emphasizes has no complementary trend following the line PX. The iron enrichment of the tholeiitic trend can be accounted for by olivine and orthopyroxene fractionation from a parental basalt magma (P) giving the complementary high magnesian trend. No mineral or commonly observed group of minerals can, however, be subtracted from a basaltic melt to produce the calc-alkaline trend. In other words, there is no complementary accumulative material having a composition falling along PX. Somewhere, somehow, Stanton contends, iron has been greatly reduced relative to the alkalis 'and its disappearance is the crux of the calc-alkaline problem.' For the Solomon Islands this can be demonstrated by a plot of lavas and associated intrusions (Fig. 13.10). Not only is this, the enigmatic gap, shown on a normal AFM plot, it is also shown by similar plots for manganese, zinc (Fig. 13.11), copper and several other metals. Clearly, these metals must have been removed from the magmatic melts during differentiation. Stanton argues that they have gone to form the exhalative and often associated manganese ores which must, on this argument, have a direct magmatic origin rather than having been deposited from circulating meteoric solutions. Urabe & Sato (1978) also prefer a magmatic origin in their discussion of certain Kuroko deposits in Japan. They contend that the circulating seawater model fails to explain the high salinities found in fluid inclusions, that it does not provide a heat engine to drive the circulating solutions or explain the intimate association of the deposits with well differentiated rhyolite domes.

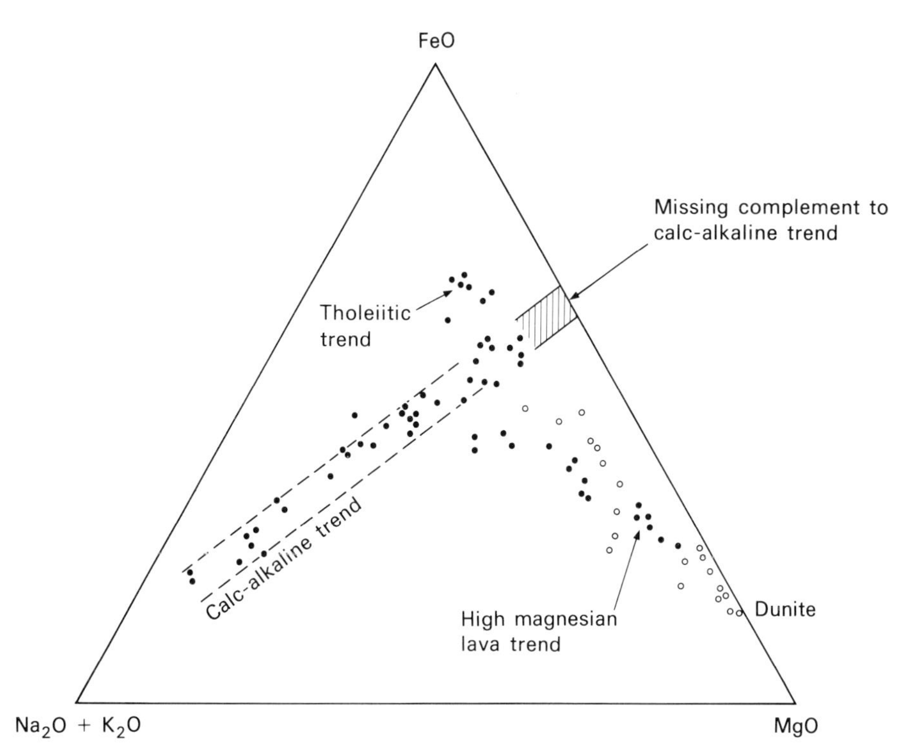

Fig. 13.10. AFM diagram with plot of lavas (solid circles) and Alpine-type peridotites (open circles) from the Solomon Islands. (Modified from Stanton 1978.)

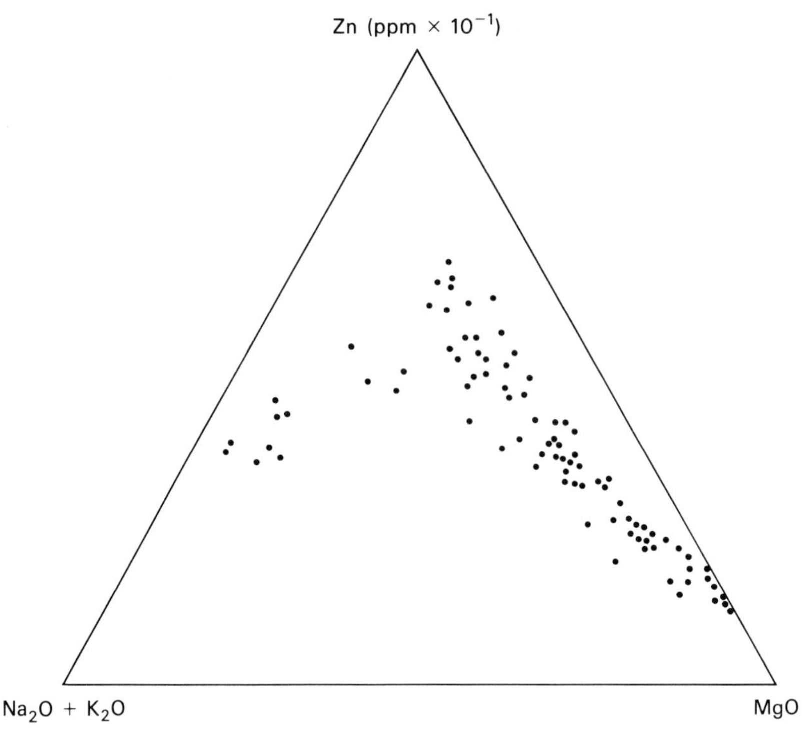

Fig. 13.11. A 'Zn' M diagram for the igneous rocks shown in Fig. 13.10. (Modified from Stanton 1978.)

Whilst the above arguments for a direct magmatic origin are very compelling, the evidence of the concomitant evolution of ore and rock types has not been ignored by those workers who prefer an origin by deposition from circulating meteoric water. This school would argue, as Graf (1977) does, that a relation should exist between the metal contents of massive sulphide deposits and the relative amounts of mafic and felsic rocks in the underlying volcanics. In a study of REE patterns in massive sulphide ores and associated rocks of the Bathurst-Newcastle District, New Brunswick, Graf found that samples high in lead sulphides have the highest europium anomalies and the strongest enrichments of heavy REE, and that these characteristics do not correlate with the abundance of any other mineral. This he argues indicates that the lead and heavy REE have been leached from a felsic rock, since heavy REE together with trace amounts of lead and zinc are preferentially enriched in plagioclase.

Other possible exhalative deposits

In addition to the sulphide deposits discussed above, there are a number of other deposits and groups of deposits which are of stratiform type, sometimes occur in a volcanic environment and may be of exhalative origin. Many workers have suggested that the Algoma type iron deposits found in Archaean greenstone belts are exhalative in origin. These iron-rich exhalites often carry gold and the spatial relationship of some gold deposits to these exhalites suggests that part of this gold was reconcentrated into veins.

As mentioned in Chapter 2, Solomon has suggested that stratiform oxide ores such as the magnetite-hematite-apatite ores of Kiruna and Gällivare in northern Sweden are oxide end members of the massive sulphide group. The famous Kiruna iron ores occur above altered acidic pyroclastics which are veined with magnetite. They have in the past been considered to be magmatic segregation deposits and recently Frietsch (1978) has again advanced a magmatic theory of origin for these and the similar ores of Missouri.

Other possible exhalative deposits include the Rexspar uranium deposit, British Columbia (Preto 1978) and certain uranium deposits in Labrador (Gandhi 1978), the stratiform tungsten deposits of Austria (Höll & Maucher 1976), some tin deposits of the Erzgebirge (Baumann 1970) and the famous Almaden mercury deposits of Spain (Saupe 1973).

14

The Vein Association

Vein, manto and pipe deposits have already received considerable attention in some earlier chapters. Their morphology and nature were described in Chapter 2 when such subjects as pinch-and-swell structure, ribbon ore shoots, mineralization of dilatant zones along faults, vein systems and orebody boundaries were outlined. The reader may find it convenient to review what is written on these subjects on pages 6-9. Similarly in Chapter 3 in a discussion of precipitation from aqueous solutions that characteristic texture of veins, crustiform banding, was described, as were fluid inclusions and wall rock alteration whose relationship to veins is most marked. It is generally agreed that vein filling minerals were deposited from hydrothermal solutions, and in Chapter 4 some consideration was given to the origin and nature of hydrothermal solutions including metamorphic processes such as lateral secretion. The relevant sections are on pages 36-40 and 40-42. In Chapter 5, paragenetic sequence and zoning were discussed with special reference to vein deposits, see pages 57-60.

If this chapter had been written thirty or so years ago it would probably have been the largest chapter in this book. However, since the 1940s the importance of veins has been steadily diminishing. The reasons for this are twofold. Firstly, a number of deposits formerly thought to belong to this association have been recognized as belonging to other ore classes. For example the Horne orebody of Noranda, Quebec was, for decades, believed to be a hydrothermal replacement pipe. It is now considered to be a volcanic massive sulphide deposit. The second and more important reason is economic. Taking copper as an example, only those ores containing more than 3% Cu were economically workable until the nineteenth century. Such high metal levels were generally only reached in hydrothermal veins, so exploration and mining concentrated on this class of orebody. Today, any rock containing 0.4% Cu can be economically exploited by large-scale mining methods and consequently veins do not hold their pre-eminent position any more. They are, however, still important for their production of tin, uranium and a number of other metals and industrial minerals such as fluorspar and baryte. As a result of their past importance, studies of vein ores have had a very profound influence on theories of ore genesis almost up to the present day, an influence out of all proportion to their true value.

Vein deposits and the allied, but less frequent, tubular orebodies show a great variation in all their properties. For example, thickness can vary from a few millimetres to more than a hundred metres and as far as environments are concerned, veins can be found in practically all rock types and situations though they are often grouped around plutonic intrusions as in Cornwall, England. Mineralogically they can vary from monomineralic to a mineral collector's paradise such as the native silver-cobalt-nickel-arsenic-uranium association of the Erzgebirge and Great Bear Lake, Northwest Territories.

Kinds of veins

The majority of veins have probably formed from uprising hydrothermal solutions which precipitated metals under environments extending from high temperature-high pressure near magmatic, to near surface low temperature-low pressure conditions. A few veins are pegmatitic and represent end stage magmatic activity, a few are clearly volcanic sublimates.

Mineralogically, gangue minerals can be the dominant constituents as in auriferous quartz veins. Quartz and calcite are the commonest, with quartz being dominant when the host rocks are silicates and calcite with carbonate host rocks. This suggests derivation of gangue material from the wall rocks. Sulphides are most commonly the important metallic minerals but in the case of tin and uranium oxides are predominant. Here and there, native metals are abundant especially in gold- and silver-bearing veins.

Zoning

This important property has already been considered. It is so commonly developed within single veins or groups of veins (district zoning) that some further mention of it must be made here. The base-metal veins of Cornwall show some of the best and earliest studied examples of orebody and district zoning (Fig. 5.6 and Table 5.1), which has had important economic implications. For example after the recognition of zoning in this tinfield, probably some time before de la Beche wrote about it in 1839, a number of abandoned copper mines were deepened and the delighted owners found tin mineralization beneath the copper zone. This tin had not been discovered earlier because there can be a barren zone of as much as 100 m or so between the tin and copper zones. On the other hand, as at the old Dolcoath Mine, there may be an overlap and a mine which commenced life as a copper producer could in time become a copper-tin producer and lastly a tin producer. Such was the history of this the richest individual tin vein that has been worked anywhere in the world. In all, Dolcoath produced 80 000 t Sn and 350 000 t Cu. It is an interesting exercise to calculate the value of this metal production at present day prices! As can be seen from Table 5.1, the zoning generally takes the form of a progressive change in composition with depth. Gradual changes in metallic minerals are accompanied by somewhat less pronounced changes in the gangue minerals with quartz being present at all levels. This zoning has been attributed to a slow fall in temperature with increasing distance from the granite intrusions though the zonal boundaries, if they are isothermal surfaces, are not parallel to the granite contacts. This discrepancy is, however, simply due to their adopting a compromise attitude between that of the source of heat and the cooling surface—ground level.

Some examples

THE LLALLAGUA TIN DEPOSITS, BOLIVIA

Bolivia possesses the greatest known reserves of tin outside the countries of south-east Asia. Most of her reserves are in vein and disseminated deposits. There is a

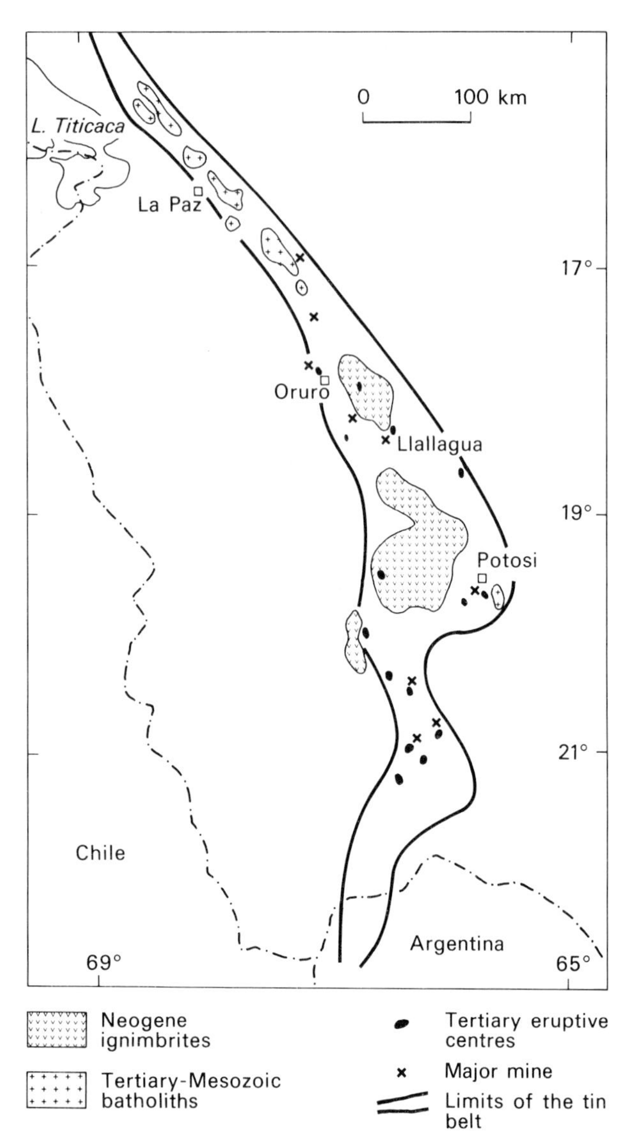

Fig. 14.1. The Bolivian Tin Belt. (After Grant *et al.* 1977.)

long history of mining dating back to the Spanish colonial days of the sixteenth century when most of the important tin deposits were discovered and were mined initially for the fabulously rich silver ores that had formed, partly by supergene enrichment processes, in the upper parts of the vein systems. Some of these deposits are still being mined, mainly for tin ores beneath the silver-rich zones.

The Bolivian tinfield extends along the Andean ranges east of the high plateau of Bolivia from north-east of Lake Titicaca to the Argentine border (Fig. 14.1). North-west of Oruro the deposits are mainly tin and tungsten veins associated with granodioritic batholiths. The batholiths range in age from Triassic to Miocene with the Miocene ones being best mineralized.

South of Oruro there is a tin-silver association spatially related to high level subvolcanic intrusions. At some of these volcanic centres both the intrusives and the coeval volcanics are preserved, at others erosion has removed the volcanics completely leaving only the intrusives. This is the case at Llallagua, the world's largest tin mine working primary tin deposits. This mine is estimated to have produced over 500 000 t of tin since the beginning of this century. The mine occurs in the Salvadora Stock which occupies a volcanic neck cutting the core of an anticline in Palaeozoic rocks. The stock is made up of xenolithic and highly brecciated porphyry. It narrows with depth. The original texture and rock composition (probably quartz-latite) are obscured by pervasive alteration. The ubiquitous brecciation is important because it has produced an increased permeability of the

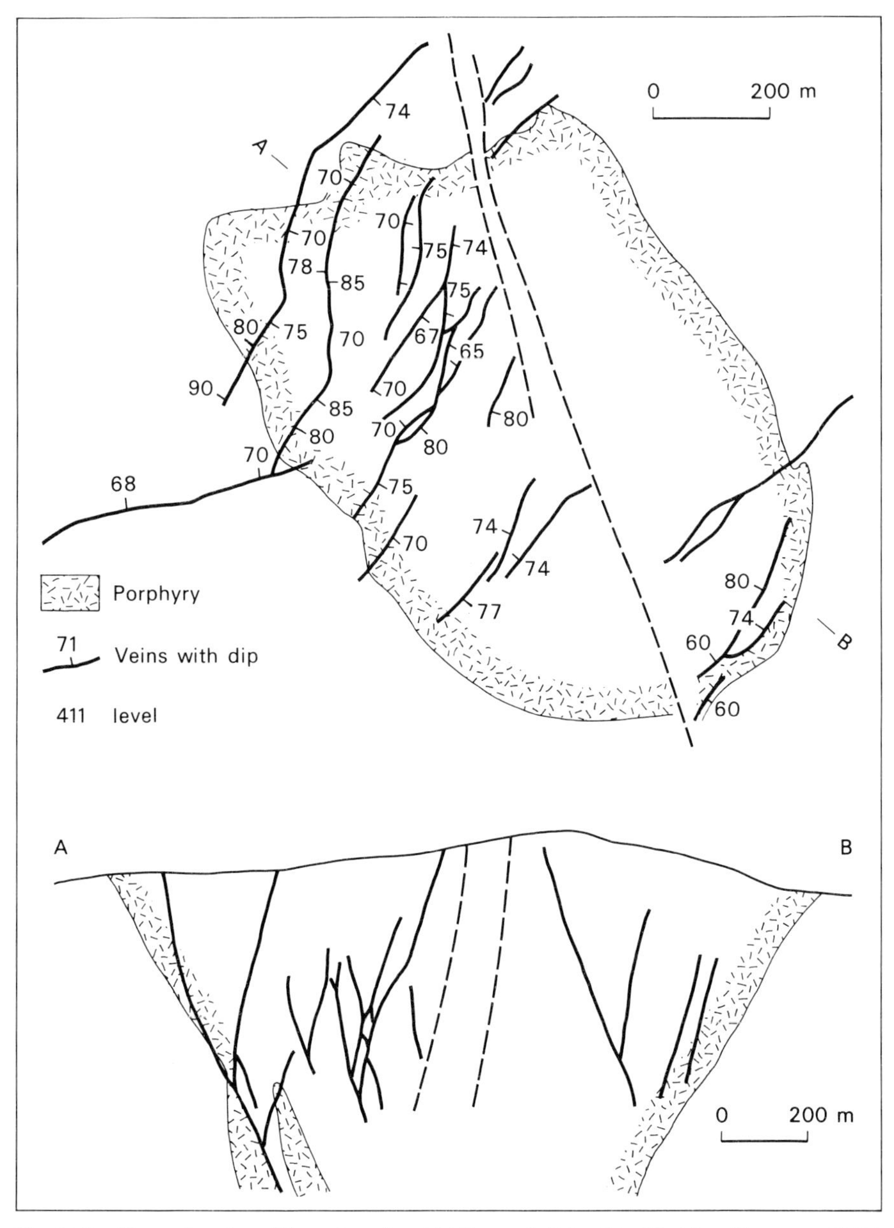

Fig. 14.2. Plan and section of the major veins at Llallagua, Bolivia. (Modified from Turneaure 1960.)

stock which has given rise to alteration and mineralization independent of the geometry of the later vein systems. This dispersed mineralization may be of high enough grade to permit bulk mining methods to be used, i.e. it may be a porphyry tin deposit (Grant *et al.* 1977). The alteration has produced a host rock consisting of primary quartz, tourmaline, sericite and secondary quartz.

There is a network of veins in and around the stock some of which are shown on Fig. 14.2. The major veins trend about 030° and appear to form part of a conjugate system of normal faults. They are typified by the San Jose type which generally has a dip of 45-80°, a good width and strike persistence. The average width is about 0.6 m but widths of up to 1.8 m are known. These are the richest veins and can contain up to 1 m of solid cassiterite. Clay gouge is very common. The Serrano vein type is much thinner (average 0.3 m), nearly vertical and impersistent. These veins can be as richly mineralized as the San Jose type, but with such narrow veins dilution with country rock material occurs during mining. There is little or no clay gouge. These veins may have formed in vertical tension gashes associated with the normal faulting.

The first stage of mineralization consisted of the formation of crusts of quartz followed by bismuthinite and cassiterite and many high grade veins are composed almost entirely of these three minerals. Wolframite and tourmaline also belong to this early stage (Turneaure 1960). This was followed by a stage of sulphide mineralization principally pyrrhotite and franckeite, the pyrrhotite being later largely replaced by pyrite, marcasite and siderite. Arsenopyrite, sphalerite, stannite and rare chalcopyrite were also formed during this later stage of sulphide deposition. The discerning reader will have noted that in this vein system there is present, over a very restricted vertical range, a host of minerals which in Cornwall are developed in a zonal sequence spread out vertically over hundreds of metres. The concentration of low and high temperature minerals into the same 'zone' is called telescoping and is thought to be due to the existence of high temperatures in near-surface host rocks induced by the volcanic activity and a blanket of hot volcanic deposits. Fluid inclusion work by Grant *et al.* (1977) has shown that the pervasive alteration and early vein growth, including cassiterite deposition, took place at temperatures (uncorrected for pressure) of about 400-350°C. The temperature dropped to 300°C and lower during the sulphide deposition.

BUTTE, MONTANA

This is one of the world's most famous vein mining districts. From 1880-1964 Butte produced 300 Mt of ore yielding 7.3 Mt Cu, 2.2 Mt Zn, 1.7 Mt Mn, 0.3 Mt Pb, 20 000 000 kg Ag, 78 000 kg Au, together with significant amounts of bismuth, cadmium, selenium, tellurium and sulphuric acid. Because of this wealth of mineral production from a very small area (little more than 6 x 3 km) with more than a score of mines Butte has been aptly called the richest hill on earth, for the monetary value of its production has only been exceeded by the much larger Witwatersrand Goldfield of South Africa. Cut-and-fill stoping of the veins was the main mining method up to 1950, then it was joined by the block caving of veined ground and in 1955 large-scale open pit working of low grade porphyry-type mineralization commenced. Reserves are still extensive, of the order of 10 Mt of high grade vein copper and silver ore and 500 Mt of low grade copper mineralization.

The Butte field is in the south-western corner of the Cretaceous Boulder

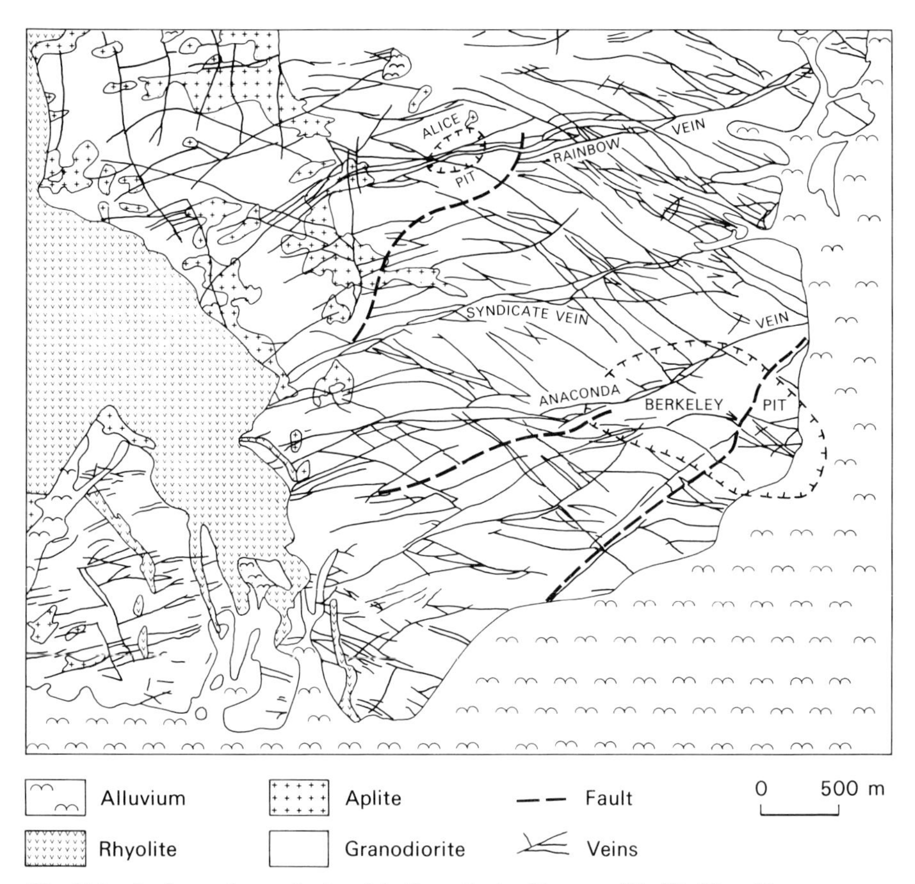

Fig. 14.3. Surface geology and veins of the Butte district, Montana. (Modified from Meyer *et al.* 1968.)

Batholith. Soon after the emplacement of the intrusion large parts of the area were covered by rhyolitic and dacitic eruptions (Fig. 14.3). The veins occur in a granodiorite which has given a radiometric age of 78 Ma. There were two mineralization stages: pre-main and main. The pre-main stage consists of small quartz veins carrying molybdenite and chalcopyrite found in the deeper central parts of the mineralized zone. They are bordered by alteration envelopes carrying potash feldspar, biotite and sericite. The biotite has been dated at 63 Ma.

The main mineralization stage occurs in several vein systems of which the most important are the easterly trending Anaconda and the later north-westerly trending Blue veins. The Anaconda veins are the major producers in the western third of the mineralized zone and also in the eastern third where they divide into myriads of closely spaced south-easterly trending minor veins. This is called horse-tailing and gives rise to porphyry-type mineralization suitable for mass-mining methods. The Anaconda veins are the largest and most productive, averaging 6-10 m in thickness with local ore pods up to 30 m thick. The Blue veins usually offset the Anaconda veins with a sinistral tear movement and sometimes ore has been dragged from the Anaconda into the Blue veins by this fault movement. Individual oreshoots persist along strike and in depth for hundreds of metres.

All the veins contain similar mineralization and this is strongly zoned (Fig. 14.4).

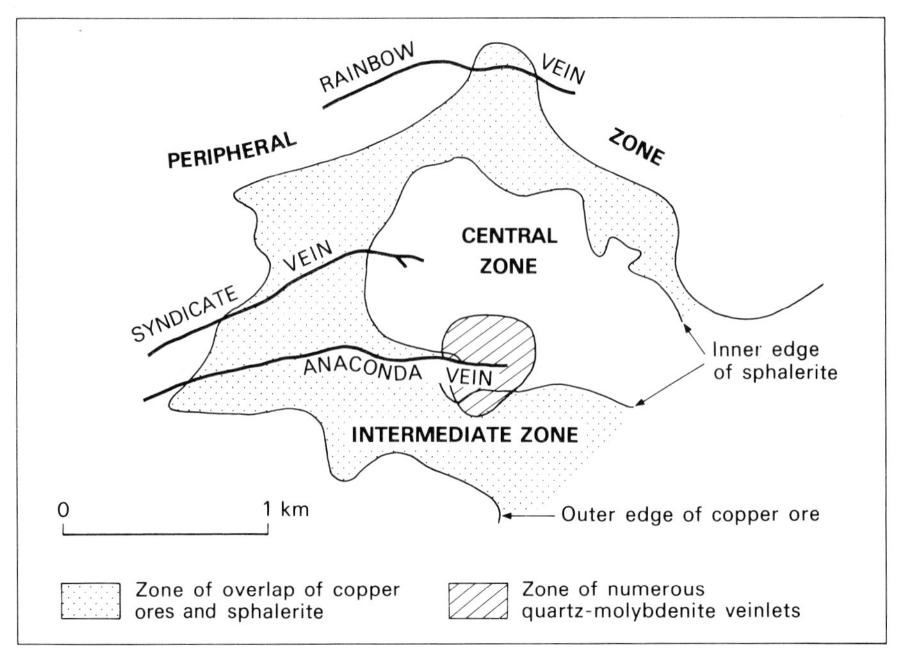

Fig. 14.4. Zoning on the 2800 ft level at Butte, Montana. (Data from Meyer *et al.* 1968.)

There is a Central Zone of copper mineralization which at depth contains the pre-main stage quartz-molybdenite veins and which is particularly rich in chalcocite-enargite ore. This gradually gives way outwards to ore dominated by chalcopyrite and containing minor amounts of sphalerite—the Intermediate Zone. The Peripheral Zone is principally sphalerite-rhodochrosite mineralization with small quantities of silver. All the veins are bordered by zones of alteration. This usually consists of sericitization next to the vein followed outwards by intermediate argillic alteration and then by propylitization. In deeper parts of Butte, advanced argillic alteration is present next to the veins. Sericite produced by this alteration dates at 58 Ma, distinctly later than the biotite of the quartz-molybdenite mineralization. The zoning suggests that the main mineralization was effected by hydrothermal solutions which passed upwards and outwards during a long period of time through a steadily evolving fracture system.

15

Strata-bound Deposits

Introduction

The term strata-bound is applied, irrespective of their morphology, to those deposits which are restricted to a fairly limited stratigraphical range within the strata of a particular region. For example, the vein, flat and pipe lead-zinc-fluorite-baryte deposits of the Pennine orefields of Britain are restricted to the Lower Carboniferous and are therefore spoken of as being strata-bound. To take a very different example, the stratiform deposits of the Zambian Copperbelt are all developed at about the same stratigraphic horizon in the Roan Series (Chapter 13) and these too may be described as strata-bound. Clearly, stratiform deposits can be strata-bound but strata-bound ores are not necessarily stratiform. Two important associations will be dealt with here: carbonate-hosted base metal deposits and sandstone-uranium-vanadium-base metal deposits.

Carbonate-hosted base metal deposits

These are important producers of lead and zinc and also, sometimes principally, of fluorite and baryte. Copper is important in some fields notably that of Central Ireland.

DISTRIBUTION IN SPACE AND TIME

Most of the lead and zinc produced in Europe and the United States comes from this type of deposit. In Europe there are important fields in Central Ireland, the Alps, southern Poland and the Pennines of England. In the United States there are the famous Appalachian, Tri-State (south-west Missouri, north-east Oklahoma and south-east Kansas), south-east Missouri and Upper Mississippi districts. There are also important fields in north Africa (Tunisia and Algeria) and Canada.

There are very few deposits of this type in the Precambrian. Substantial deposits first appear in the Ordovician of south-east Missouri and important deposits occur in all systems, except the Silurian, up to the Cretaceous as can be seen from the following examples:

Ordovician: South-east Missouri (Old and New Lead Belts);
Devonian: Pine Point, Northwest Territories, Canada;
Carboniferous: Central Ireland; British Pennines; Tri-State;
Permian: Trento Valley, Italy;
Triassic: Eastern Alps (Austria and northern Yugoslavia);
Jurassic: Southern Poland (also in Triassic and Devonian);
Cretaceous: northern Algeria and Tunisia.

ENVIRONMENT

We are not concerned in this group of deposits with skarns and the skarn environment, but with deposits which occur mainly in dolomites and, to a lesser extent, in limestones. The most important feature is the presence of a thick carbonate sequence. Thin carbonate layers in shales seldom contain important deposits of this type (Sangster 1976). The fauna and lithologies of the limestone hosts show that they were mostly formed in shallow water near shore environments of warm seas, and a plot of major carbonate-hosted deposits on palaeolatitude maps shows a grouping of these deposits in low latitudes (Dunsmore & Shearman 1977). The warmer climates of low latitudes encourage the development of reefs and so the frequent, but by no means universal, association of these deposits with reefs (e.g. south-east Missouri) and carbonate mudbanks (e.g. Ireland) is not surprising. The occurrence of reefs and carbonate mudbanks is related to ancient shorelines and sea bed topographies. Nowadays, along the shorelines where carbonate deposition is occurring we often find arid zones, as in the Persian Gulf, with desert prograding supratidal flats or coastal sabkhas. There, gypsum and anhydrite are precipitated from marine-derived groundwaters to form evaporites. These may be of considerable significance. The isotopic composition of the sulphur or sulphides from a number of carbonate-hosted deposits suggests origination from sea water sulphate, particularly sea water of the same age as the limestone country rocks. Sulphate evaporites are known to be interbedded with the limestones in relatively close regional proximity to many carbonate-hosted deposits (Dunsmore & Shearman 1977). Thus, as Stanton (1972) has emphasized, the primary regional control of such deposits is palaeogeographical.

Environments such as those described above developed in the past along the margins of marine basins which formed in stable cratonic areas such as the Devonian Elk Point Basin of western Canada in which the Pine Point deposits occur. Carbonate-hosted lead-zinc deposits also occur in a very different environment—in the failed arms (aulacogens) of the triple junctions of rifted continental areas as in the Benue Trough and the Amazon Rift Zone, and on the flanks of embryonic oceans as down the flanks of the Red Sea. In this environment too there is an important development of evaporites. A negative but important point is that they are remote from post-host rock igneous intrusives which might be the source for mineralizing solutions.

Returning to the cratonic basin-shelf sea environment we may note other regional controls which are well exemplified by the Mississippi Valley region and the British deposits. In these regions the orefields are present in positive areas of shallow water sedimentation separated from each other by shale-rich basins. Such positive areas in the British Isles are often underlain by older granitic masses, and Evans & Maroof (1976) suggested that these very competent rocks fractured easily to produce channelways for uprising solutions which, on reaching the overlying limestones, gave rise to the mineralization. In addition, a large number of deposits are clearly related spatially to faults sometimes of a regional character (Figs 2.10, 5.7) up which the ore solutions may have passed.

Sangster (1976) has divided carbonate-hosted base metal deposits into two major types: (1) Mississippi Valley, and (2) Alpine. Other workers do not make this distinction and refer to all low temperature carbonate-hosted deposits as being of Mississippi Valley-type. Sangster contends that the first type, which are strata-bound,

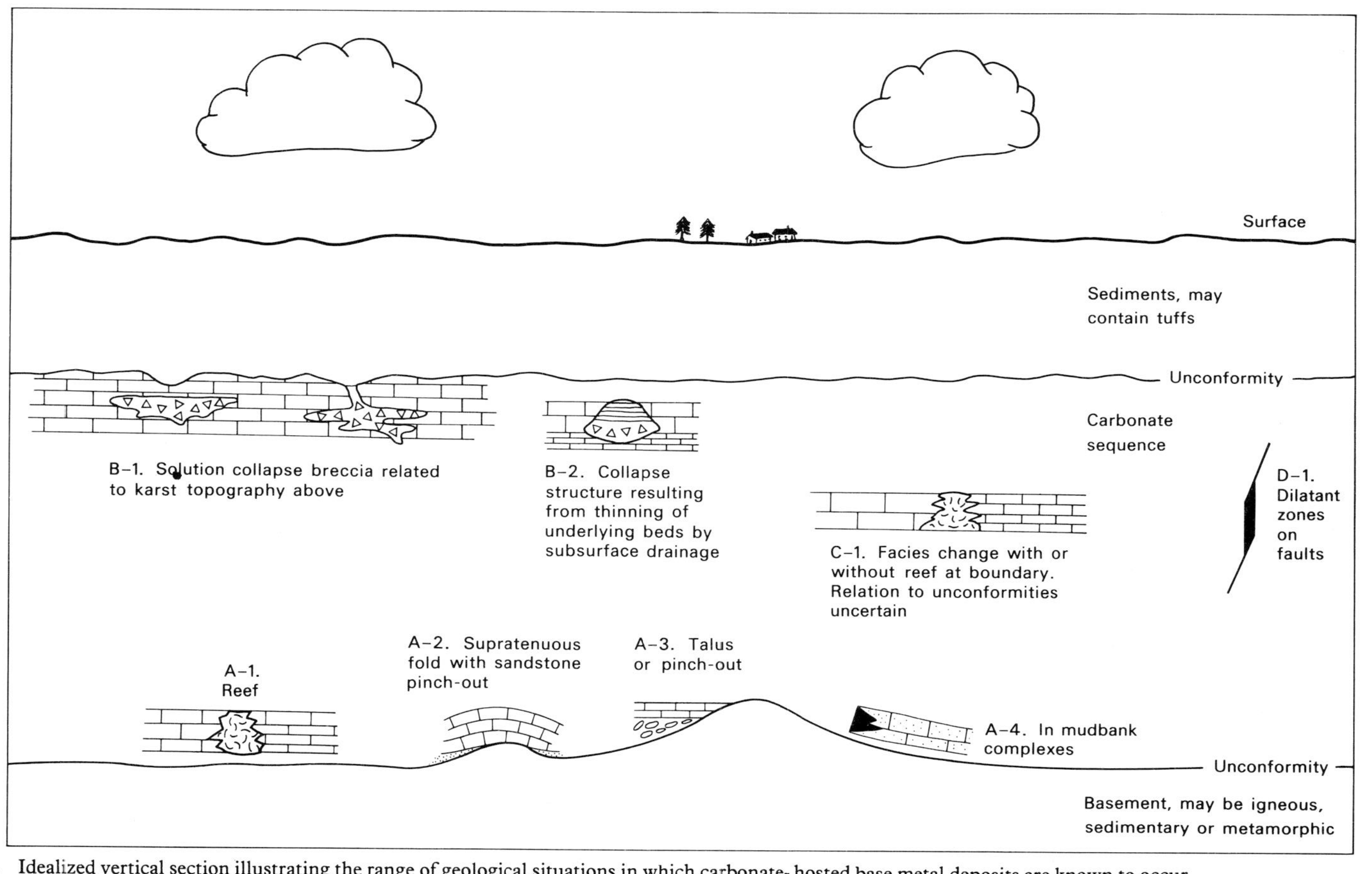

Fig. 15.1. Idealized vertical section illustrating the range of geological situations in which carbonate- hosted base metal deposits are known to occur. (Modified from Callahan 1967.)

were emplaced after lithification of the host rocks and were largely controlled by pre-ore structures, whilst he considers the Alpine-type to be stratiform and synsedimentary. The author of this book feels that this is a difficult distinction to sustain. For example, Sangster classifies the Irish deposits as being Alpine in type although they show a complete spectrum from thoroughly epigenetic types through to stratiform perhaps synsedimentary orebodies (Evans 1976a), and thus what seems to be a single orefield consisting of related deposits would have to be split into two different types. However, because this distinction is made by some workers on a global basis, the author has avoided using the term Mississippi Valley-type for all carbonate-hosted base metal deposits.

OREBODY TYPES AND SITUATIONS

As has been made clear by the above discussion, the orebodies are very variable in type. In the British Pennines, vein orebodies with ribbon ore shoots occupying normal faults are the main deposit type in the northern field (Fig. 2.4). In the southern Pennines, veins are again the most important orebodies but there they occupy tear (wrench) faults. The orebodies in the Tri-State field are in solution and collapse structures, caves and underground channelways connected with karst topography. In Ireland, the orebodies vary from stockwork brecciation zones to stratiform deposits (Fig. 2.10). At Pine Point, the ores are in interconnected small-scale solution cavities which Dunsmore (1973) has suggested are the result of the dissolution of carbonate rocks by corrosive fluids generated by a reaction between petroleum and sulphate ion. Some of the geological situations in which these deposits occur are shown in Fig. 15.1. They may be listed as follows:

(1) Above unconformities in environments such as reef and facies changes (A-1), supratenuous folds (A-2); above the pinch-outs of permeable channelway horizons (A-3); above or in mudbank complexes (A-4);

(2) Below unconformities in solution-formed open spaces (caves, etc.) related to a karst topography predating the unconformity (B-1), or in collapse structures formed by the dissolution of underlying beds by subsurface drainage (B-2);

(3) At a facies change in a formation, or between basins of deposition (C-1);

(4) In regional fracture systems (D-1).

GRADE, MINERALOGY AND ISOTOPIC CHARACTERISTICS

Average ore grades range mainly from 3-10% combined Pb+Zn with individual orebodies running up to 50%. Tonnages generally range from a few tens of thousands up to 20 Mt. As an example, we can look at the Navan Mine in Ireland, the largest zinc mine in Europe. This has a number of closely spaced orebodies whose proven, probable and possible reserves total 62 Mt grading 12% Zn+Pb. Zn/Pb = 5/1, the cut-off grade is 4% and the annual production of ore is 2.25 Mt.

The characteristic minerals of this ore association are galena, sphalerite, fluorite and baryte in different ratios to one another varying from field to field. Pyrite and especially marcasite may be common and chalcopyrite is important in a few deposits. Calcite, dolomite, other carbonates and various forms of silica usually constitute the main gangue material. Colloform textures are common in some ores. High trace amounts of nickel seem to be characteristic of these ores (Ixer & Townley 1979).

Sulphur isotopic abundances have been studied in both the sulphide and

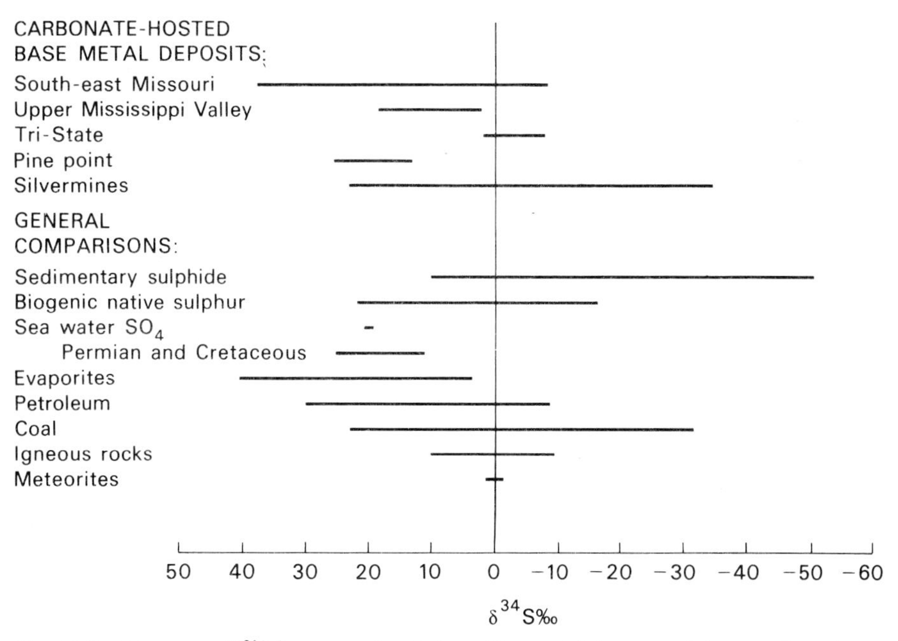

Fig. 15.2. The range of $\delta^{34}S$ for some carbonate-hosted base metal deposits and the range for major sources of sulphur that could have contributed to the ore deposits. (Modified from Heyl *et al.* 1974.)

sulphate minerals. These tend to vary from field to field (Fig. 15.2) but generally show a range of positive $\delta^{34}S$ values. This range may be explained in terms of fractionation as a function of mineral species, temperature or chemical environment or by the mixing of sulphur from different sources (Heyl *et al.* 1974). These authors have suggested that a comparison with values for crustal rocks (Fig. 15.2) indicates a crustal source for the sulphur of these deposits. Thus, the Pine Point values suggest that the sulphur was derived from marine evaporites (Rye & Ohmoto 1974). Silvermines, however, appears to be a notable exception, for sulphur isotopic studies have suggested a meteoric-type (mantle?) source for all the sulphur (Greig *et al.* 1971), though Coomer & Robinson (1976) have suggested that some of the sulphur was derived from sea water.

Like sulphur, lead isotope abundances are variable in nature. This variation results from the addition to existing lead of varying amounts of ^{206}Pb, ^{207}Pb and ^{208}Pb produced by the radioactive decay of uranium and thorium. Two distinct categories of lead, ordinary lead and anomalous lead, have been recognized. Ordinary lead has isotopic ratios which increase steadily with time so long as it remains in uniform source rocks (probably the mantle) in contact with constant amounts of uranium and thorium. Once it has been removed from its source and separated from the elements producing radiogenic lead ($^{206-208}Pb$) its isotopic composition is fixed and if a mathematical model is assumed for the rate of addition of radiogenic leads in the source region then a model lead age can be calculated. Generally, such ages are in reasonable agreement with other radiometric age determinations, but sometimes they are grossly incorrect and such leads are defined as anomalous. Some give negative ages, i.e. the model lead age says that they have not yet been formed! These leads must have had a more complicated history, presumably within the crust, during which they acquired extra amounts of radiogenic lead. They are sometimes called J-type leads after Joplin, Missouri from which the first examples to be found were collected. Other leads, on a simple mathematical model, are older than the rocks in which they occur. These could be leads which were first removed from the mantle then 'stored' in some older rocks before being remobilized and redeposited in younger rocks.

Leads from the various fields of the Mississippi Valley have been found to be notably enriched in radiogenic lead (i.e. having $^{206}Pb/^{204}Pb$ ratios of twenty or greater) compared to ordinary lead. They are all J-type leads. The strange thing is that although J-type lead is ubiquitous in the Mississippi Valley fields, most similar carbonate-hosted deposits such as those of Pine Point, the British Pennines, central Ireland and southern Poland contain ordinary lead. These facts suggest that the Mississippi Valley lead was derived from a crustal source relatively high in uranium and thorium which could have provided it with anomalous amounts of radiogenic lead. A highly probable source of this nature would be the Precambrian basement (Heyl *et al.* 1974) because of the conspicuous similarity between the source age of the lead, as determined from the slope of the lead-lead isochron, and the accepted age of the basement rocks (1450 Ma) in the Upper Mississippi Valley

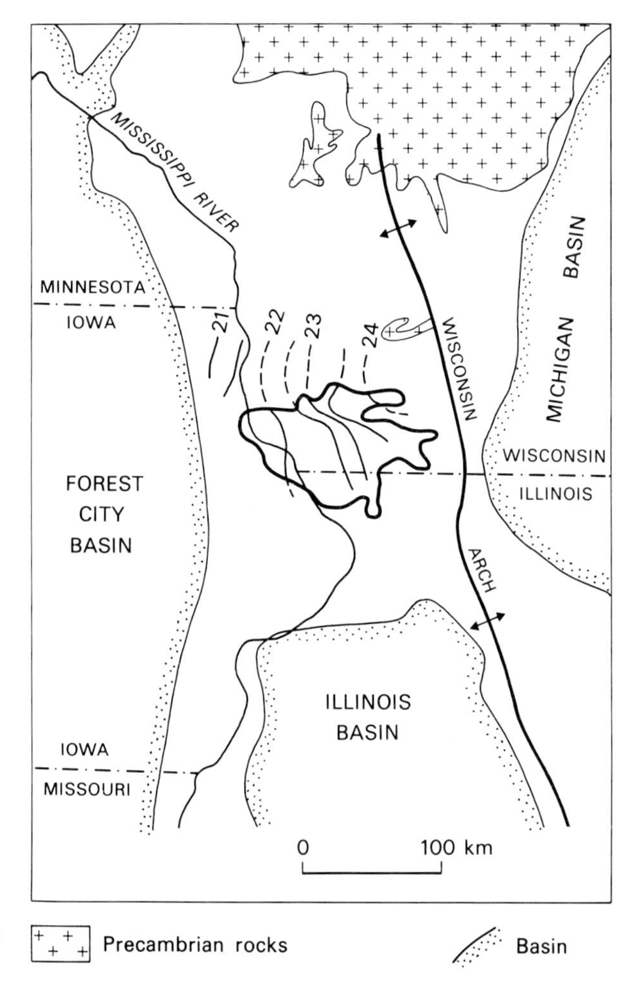

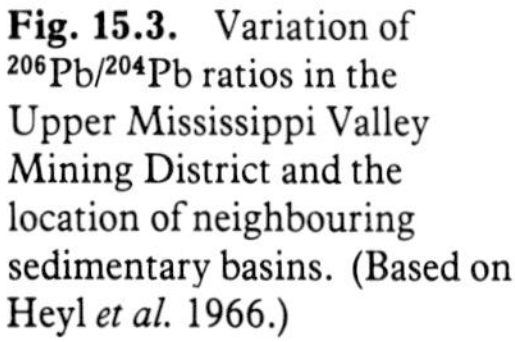
Fig. 15.3. Variation of $^{206}Pb/^{204}Pb$ ratios in the Upper Mississippi Valley Mining District and the location of neighbouring sedimentary basins. (Based on Heyl *et al.* 1966.)

district. On the other hand work in the south-east Missouri district suggests that all the lead may have been derived by hydrothermal solutions passing through the permeable Cambrian Lamotte Sandstone.

The Upper Mississippi Valley district not only contains the most radiogenic lead yet known from a carbonate-hosted deposit anywhere in the world, it also shows an isotopic zoning (Fig. 15.3). The lead becomes progressively more radiogenic from west to east. Heyl *et al.* (1966) have pointed out that this could be produced by addition of highly radiogenic lead leached from the basement to more normal lead carried by basinal brines moving up dip into the district from the neighbouring Forest City and Illinois Basins (Fig. 15.3).

The fact that the lead in other carbonate-hosted ore districts (Pine Point, etc.) is ordinary lead has caused most workers to postulate a deep-seated uniform source (e.g. Greig *et al.* 1971 for Ireland), and it has been pointed out that there are a number of profound faults in the nearby Canadian Shield which, when projected south-westwards, pass under the ore zone at Pine Point.

ORIGIN

There is little doubt that the majority of deposits of this class have been formed from epigenetic hydrothermal solutions. A few show syngenetic features and the occasional deposit shows evidence of both epigenetic and syngenetic deposition, e.g. the Mogul Mine at Silvermines, Ireland, (Fig. 2.10) where the lower ore along the fault is epigenetic and yields meteoritic values for the sulphur isotopic ratios but the upper orebody appears to be syngenetic and to mark the time of ore formation. Some of the mechanisms of transportation and deposition of lead and zinc in hydrothermal solutions have been discussed on pages 36-40. The source(s) of these solutions and their metallic constituents is very problematical. For most fields, lead isotopic studies suggest a deep source for the metals but in some cases basinal brines may have played an important role. Jackson & Beales (1967) have argued strongly for such a mechanism and this may have been the case for the Upper Mississippi Valley district. Again the source of sulphur may have been mainly or wholly deep-seated (Ireland?) or from marine evaporites (Pine Point?). Clearly, we still have much to learn about the origin of this class of deposit.

Sandstone-uranium-vanadium-base metal deposits

These deposits are found in terrestrial sediments, frequently fluviatile, which were generally laid down under arid conditions. As a result, the host rocks are often red in colour and for this reason copper deposits of this type are commonly referred to as 'red bed coppers.' Uranium-rich examples are called Colorado Plateau-type, carnotite-type, sandstone-type, Wyoming roll-front-type, Wyoming geochemical cell uranium-type or western states-type. The last term is advocated as being the best by Rackley (1976), but all are in common use.

In deposits of this type one or two metals are present in economic amounts, whilst the others may be present in minor or trace quantities. Thus copper mineralization (with chalcocite, bornite and covellite) is widespread in red bed successions though it is not often up to ore grade (Chapter 2, page 15). The same applies to silver and lead-zinc mineralization. Uranium mineralization ($\pm$vanadium) may be accompanied by trace amounts of the above metals but usually occurs as separate deposits.

WESTERN STATES-TYPE URANIUM DEPOSITS

Uranium deposits of this general type are widespread in the United States and they have provided over 90% of its domestic production of uranium and vanadium. From the global point of view they probably constitute one fourth of the non-communist world's reserves and they are now known in many parts of the world, e.g. South Australia (Brunt 1978). In the USA, these deposits are well developed in the Colorado Plateau region and in Wyoming (Fig. 15.4).

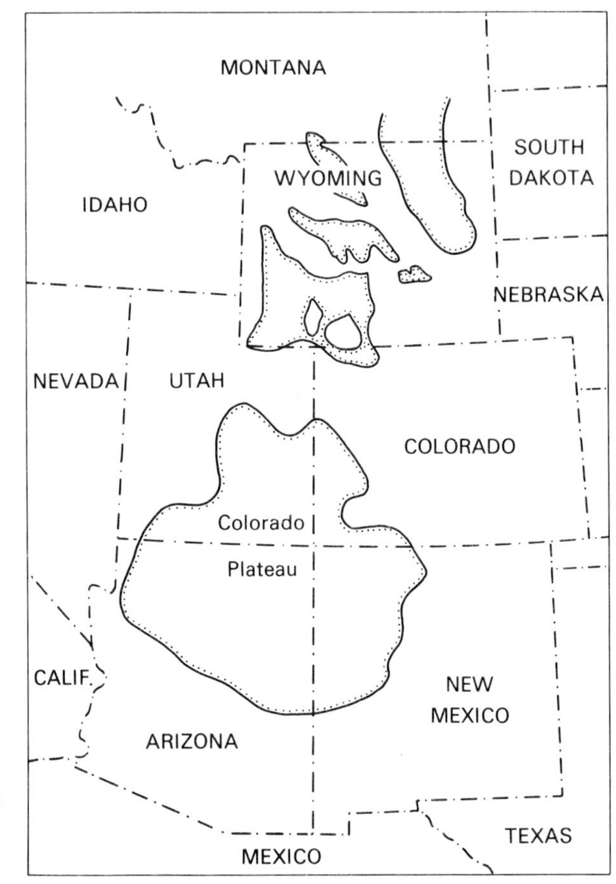

Fig. 15.4. Map showing the Colorado Plateau and Wyoming basins. (After Rackley 1976.)

Metals occurring in these deposits in significant quantities are: uranium, vanadium, copper, silver, selenium and molybdenum. A deposit may contain any one or more of these metals in almost any combination except vanadium and copper which are usually mutually exclusive. Amounts of uranium, vanadium and copper vary enormously within and between deposits and many orebodies fluctuate so much in grade that a single overall average figure is not informative. Generally grades vary from 0.1-1% U_3O_8, but can be locally much higher with such phenomena as whole tree trunks entirely replaced by uranium minerals. Most of the orebodies are similar. Small irregular pods are common and are sporadically distributed within a favourable rock unit. The larger deposits form mantos hundreds of metres long, about a hundred metres wide and a few metres thick. They can be mined by open pit or underground methods whilst the smaller bodies can be exploited by *in situ* leaching. The elongate orebodies follow buried stream courses or lenses of conglomeratic material. The deposits are epigenetic, in the sense that they were formed in their present position after the host sediment was deposited—how much later is very debatable. The typical orebody represents an addition of less than 1% of ore minerals which are accommodated in pore spaces where they form thin coatings on the detrital grains, whilst, in the case of high grade deposits, they may entirely fill the pore spaces. The disseminated form and microscopic size of the ore minerals increases the susceptibility to subsequent oxidation and remobilization by both alteration and weathering. The principal primary uranium minerals are pitchblende and coffinite $[(USiO_4)_{1-x}(OH)_{4x}]$. Vanadium, if present, is generally in the form of roscoelite (vanadium mica) and montroseite $[VO(OH)]$.

Sedimentological studies have shown that the usual immediate host rocks of these ores are fossil stream channel deposits. These consist of linear formations of permeable sandstone and conglomeratic sandstone enclosed by relatively impermeable rocks—shales, mudstones, etc. (Figs 15.5-7). During deposition, climatic

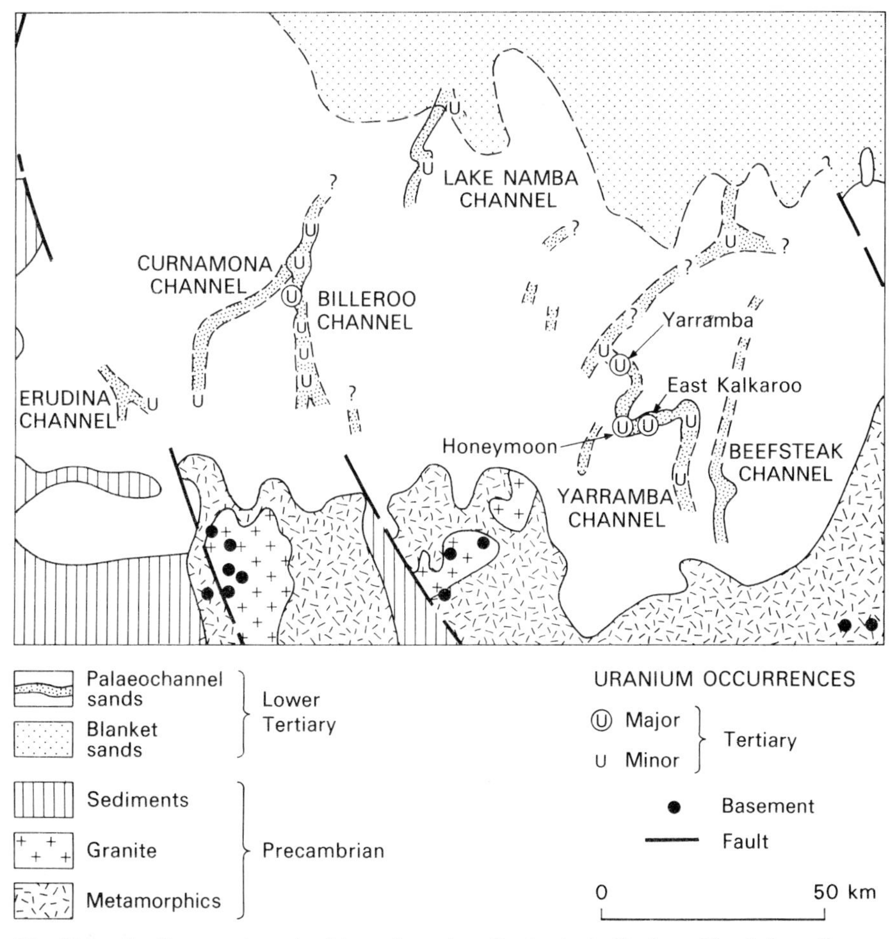

Fig. 15.5. Fossil stream channels with uranium mineralization in the Tertiary of South Australia. Note the many uranium occurrences in the basement which may have been the source of the uranium. (Redrawn from Brunt 1978.)

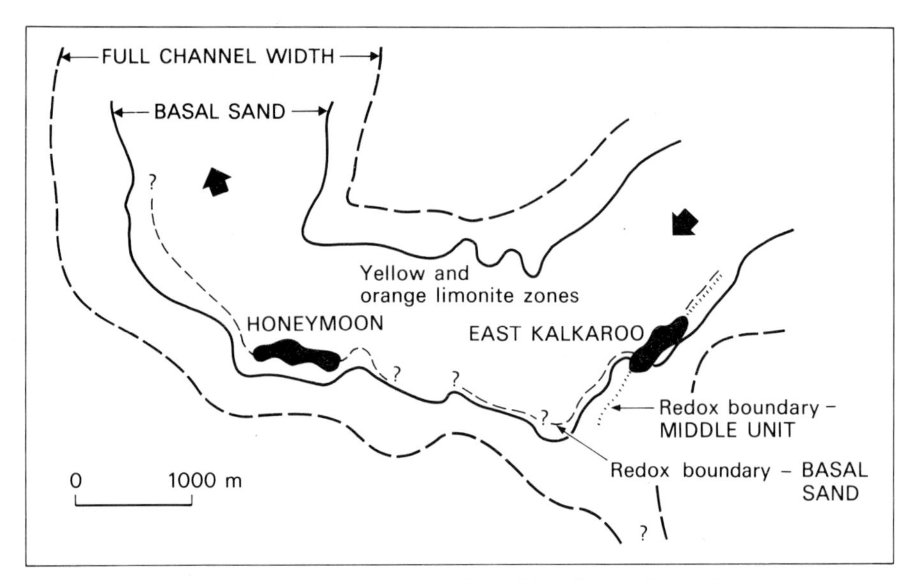

Fig. 15.6. Part of the Yarramba channel showing the position of two major uranium occurrences. Both deposits occur in embayments along the same channel bank. Patchy ore grade mineralization is also present along the redox boundaries away from the deposits. The opposite bank is barren as the yellow limonitic oxidation zone extends up to the pinchout of the sand units. (Redrawn from Brunt 1978.)

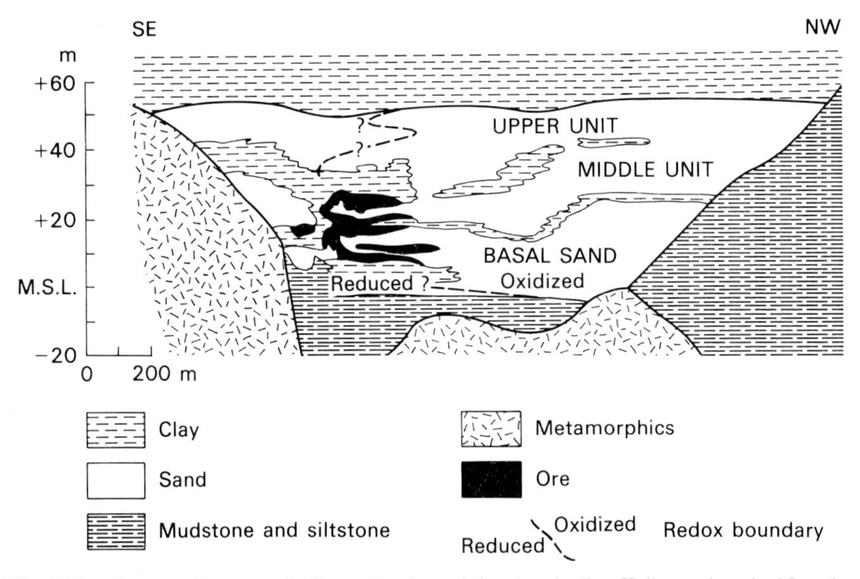

Fig. 15.7. Cross-section across the Yarramba channel showing the East Kalkaroo deposit. Note the roll-front configuration of the ore. (Redrawn from Brunt 1978.)

conditions were warm to hot and seasonably humid. Abundant vegetation grew in the depositional area and animals burrowed and mixed dead vegetation with the sediment. Frequent reworking by the streams incorporated sufficient organic material into the sediment to produce reducing conditions when it decayed. The sands deposited under these conditions were organic-rich, pyritic, light to dark grey and the associated clays were light to medium grey or green, pyritic and commonly carbonaceous. Petrographic studies show that the sediment was often derived from a granitic source area. During weathering of the granite its trace content of uranium would be oxidized to the hexavalent state and taken into solution. This uranium would migrate through the basin of deposition to be lost in the sea unless it came into contact with reducing conditions in the organic-rich sediment a short distance beneath the sediment-water interface, in which case it would enrich the sediment (Rackley 1976). In some areas, acid tuffs rather than granites appear to have been the source rocks for the uranium. As the oxygen-rich waters encroached upon the reducing environment an irregular tongue-shaped zone of oxidized rock was formed. The interface, or redox boundary, between the oxidized and reduced rocks has, in cross-section, the shape of a crudely crescentic envelope or roll, the leading edge of which cuts across the host strata (Fig. 15.8) and points down dip towards reduced ground that still contains authigenic iron disulphides. The reduced ground is generally grey in colour, the oxidized ground is drab yellow to orange or red due to the development of limonite and hematite by the alteration of the sulphides. Upon encountering reducing conditions, the uranium became reduced to the insoluble tetravalent state and was precipitated. Continuous or episodic introduction of oxygenated groundwater resulted in continuous or episodic solution and redeposition of uranium and migration of the redox interface down the palaeoslope. This process can lead to ore grade accumulation at or near the concave edge of the roll and, to a lesser extent, in reduced rock near the upper and lower limbs of the roll. Later reduction or oxidation of the ore beds may materially alter the form and mineralogy of the orebodies and obscure the primary redox relationships.

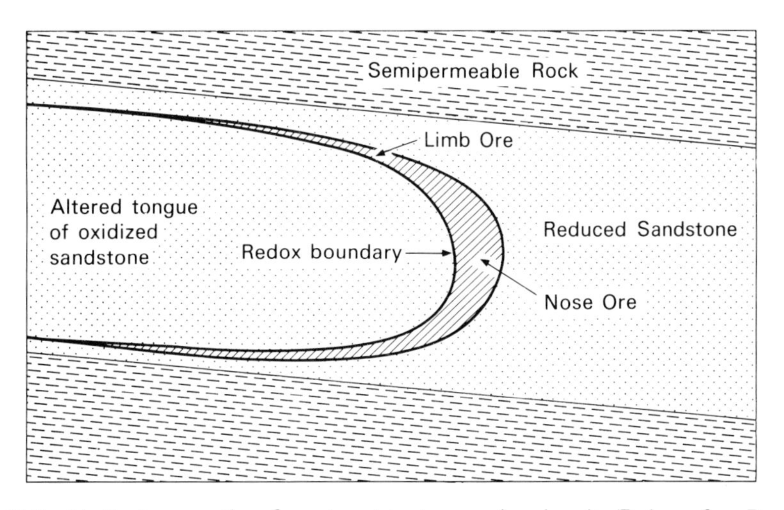

Fig. 15.8. Idealized cross section of a western states-type uranium deposit. (Redrawn from Reynolds & Goldhaber 1978.)

16

Sedimentary Deposits

Broadly speaking, sediments can be divided into two great groups, allochthonous deposits and autochthonous deposits. The allochthonous deposits are those which were transported into the environment in which they are deposited. They include the terrigenous (clastic) and pyroclastic classes. The autochthonous sediments are those which form within the environment in which they are deposited. They include the chemical, organic and residual classes. Table 16.1 shows the relationships between these various terms.

Table 16.1. A classification of sedimentary rocks.

Group	Class
I. Allochthonous sediments	(a) *Terrigenous deposits* - clays, siliclastic sands and conglomerates (b) *Pyroclastic deposits* - tuffs, lapillituffs, agglomerates volcanic breccias.
II. Autochthonous sediments	(c) *Chemical precipitates* - carbonates, evaporites, cherts, ironstones, phosphates (d) *Organic deposits* - coal, lignite, oil shales (e) *Residual deposits* - laterites, bauxites

Some sediments are sufficiently rich in elements of economic interest to form orebodies. Examples from both sedimentary groups will be described in this chapter. The student must realize, however, that the total range of sedimentary material of economic importance is much greater than can be included in this small volume. The residual deposits will be dealt with in the next chapter together with other ores in whose formation weathering has played an important role.

Allochthonous deposits

Allochthonous sediments of economic interest are usually referred to by ore geologists as mechanical accumulations or placer deposits. They belong to the terrigenous class formed by the ordinary sedimentary processes which concentrate heavy minerals. Usually this natural gravity separation is accomplished by moving water, though concentration in solid and gaseous mediums may also occur. The dense or heavy minerals so concentrated must first be freed from their source rock and must possess a high density, chemical resistance to weathering and mechanical durability. Placer minerals having these properties in varying degrees include: cassiterite, chromite, columbite, copper, garnet, gold, ilmenite, magnetite, monazite, platinum, ruby, rutile, sapphire, xenotime and zircon. Since sulphides readily break up and decompose they are rarely concentrated into placers. There are, however, some notable Precambrian exceptions (perhaps due to a non-oxidizing atmosphere) and a few small recent examples.

Placer deposits have formed throughout geological time, but most are of Tertiary and Recent age. The majority of placer deposits are small and often ephemeral as they form on the earth's surface usually at or above the local base level, so that many are removed by erosion before they can be buried. Most placer deposits are low grade, but can be exploited because they are loose easily worked materials which require no crushing and for which relatively cheap semi-mobile separating plants can be used. Mining usually takes the form of dredging which is about the cheapest of all mining methods. Older placers are likely to be lithified, tilted and partially or wholly buried beneath other lithified rocks. This means that exploitation costs are much higher and then the deposits, to be economic, must contain unusually valuable minerals (e.g. gold) or be of high grade.

Placers can be classified in various ways, in this book the simple genetic classification shown in Table 16.2 will be used.

Table 16.2. A classification of placer deposits.

Mode of origin	Class
Accumulation *in situ* during weathering	(a) *Residual placers*
Concentration in a moving solid medium	(b) *Eluvial placers*
Concentration in a moving liquid medium (water)	(c) *Stream or alluvial placers*
	(d) *Beach placers*
Concentration in a moving gaseous medium (air)	(e) *Aeolian placers*

RESIDUAL PLACERS

These accumulate immediately above a bedrock source (e.g. gold or cassiterite vein) by the chemical decay and removal of the lighter rock materials. They may grade downwards into weathered veins as in some tin areas of Shaba. In residual placers chemically resistant light minerals (e.g. beryl) may also occur. Residual placers only form where the ground surface is fairly flat, when a slope is present, creep will occur and eluvial placers will be generated (Fig. 16.1).

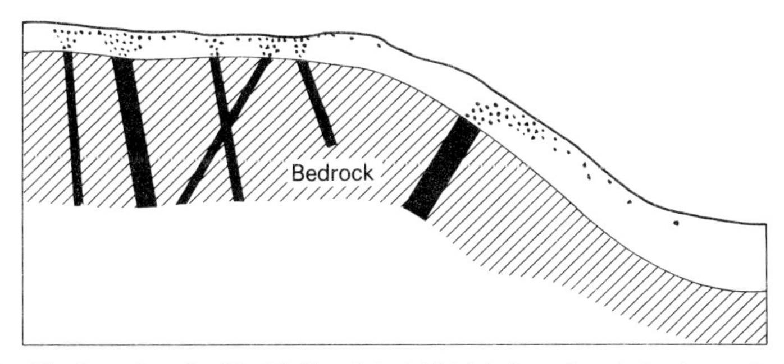

Fig. 16.1. The formation of residual (left) and eluvial (right) placer deposits by the weathering of cassiterite veins.

ELUVIAL PLACERS

These are formed upon hill slopes from minerals released from a nearby source rock. The heavies collect above and just downslope of the source. The lighter non-resistant minerals are dissolved or swept downhill by rain wash or are blown away by the wind. This produces a partial concentration by reduction in volume, a process which continues with further downslope creep. Obviously to yield a workable deposit this incomplete process of concentration requires a rich source. In some areas with eluvial placers, the economic material has accumulated in pockets in the bedrock surface, e.g. cassiterite in potholes and sinkholes in marble in Malaysia.

STREAM OR ALLUVIAL PLACERS

These were once the most important type of placer deposit. Primitive mining made great use of such deposits. The ease of extraction made them eagerly sought after in early as well as in recent times and they have been the cause of some of the world's greatest gold and diamond rushes.

Our understanding of the exact mechanisms by which concentrations of heavy minerals are formed in stream channels is still incomplete. Rubey (1933) considered fall velocity to be the most important segregating mechanism. Rittenhouse (1943) used the concept of hydraulic equivalence to explain heavy mineral concentrations. Brady & Jobson (1973) showed that fall velocities are of little importance. They found that bed configuration and grain density are the most important factors.

It is well known that the heavy mineral fraction of a sediment is much finer grained than the light fraction (Selley 1976). There are several reasons for this. Firstly, many heavy minerals occur in much smaller grains than do quartz and feldspar in the igneous and metamorphic rocks from which they are derived. Secondly, the sorting and composition of a sediment is controlled by both the density and size of the particles, this is known as their hydraulic ratio. Thus a large quartz grain requires the same current velocity to move it as a small heavy mineral. Clearly, if we have a very rapid flow all grains of sand grade will be in motion. With a slackening of velocity, the first materials to be deposited will be large heavies then smaller heavies plus large grains of lighter minerals. If the velocity of the transporting current does not drop any further then a heavy mineral concentration will be built up. For this reason such concentrations are developed when we have irregular flow and this may occur in a number of situations—always provided a source rock is present in the catchment area.

The first example is that of emergence from a canyon. In the canyon itself net deposition is zero. As the stream widens and the gradient decreases at the canyon exit any heavies will tend to be deposited and lighter minerals will be winnowed away. Again, where we have fast-moving water passing over projections in the stream bed the progress of heavy minerals may be arrested (Fig. 16.2). Waterfalls and potholes form other sites of accumulation (Fig. 16.3) and the confluence of a swift tributary with a slower master stream is often another site of concentration (Fig. 16.4). Most important of all, however, is deposition in rapidly flowing meandering streams. The faster water is on the outside curve of meanders and slack water is opposite. The junction of the two, where point bars form, is a favourable site for deposition of heavies. With lateral migration of the meander (Fig. 16.5) a

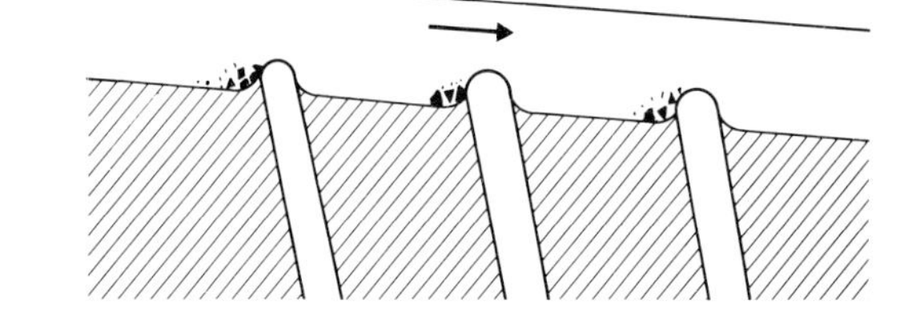

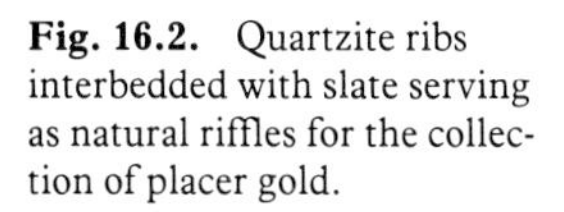

Fig. 16.2. Quartzite ribs interbedded with slate serving as natural riffles for the collection of placer gold.

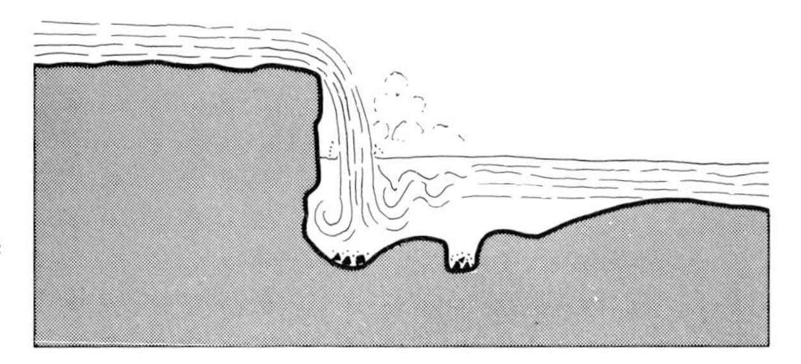

Fig. 16.3. Plunge pools at the foot of waterfalls and potholes can be sites of heavy mineral accumulations.

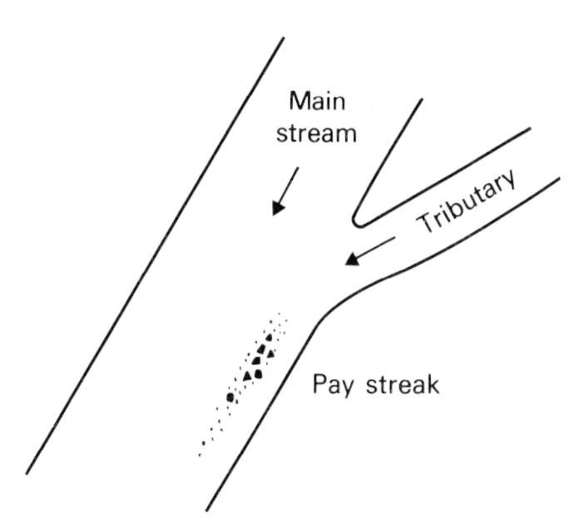

Fig. 16.4. A pay streak may be formed where a fast-flowing tributary enters a master stream.

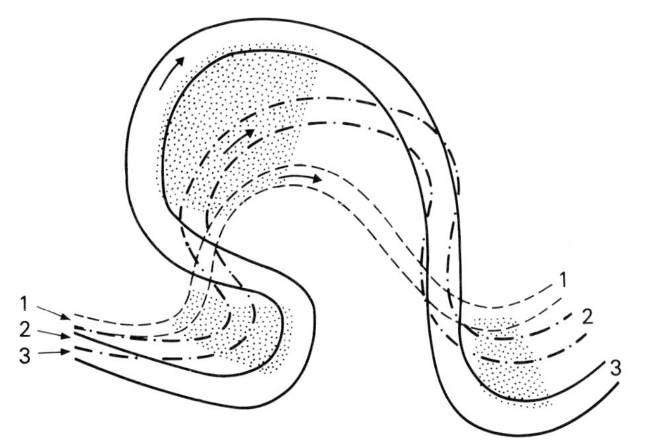

Fig. 16.5. Formation of pay streaks (dotted) in a rapidly flowing meandering stream with migrating meanders.
1 - original position of stream;
2 - intermediate position;
3 - present position. Note that pay streaks are extended laterally and downstream. Arrows indicate direction of water flow.

pay streak is built up which becomes covered with alluvium and eventually lies at some distance from the present stream channel.

Obviously, placer deposits do not form in the meanders of old age rivers because current flow is too sluggish to transport heavy minerals. In the upper reaches, current flow may be too rapid and there may also be a lack of source material. The middle reaches are most likely to contain placer deposits where we have well graded streams in which a balance has been achieved between erosion, transportation and deposition. Gradients measured on a number of placer gold and tin deposits average out at a little under 1 in 175.

BEACH PLACERS

The most important minerals of beach placers are: cassiterite, diamond, gold, ilmenite, magnetite, monazite, rutile, xenotime and zircon. Examples include the gold placer of Nome, Alaska, diamonds of Namibia, ilmenite-monazite-rutile sands of Travencore and Quilon, India, rutile-zircon-ilmenite sands of eastern and western Australia, and magnetite sands of North Island, New Zealand. Of course a source or sources of the heavies must be present. These may be coastal rocks, or veins cropping out along the coast or in the sea bed, or rivers or older placer deposits being reworked by the sea. Recent marine placers occur at different topographical levels due to Pleistocene sea level changes (Fig. 16.6). The optimum zone for heavy mineral separation to take place is the tidal zone of an unsheltered beach. Concentration may also occur on wave-cut terraces. Some raised beaches formed during high Pleistocene sea levels contain placer ores such as at Nome, Alaska.

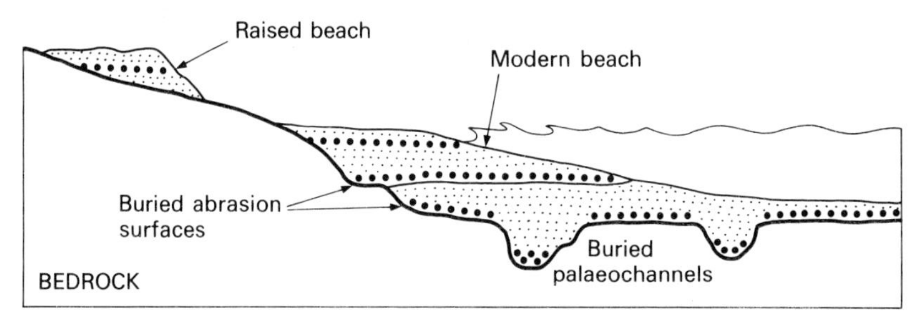

Fig. 16.6. Sketch section to illustrate some sites of beach placer deposits. Placers shown by heavy stipple. (After Selley 1976.)

Important beach placers stretch for about 900 km along the eastern coast of Australia. These are particularly important for their rutile and zircon production. They occur in Quaternary sediments which form a coastal strip up to 13 km wide. The sediments are usually 30-40 m thick. Placer deposits occur along the present day beaches and in the Pleistocene sands behind them (Fig. 2.13). These stabilized sands are characterized by low arcuate ridges which probably outline the shape of former bays. As the thickest heavy mineral accumulations are usually adjacent to the southern side of headlands, a reconstruction of the palaeogeography is important during exploration. Thus at Crowdy Head (Fig. 2.13) the deposit between A and B appears to be related to an old headland, now a bedrock outcrop at B. The point A also appears to mark the site of a former headland and further mineralization may be present to the south-west of this point.

Beach placers are formed along shorelines by the concentrating agency of waves and shore currents. Waves throw up materials on to the beach, the backwash carrying out the lighter materials which are moved away by longshore drift. The larger and heavier particles are thus concentrated on the beaches. Heavy mineral accumulations can be seen on present day beaches to have sharp bases and to form discrete laminae. They are especially well developed during storm wave action. Inverse grading is present in these laminae. At the base there is a fine-grained and/or heavy mineral-rich layer which grades upwards into coarser and/or heavy mineral-poor sands. These laminations develop during the backwash phase of wave action (Winward 1975). The breaking wave moves sand into suspension and carries it beachward, as its velocity drops its load is deposited. Then the water flow reverses and a surface layer of sand is disturbed and becomes a high particle density bed flow. During such a bed flow the smaller and denser particles sink to the bottom of the flow producing the reverse grading and also helping to concentrate the heavies. The heavy mineral-poor sand is thus closest to the surface waiting to be removed by the next wave. A considerable tidal variation is also important in that it exposes a wider strip of beach to wave action. This may lead to the abandonment of heavy mineral accumulations at the high water mark, where they may be covered and preserved from erosion by seaward advancing aeolian deposits.

Thus beaches on which heavy mineral accumulations are forming today include many upon which trade winds impinge obliquely and ocean currents parallel the coast, these two factors favouring longshore drift. In addition, these beaches face large areas of ocean and so are subjected to fierce storms and large waves. Such situations are found along the eastern and western coasts of Africa and Australia where various important heavy mineral concentrations occur. But just how do such ephemeral deposits become preserved? The answer is still the subject of considerable debate. Along the present day coast of New South Wales, heavy mineral accumulations formed by storm action are rarely preserved. They are reworked and redeposited in diluted form, possibly because these beaches are now in a stable or slightly erosive stage. In the mining operations inland from the foredune of this area the Holocene and Pleistocene deposits are seen to form overlapping layers separated by heavy mineral-poor quartz sand. These layers dip south-eastwards towards the prevailing winds. This suggests that for the preservation of heavy mineral deposits either the shoreline must prograde because of a previously more abundant sediment supply than at present, or the sea level must fall to remove the heavies from the sphere of wave activity.

FOSSIL PLACER DEPOSITS

The most outstanding examples are the gold-uranium-bearing conglomerates of the early Proterozoic. The principal deposits occur in the Witwatersrand Goldfield of South Africa, the Blind River area along the north shore of Lake Huron in Canada (only trace gold) and at Serra de Jacobina, Bahia, Brazil. Other occurrences are known in West Africa and Western Australia. The host rocks are oligomictic conglomerates (vein quartz pebbles) having a matrix rich in pyrite, sericite and quartz. The gold and uranium minerals (principally uraninite) occur in the matrix together with a host of other detrital minerals.

In the Witwatersrand Goldfield (Fig. 16.7) the orebodies appear to have been formed around the periphery of an intermontane, intracratonic lake or shallow water inland sea at and near entry points where sediment was introduced into the

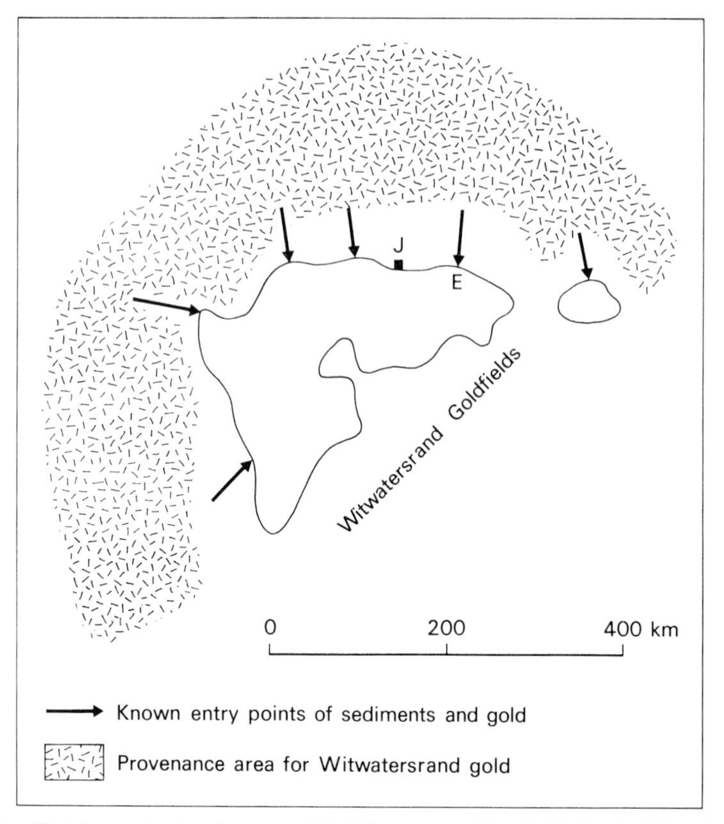

Fig. 16.7. Sketch map showing the extent of the Witwatersrand Goldfields, South Africa, the known entry points of sediment into the basin and the probable extent of the source area. J indicates the location of Johannesburg and E the East Rand Goldfield (Fig. 2.14). (After Anhaeusser 1976.)

basin. Deposition took place along the interface between river systems which brought the sediments and heavies from source areas to the north and west and a lacustrine littoral system that reworked the material (Pretorius 1975). The individual mineralized areas formed as fluvial fans (Fig. 2.14) which were built up at the entry points. Each fan was the result of sediment deposition at a river mouth that discharged through a canyon and flowed across a relatively narrow piedmont plain before entering the basin.

The Blind River area has also been closely studied and here the uranium deposits appear to have been laid down in a fluviatile or deltaic environment, perhaps during a wet period preceding an ice age. Unlike the Witwatersrand, the host conglomerates are now at the base of the enclosing arenaceous succession and they appear to occupy valleys eroded in the softer greenstones of the metamorphic basement (Roscoe 1968).

The presence of large amounts of pyrite and uraninite in these old conglomerates is a problem. At the present day detrital pyrite is uncommon, but known. Detrital uraninite is unknown as it is very susceptible to oxidation. Stanton (1972) has suggested that the most probable explanation is that of very rapid accumulation and burial of the whole clastic sequence not far from its erosional source at a time in earth history when the oxygen content of the atmosphere and the prevailing surface temperatures were low.

Autochthonous deposits

In this section we will be concerned with bedded iron and manganese deposits. The iron deposits can be conveniently divided into the Precambrian Banded Iron Formations (BIF) and the Phanerozoic ironstones.

BANDED IRON FORMATIONS (BIF)

These form one of the earth's great mineral treasures. Besides the term BIF, these rocks are known in different continents under the terms itabirite, jaspillite, hematite-quartzite and specularite. They occur in stratigraphical units hundreds of metres thick and hundreds or even thousands of kilometres in lateral extent. Substantial parts of these iron formations are directly usable as a low grade iron ore (e.g. taconite) and other parts have been the protores for higher grade deposits (Chapter 17). Compared with the present enormous demand for iron ore, now approaching 10^9 t yr^{-1}, the reserves of mineable ore in the banded iron formations are very large indeed (James & Sims 1973). An extraordinary fact emerging from recent studies is that the great bulk of iron formations of the world was laid down in the very short time interval of 2600-1800 Ma ago (Goldich 1973). The amount of iron laid down during this period, and still preserved, is enormous—at least 10^{14} t and possibly 10^{15} t. BIF are not restricted to this period, older and younger examples are known, but the total amount of iron in them is far outweighed by that deposited during this short time interval and now represented by the BIF of Labrador, the Lake Superior region of North America, Krivoi Rog and Kursk, USSR and the Hamersley Group of Western Australia.

Banded iron formation is characterized by its fine layering. The layers are generally 0.5-3 cm thick and in turn they are commonly laminated on a scale of millimetres or fractions of a millimetre. The layering consists of silica layers (in the form of chert or better crystallized silica) alternating with layers of iron minerals. The simplest and commonest BIF consists of alternating hematite and chert layers. Note that the content of alumina is less than one per cent, this contrasts with Phanerozoic ironstones which normally carry several per cent of this oxide. James (1954) identified four important facies of BIF.

(a) *Oxide facies.* This is the most important facies. It can be divided into the hematite and magnetite subfacies according to which iron oxide is dominant. There is a complete gradation between the two subfacies. Hematite in least altered BIF takes the form of fine-grained grey or bluish specularite. An oolitic texture is common in some examples, suggesting a shallow water origin, but in others the hematite may have the form of structureless granules. Carbonates (calcite, dolomite and ankerite rather than siderite) may be present. The 'chert' varies from fine-grained cryptocrystalline material to mosaics of intergrown quartz grains. In the much less common magnetite subfacies layers of magnetite alternate with iron silicate or carbonate and cherty layers. Oxide facies BIF typically averages 30-35% Fe. These rocks are mineable provided they are amenable to beneficiation by magnetic or gravity separation of the iron minerals.

(b) *Carbonate facies.* This commonly consists of interbanded chert and siderite in about equal proportions. It may grade through magnetite-siderite-quartz rock into the oxide facies, or, by the addition of pyrite, into the sulphide facies. The siderite lacks oolitic or granular texture and appears to have accumulated as a fine mud below the level of wave action.

(c) *Silicate facies.* Iron silicate minerals are generally associated with magnetite, siderite and chert. These form layers alternating with each other. This mineralogy suggests that the silicate facies forms in an environment common to parts of the oxide and carbonate facies. However, of all the facies of BIF, the depositional environment for iron silicates is least understood. This is principally because of the number and complexity of these minerals and the fact that primary iron silicates are difficult to distinguish from low rank metamorphic silicates. Probable primary iron silicates include greenalite, chamosite and glauconite, some minnesotaite and probably stilpnomelane. Most of the iron in these minerals is in the ferrous rather than the ferric state which, like the presence of siderite, suggests a reducing environment. P_{CO_2} may be important, a high value leading to siderite deposition, a lower one to iron silicate formation (Gross 1970). Carbonate and silicate facies BIF typically run 25-30% Fe, which is too low to be of economic interest.

(d) *Sulphide facies.* This consists of pyritic carbonaceous argillites—thinly banded rocks with organic matter plus carbon making up 7-8%. The main sulphide is pyrite which can be so fine-grained that its presence may be overlooked in hand specimen unless the rock is polished. The normal pyrite content is around 37%. The banding is due to the concentration of pyrite into certain layers. This facies clearly formed under anaerobic conditions. Its high sulphur content precludes its exploitation as an iron ore. It has however been mined for its sulphur content at Chvaletice in Czechoslovakia.

Precambrian BIF can be divided into two principal types (Gross 1970).

(a) *Algoma type.* This type is characteristic of the Archaean greenstone belts where it finds its most widespread development. It also occurs, however, in younger rocks including the Phanerozoic. It shows a greywacke-volcanic association suggesting a geosynclinal environment. The oxide, carbonate and sulphide facies are present, with iron silicates often appearing in the carbonate facies. Algoma type BIF generally ranges from a few centimetres to a hundred or so metres in thickness and is rarely more than a few kilometres in strike length. Exceptions to this observation occur in Western Australia where the main deposits of economic importance can be found. Oolitic and granular textures are absent or inconspicuous and the typical texture is a streaky lamination. A close relationship in time and space to volcanic rocks hints at a volcanic source of the iron and many regard deposits of this type as being exhalative in origin. Goodwin (1973) in a study of this deposit type in the Canadian Shield showed that facies analysis was a powerful tool in elucidating the palaeogeography and could be used to outline a large number of Archaean basins. His section across the Michipicotin Basin is shown in Fig. 16.8.

(b) *Superior type.* These are thinly banded rocks mostly belonging to the oxide, carbonate and silicate facies. They are usually free of clastic material. The rhythmical banding of iron-rich and iron-poor cherty layers, which normally range in thickness from a centimetre or so up to a metre, is a prominent feature. This distinctive feature allows correlation of BIF over considerable distances. Individual parts of the main Dales Gorge Member of the Hamersley Brockman BIF of Western Australia can be correlated at the 2.5 cm scale over about 50 000 km^2 (Trendall & Blockley 1970), and correlations of varves within chert bands can be made on a microscopic scale over 300 km (Trendall 1968).

Superior BIF is stratigraphically closely associated with quartzite and black carbonaceous shale and usually also with conglomerate, dolomite, massive chert,

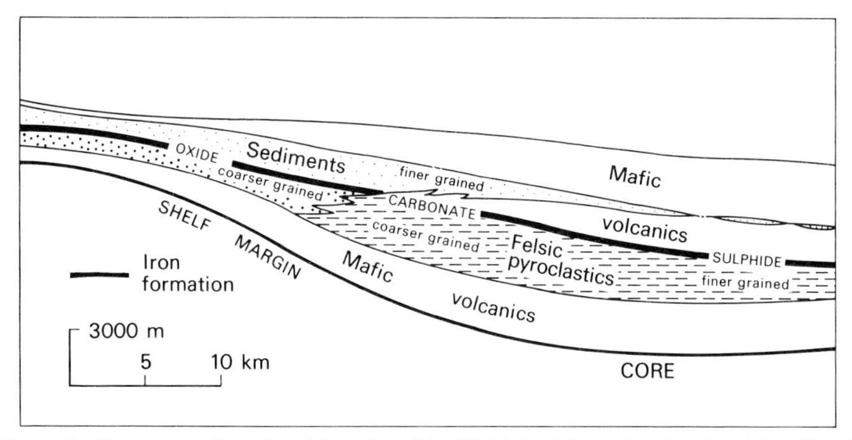

Fig. 16.8. Reconstructed stratigraphic section of the Michipicotin basin showing the relationship of the oxide, carbonate and sulphide facies of banded iron formation to the configuration of the basin and the associated rock-types. (After Goodwin 1973.)

chert breccia and argillite. Volcanic rocks are not always directly associated with this BIF, but they are nearly always present somewhere in the stratigraphic column. Superior type BIF may extend for hundreds of kilometres along strike. It may thicken from a few tens of metres to several hundred metres. The successions in which these BIF occur usually lie unconformably on highly metamorphosed basement rocks. The BIF are, as a rule, in the lower part of the succession. In some places they are separated from the basement rocks by only a metre or so of quartzite, grit and shale and in some parts of the Gunflint Range, Minnesota, they rest directly on the basement rocks.

The development of superior BIF reached its acme during the early Proterozoic, and Ronov (1964) has calculated that BIF accounts for 15% of the total thickness of sedimentary rocks of this age. Stratigraphic studies show that BIF frequently extended right around early Proterozoic sedimentary basins and Gross (1965) suggested that BIF was once present around the entire shoreline of the Ungava craton for a distance of more than 3200 km.

The associated rock sequences and sedimentary structures indicate that Superior BIF formed in fairly shallow water on continental shelves or along the margins of shelves and miogeosynclinal basins. There is general agreement that they are chemically precipitated sediments but there is no agreement on the source of the iron. One school considers that this was derived by erosion from nearby landmasses, another that it is of exhalative origin. One major drawback to the terrestrial derivation is that if large amounts of iron and silica were transported from the continents, large quantities of aluminous material must either have been left behind or transported and dispersed in the sea with, or not far from, the iron deposits. No such residual bauxites or aluminous sediments have been discovered. Trendall (1973) has suggested that the rhythmic microbands of the Hamersley Group so closely resemble evaporitic varves that a common origin is probable. He suggested that the banding originated by the annual accumulation of iron-rich precipitates whose deposition was triggered by evaporation from a partially enclosed basin with an average water depth of about 200 m. Many authors, e.g. Laberge (1973), Trudinger (1976), have discussed the role that micro-organisms may have played in the precipitation of the iron and the chert.

The Lake Superior region

For an example of BIF, we can look briefly at the deposits in the United States to the west and south of Lake Superior (Bayley & James 1973). This is one of the greatest iron ore districts of the world. The western part, which is shown in Fig. 16.9, can be divided into three major units: a basement complex (> 2600 Ma old); a thick sequence of weakly to strongly metamorphosed sedimentary and volcanic rocks—the Marquette Range Supergroup—and later Precambrian (Keweenawan) volcanics and sediments.

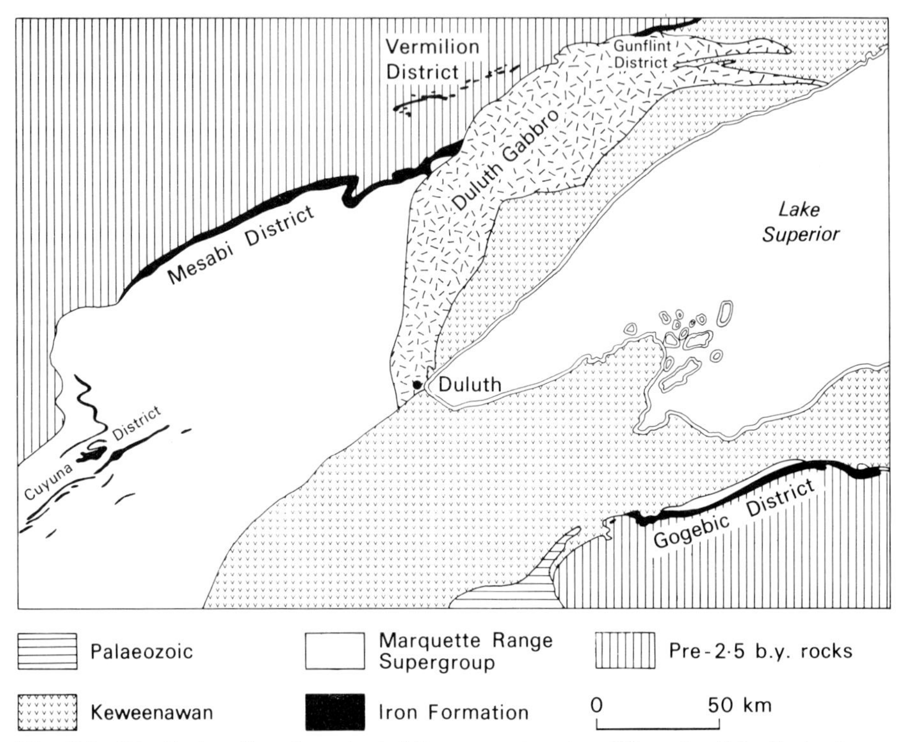

Fig. 16.9. Distribution of iron formation in Minnesota and northern Wisconsin. (After Bayley & James 1973.)

BIF is mainly developed in the Marquette Range Supergroup, but in the Vermilion district it is present in the basement. In this district there is a great thickness of mafic to intermediate volcanic rocks and sediments. BIF, mainly of oxide facies, occurs at many horizons as generally thin units rarely more than ten metres thick. One iron formation (the Soudan) is much thicker and it has been extensively mined.

The remaining iron ore of this region comes from the Menominee Group of the Marquette Range Supergroup. All the iron formations of this group in the different districts are of approximately the same age. The Marquette Range Supergroup shows a complete transition from a stable craton to eugeosynclinal conditions. Clastic rocks were first laid down on the bevelled basement. Most of these, however, were removed by later erosion and in many places the Menominee Group rests directly on the basement. Iron formation is the principal rock-type of this group. Despite the approximate stratigraphic equivalence of the major iron formations, they differ greatly from one district to another in thickness, stratigraphic detail and facies type. They appear to have been deposited either in separate basins or in isolated parts of the same basin. The only evidence of contemporaneous volcanism is the occurrence of small lava flows in the Gunflint and Gogebic districts. It has been suggested apropos this region that volcanism appears to have been detrimental rather than conducive to iron concentration.

The same iron formation appears in the Mesabi and Gunflint districts. It is 100-270 m thick and consists of alternating units of dark, non-granular, laminated rock and cherty, granule-bearing irregularly to thickly bedded rock. The granules are mineralogically complex containing widely different proportions of iron silicates, chert and magnetite; some are rimmed with hematite. The iron formation of the Cuyuna district consists principally of two facies, thin bedded and thick bedded which differ in mineralogy and texture. The first is evenly layered and laminated, the layers carrying varying proportions of chert, siderite, magnetite, stilpnomelane, minnesotaite and chlorite, while the second contains evenly bedded and wavy bedded rock in which chert and iron minerals alternately dominate in layers 2-30 cm thick. Granules and oolites are present. In the Gogebic district the iron formation is 150-310 m thick and consists of an alternation of wavy to irregularly bedded rocks characterized by granule and oolitic textures. The iron in the irregularly bedded rocks is principally in the form of magnetite and iron silicates, and granule textures are common. The evenly bedded iron formation is mineralogically complex consisting of chert, siderite, iron silicates and magnetite. Each mineral may dominate a given layer and may be accompanied by one or more of the other minerals.

PHANEROZOIC IRONSTONES

These are usually classified into two types, Clinton and Minette.

(a) *Clinton type.* This forms deep red to purple massive beds of oolitic hematite-chamosite-siderite rock. The iron content is about 40-50% and they are higher in Al and P than BIF. They also differ from BIF in the absence of chert bands, the silica being mainly present in iron silicate minerals with small amounts as clastic quartz grains. Clinton ironstones are associated with carbonaceous shale, sandy shales, dolomite and limestone. They form lenticular beds usually 2-3 m thick and never greater than 13 m. Individual beds may thicken and thin within short distances to form orebodies filling channels formed by current and wave action. Deposits of this type occur along one horizon or within a restricted part of the stratigraphic succession for hundreds of kilometres in a tectonic belt, but only locally may the lenses be thick enough to be mined. For example, Clinton ironstones are present in the Silurian almost continuously from northern New York State to Alabama.

This type of ironstone appears to have formed in shallow water along the margins of continents, on continental shelves or in shallow parts of miogeosynclines. It is common in rocks of Cambrian to Devonian age. One of the best examples is the Ordovician Wabana ore of Newfoundland (Gross 1970).

(b) *Minette type.* These are the most common and widespread ironstones. They differ principally from the Clinton type in being brown to dark green in colour, in having individual beds which are less massive and in the iron oxide mineral being limonite rather than hematite. The principal minerals are siderite and chamosite or another iron chlorite, the chamosite often being oolitic. A high proportion of fine-grained clastic sediment and organic material may be present. Iron content is around 30%, lime runs 5-20% and silica is usually above 20%. The high lime content forms one contrast with BIF and often results in these ironstones being self-fluxing ores.

Minette ironstones are particularly widespread and important in the Mesozoic of Europe, examples being the ironstones of the English Midlands, the minette ores of Lorraine and Luxembourg, the Salzgitter ores of Saxony and the iron ores of the Peace River area, Alberta. These ironstones are usually interbedded with black carbonaceous shale, mudstones, sandy shales and limestones which formed in shallow marine or brackish water in shallow basins of deposition. Cyclic sedimentation is sometimes present. Unlike the BIF, neither the Minette nor the Clinton ironstones show a separation into oxide, carbonate and silicate facies. Instead the minerals are intimately mixed often in the same oolite. These low grade ores are not amenable to beneficiation and the future of most operations is now in doubt.

SEDIMENTARY MANGANESE DEPOSITS

Sedimentary manganese deposits and their metamorphosed equivalents produce the bulk of the world's output of manganese. Residual deposits are the other main source. The USSR is the world leader in manganese production, and in 1977 produced 8 500 000 t, i.e. about one third of world production. Approximately 75% of this came from the Nikopol Basin in the Ukraine and much of the remainder from the Chiatura Basin in Georgia. The other important producers are the Republic of South Africa, Gabon, India, Australia and Brazil, though the last named has been cutting down production to conserve its resources for its own steel industry.

The geochemistry of iron and manganese is very similar and the two elements might be expected to move and be precipitated together. This is indeed the case *in some* Precambrian deposits, for example in the Cuyuna District (Fig. 16.9) manganese is abundant in some of the iron formations and forms over 20% of some of the ores. In other areas there seems to have been a complete separation of manganese and iron during weathering, transportation and deposition so that many iron ores are virtually free of manganese and many manganese ores contain no more than a trace of iron. The mechanism of this separation is still unknown. Stanton (1972) and Roy (1976) discuss a number of possibilities. Firstly, there is the possibility of segregation at source due to manganese being leached more readily from source rocks because of its relatively low ionic potential. This is, however, unlikely to produce more than a few per cent difference in the Mn/Fe of the extracted material compared with the ratio in the source rocks. A second possibility arises from the observation that many hot springs produce more manganese than iron. This suggests that iron has been precipitated preferentially from these hydrothermal solutions before they reached the surface. The third possibility is segregation by differential precipitation. Chemical considerations suggest that a limited increase in pH in *some* natural situations may lead to the selective elimination of iron from iron-manganese solutions. A fourth possibility is that separation occurs during diagenesis. The development of reducing conditions will cause both the iron and the manganese to be reduced and to go into solution and move laterally and upward. When the solutions reach an oxidizing environment the two are precipitated. However, since iron will always be the last to be reduced and hence mobilized, and the first to be oxidized and hence immobilized, manganese will tend to be progressively separated in diagenetic solutions.

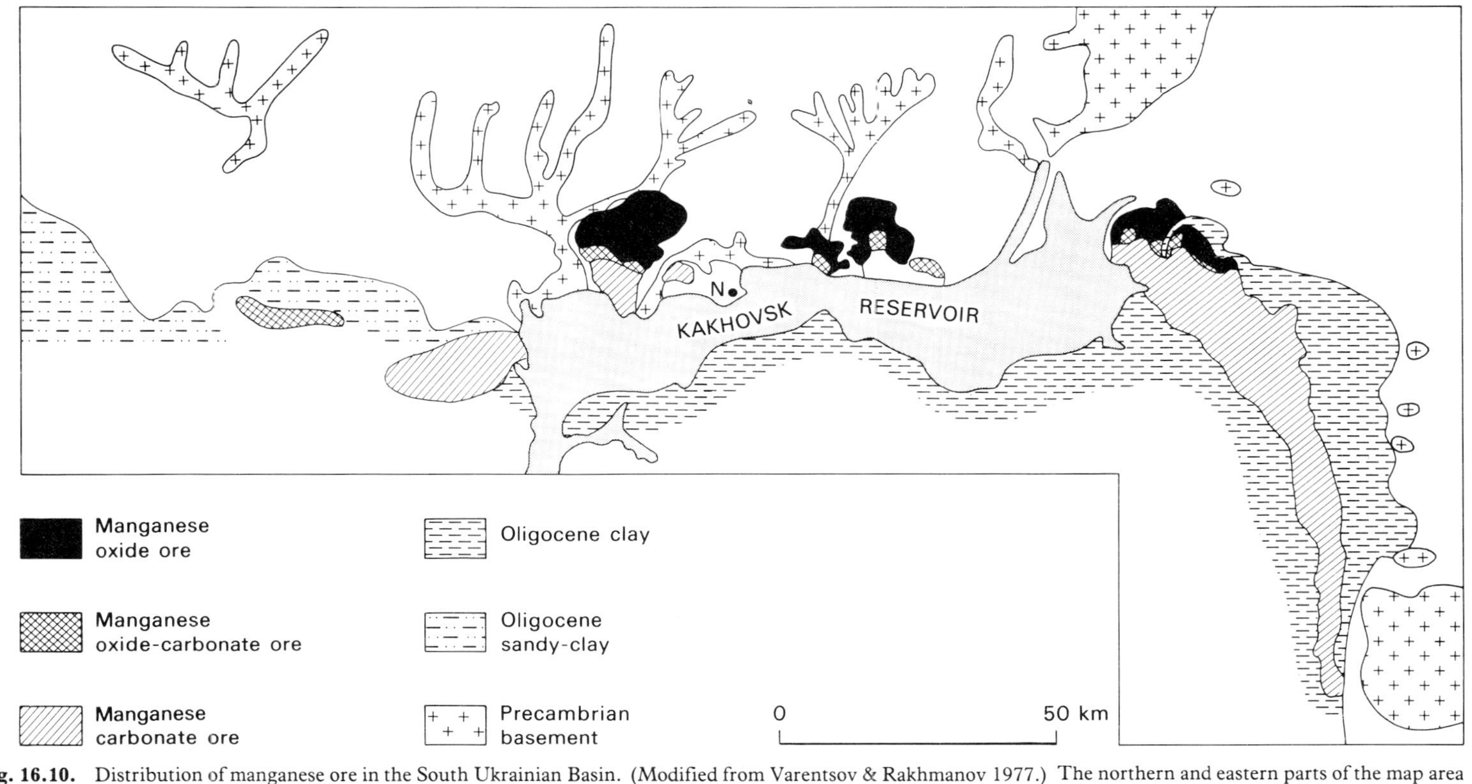

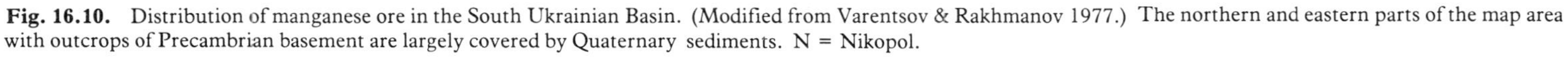

Fig. 16.10. Distribution of manganese ore in the South Ukrainian Basin. (Modified from Varentsov & Rakhmanov 1977.) The northern and eastern parts of the map area with outcrops of Precambrian basement are largely covered by Quaternary sediments. N = Nikopol.

There are two main classes of ancient sedimentary deposits: nonvolcanogenic and volcanogenic-sedimentary (Roy 1976). Among the nonvolcanogenic, the quartz-glauconite sand-clay association, which includes the chief USSR deposits, and the manganiferous carbonate association are the most important (Varentsov & Rakhmanov 1977).

(a) *The quartz-glauconite sand-clay association.* This formed in a shelf environment under estuarine and shallow marine conditions. On one side it passes into a non-ore-bearing coarse clastic succession, sometimes with coal seams, that lies between the manganese orefields and the source area for the sediments. On the other side it passes into an argillaceous sequence that marks deeper water deposition.

The largest manganese ore basin of this type is the South Ukrainian Oligocene Basin, its deposits include about 70% of the world's reserves of manganese ores. It forms a part of the vast South European Oligocene Basin which also contains the deposits of Chiatura and Mangyshlak in the USSR and Varnentsi in Bulgaria. The distribution of ore deposits is shown in Fig. 16.10.

The manganese ore forms a layer interstratified with sands, silts and clays. It is 0-4.5 m thick, averages 2-3.5 m and extends for over 250 km. There are intermittent breaks due to post-Oligocene erosion. A glauconitic sand is frequently present at the base of the ore layer which consists of irregular concretions, nodules and rounded earthy masses of manganese oxide and/or carbonate in a silty or clayey matrix. A shoreward to deeper water zoning is present (Figs 16.10 and 11). The dominant minerals of the oxide zone are pyrolusite and psilomelane. The principal carbonate minerals are manganocalcite and rhodochrosite. Progressing into a deeper water environment, the ore layer in the carbonate zone grades into green-blue clays with occasional manganese nodules. The Chiaturi deposit of Georgia shows a similar zonal pattern. The average ore grade in the Nikopol deposits is 15-25% Mn and at Chiaturi it varies from a few to 35%.

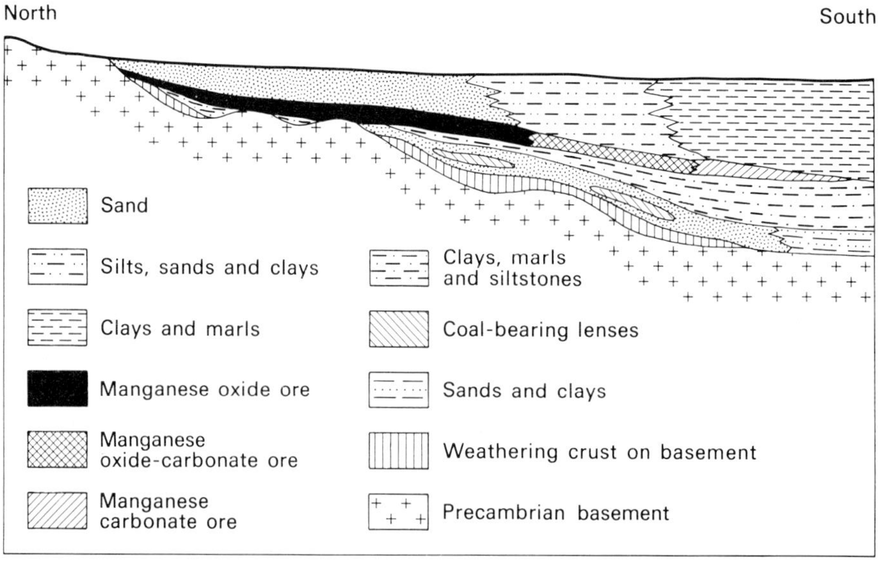

Fig. 16.11. Diagrammatic cross section through the Nikopol manganese deposits showing the zonation of the manganese ores and the transgressive nature of the sedimentary sequence with its overlap on to the Precambrian basement of the Ukrainian Platform. (After Varentsov 1964.)

The lack of associated volcanic rocks suggests that these deposits were formed by weathering and erosion from the nearby Precambrian shield which contains a number of rock-types (spilites, etc) that could have supplied an abundance of manganese. Palaeobotanical research has shown that the deposition of the ores coincided with a marked climatic change from humid subtropical to cold temperate. This change could have affected the pattern of weathering and transportation of the manganese.

(b) *Manganiferous carbonate association.* Varentsov & Rakhmanov (1977) divide this association into two types:

(1) a geosynclinal manganiferous dolomite-limestone formation, and

(2) a manganiferous limestone-dolomite formation developed on rigid cratonic blocks.

These deposits have many features in common and both are basically minor manganese-rich divisions of larger non-ore carbonate formations. Like the quartz-glauconite sand-clay association, deposits of significant size are not common. The principal deposits are in the USSR and north Africa, in particular Morocco. They vary from small lenses to quite extensive, continuous beds of manganese limestone. Often the lenses occur in swarms within a manganiferous section of the general carbonate succession.

The geosynclinal type includes the Usa deposit in the Lower Cambrian of south-western Siberia. There, 1000 m of limestones and dolomites underlie the ore horizon. This is succeeded conformably by interstratified manganese limestones, weakly manganous limestones and slates. These carry manganocalcite and calcic-rhodochrosite ores. The deposit consists of three lenses elongated parallel to the strike. The largest of these is 370 m thick. The others are 215 and 170 m thick. Most of the manganese is present as carbonates, some manganostilpnomelane is present. Ore grades vary from 5 to over 30%.

The cratonic type occur on the eroded surfaces of cratons close to uplifted areas that formed land at the time of sedimentation. Usually the succession contains three distinct units: (a) a lower terrigenous, often red bed, unit; (b) the ore-bearing unit, dominantly a carbonate formation with subordinate red clay and sometimes gypsum; (c) an upper terrigenous unit, again often red. The manganese ore occurs as layers and lenses in the central unit. The Lower Permian Ulutelyak deposit of the USSR occurs on the gently dipping slope of the south-eastern portion of the Russian Platform where it merges with the Pre-Urals Marginal Foredeep. Here the seams of carbonate manganese ore are up to 1 m thick with a grade of 12-18% Mn.

Volcanogenic-sedimentary deposits.

These are geologically the most widespread manganese deposits, but economically they are of minor importance compared with the nonvolcanogenic deposits described above. They were divided by Shatsky (see Varentsov & Rakhmanov 1977) into a greenstone association with spilite-keratophyre volcanism and a porphyry association with trachy-rhyolite volcanism. The greenstone association is the important one. It occurs in geosynclinal terrain and consists mainly of a silicate facies, sometimes carbonates are dominant but rarely oxides. A good example is the Kusimovo deposit (Fig. 16.12) which occurs in a group of deposits near Magnetogorsk in the Southern Urals. There, an intimate association exists in strongly folded rocks with basic and intermediate volcanics. The orebodies are

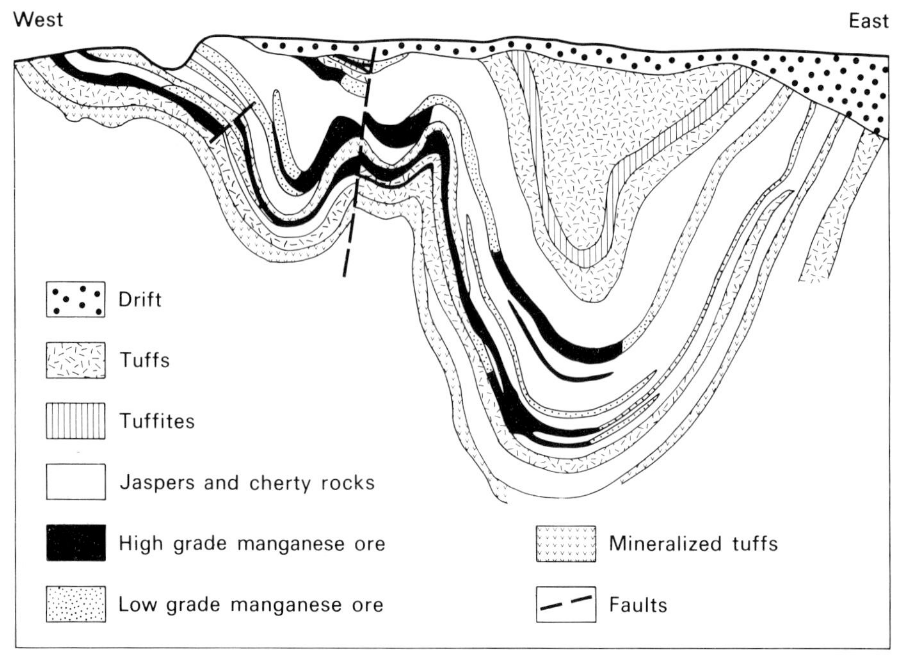

Fig. 16.12. Geological section through the Kusimovo deposits, Southern Urals. (After Varentsov & Rakhmanov 1977.)

layers or lenses which rest conformably on the underlying pyroclastics. They range from a few centimetres to five metres in thickness and are bedded, with bands of braunite alternating with cherty jasperoid layers. The manganese mineralogy of these deposits varies with the grade of metamorphism in a complicated manner. The tenor of the ores is commonly 15-25% Mn.

17

Residual Deposits and Supergene Enrichment

In the previous chapter we considered the concentration into orebodies of sedimentary material removed by mechanical or chemical processes and redeposited elsewhere. Sometimes the material left behind has been sufficiently concentrated by weathering processes and ground water action to form residual ore deposits. For the formation of extensive deposits, intense chemical weathering, such as in tropical climates having a high rainfall, is necessary. In such situations, most rocks yield a soil from which all soluble material has been dissolved. These soils are called laterites. As iron and aluminium hydroxides are amongst the most insoluble of natural substances, laterites are mainly composed of these materials and are, therefore, of no value as a source of either metal. Sometimes, however, residual deposits can be high grade deposits of one metal only.

Residual deposits of aluminium

When laterite consists of almost pure aluminium hydroxide, it is called bauxite. This is the chief ore of aluminium. Bauxite will develop on any rock with a low iron content or one from which the iron is removed during weathering. As with placer deposits, bauxites are vulnerable to erosion and most deposits are therefore post-Tertiary. Older deposits, however, are known, for example those in the Palaeozoic of the USSR. Some eroded bauxite has been redeposited to form what are called transported bauxites.

After oxygen and silicon, aluminium is the third most common element in the earth's crust, of which it forms 8.1%. Aluminium displays a marked affinity for oxygen and is not found in the native state. In weathered materials it accumulates in clay minerals or in purely aluminous ones such as gibbsite, boehmite and diaspore which are the principal minerals of bauxite. The mineralogy of bauxites depends on their age. Young bauxites are gibbsitic. With age, gibbsite gives way to boehmite and diaspore. There are certain chemical requirements that bauxites must meet to be economic, these vary according to whether the bauxite is gibbsitic or boehmitic (Table 17.1). Bauxite deposits are usually large deposits worked in

Table 17.1. Maximum and minimum contents of certain oxides in bauxites (% by wt on a dry basis).

	Gibbsitic bauxite	Boehmitic bauxite
Al_2O_3 (min.)	55.0	47.0
SiO_2 (max.)	5.0	4.0
FeO (max.)	3.0	3.0
P (max.)	1.0	1.0
Total alkalies (max.)	1.0	1.0
$MnO_2 + Cr_2O_3 + V_2O_5$ (max.)	2.0	2.0

open pits. The largest deposit is that of Sangaredi in Guinea where there is at least 180 Mt forming a plateau up to 30 m thick and averaging 60% alumina.

CLASSIFICATION OF BAUXITES

Bauxite deposits are extremely variable in their nature and geological situations. As a result, many different classifications have been put forward although here we have space to look at only three, those of Harder, Hose and Grubb (Table 17.2). Grubb's simple scheme is based on the topographical levels at which these deposits were formed.

Table 17.2. Classification of bauxite deposits.

Harder & Greig (1960)	Hose (1960)	Grubb (1973)
Surface blanket deposits	Bauxites formed on peneplains	High level or upland bauxites
Interlayered beds or lenses in stratigraphic sequences	Bauxites formed on volcanic domes or plateaux	
Pocket deposits in limestones, clays or igneous rocks		
Detrital bauxites	Bauxites formed on limestones or karstic plateaux	Low-level peneplain-type bauxites
	Sedimentary reworked bauxites	

(a) *High-level or upland bauxites.* These generally occur on volcanic or igneous source rocks forming thick blankets of up to 30 m which cap plateaux in tropical to sub-tropical climates. Examples occur in the Deccan Traps of India, southern Queensland, Ghana and Guinea. These bauxites are porous and friable and show a remarkable retention of parent rock textures. They rest directly on the parent rock with little or no intervening underclay. Bauxitization is largely controlled by joint patterns in the parent rock, with the result that chimneys and walls of bauxite often extend deep into the footwall.
(b) *Low-level peneplain-type bauxites.* These occur at low levels along tropical coastlines such as those of South America, Australia and Malaysia. They are distinguished by the development of pisolitic textures and are often boehmitic in composition. Peneplain deposits are generally less than 9 m thick and are usually separated from their parent rock by a kaolinitic underclay. They are frequently associated with detrital bauxite horizons produced by fluvial or marine activity.
(c) *Karst bauxites.* These include the oldest known bauxites—those in the lands just north of the Mediterranean which range from Devonian to mid-Miocene. Other major deposits are the Tertiary ones of Jamaica and Hispaniola. These bauxites overlie a highly irregular karstified limestone or dolomite surface. Texturally karst bauxites are quite variable. The West Indian examples are gibbsitic ores with a structureless earthy, sparsely concretionary texture. European karst bauxites, on the other hand, are generally lithified and texturally pisolitic, oolitic, fragmental or even bedded. Mineralogically they are boehmitic ores. From these and other facts Grubb contended that the West Indian bauxites have strong affinities with

upland deposits, whilst the European karst bauxites are more reminiscent of peneplain deposits.

Iron-rich laterites

Most iron-bearing laterites are too low in iron to be of economic interest. Occasionally, however, laterites derived from basic or ultrabasic rock may be sufficiently rich in iron to be workable, though in some cases other metals such as cobalt and nickel may also have been enriched to such an extent as to poison the ore. These deposits, which may be as much as 20 m thick but are usually less than 6 m, consist of nodular red, yellow or brown hematite and goethite which may carry up to 20% alumina. Deposits of this type form mantles on plateaux and are worked in Guyana, Indonesia, Cuba, the Philippines, etc.

A good example is the Conakry deposit in Guinea which is developed on dunite, the change from laterite to dunite being sharp. Most of the laterite consists of a hard crust usually about 6 m thick. The ore as shipped contains 52% Fe, 12% alumina, 1.8% silica, 0.25% phosphorus, 0.14% S, 1.8% Cr, 0.15% Ni, 0.5% TiO_2 and 11% combined water. This last figure illustrates one of the drawbacks of these ores—their high water content, which may range up to 30%. This has to be transported and then removed during smelting.

Residual deposits of nickel

The first major nickel production in the world came from nickeliferous laterites in New Caledonia where mining commenced about 1876. World nickel consumption is forecast to increase from 690 000 t in 1978 to 1.3 Mt in 1990. Some of this increase will be supplied by manganese nodules recovered from the ocean bed—perhaps 10-18% by 1990—some will come from increased production at nickel sulphide deposits, the rest must come from nickel laterites. It has been calculated that there are about 64 Mt of economically recoverable nickel in land-based deposits. Of this, about 70% occurs in lateritic deposits, although less than a half of current nickel production comes from these ores.

Residual nickel deposits are formed by the intense tropical weathering of rocks rich in trace amounts of nickel such as peridotites and serpentinites, these run about 0.25% Ni. During the lateritization of such rocks, nickel passes (temporarily) into solution but is generally quickly reprecipitated either on to iron oxide minerals in the laterite or as garnierite and other nickeliferous phyllosilicates in the weathered rock below the laterite. Cobalt too may be concentrated, but it is usually fixed in wad.

(a) *Nickel deposits of New Caledonia.* Much of New Caledonia is underlain by ultrabasic rocks many of which are strongly serpentinized. A typical environment of nickel mineralization is shown in Fig. 17.1 and a more detailed profile in Fig. 17.2. The nickel occurs in both the laterite and the weathered rock zone. In the latter it forms distinct masses, veins, veinlets or pockets rich in garnierite which occur around residual blocks of unweathered ultrabasic rock and in fissures running down into the underlying rock. The material mined is generally a mixture of the lower parts of the laterite and the weathered rock zone. Above the nickel-rich zone there are pockets of wad containing significant quantities of cobalt. Grades of up to 10% Ni were worked in the past, but today the grade is around 1.5% Ni. It has been estimated that there are 1.5 Gt of material on the island assaying a little over

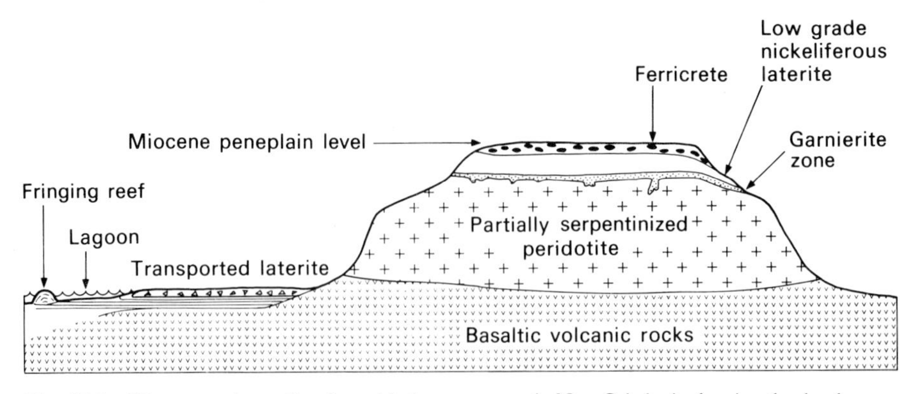

Fig. 17.1. Diagrammatic profile of a peridotite occurrence in New Caledonia showing the development of a residual nickel deposit. (After Dixon 1979.)

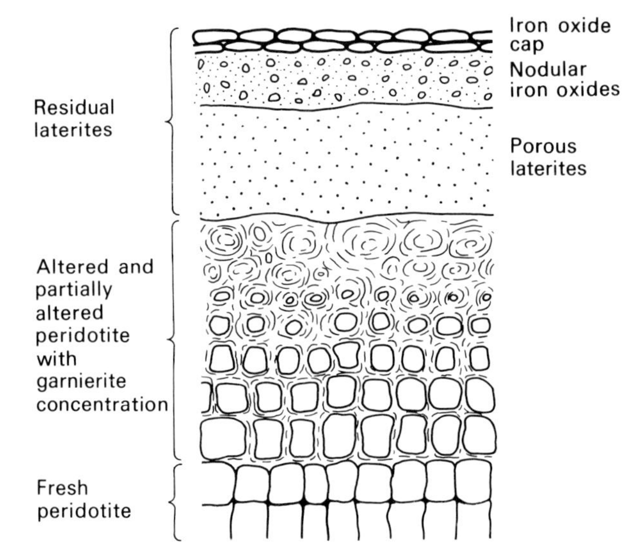

Fig. 17.2. Section through nickeliferous laterite deposits, New Caledonia. (After Chételat 1947.)

1% Ni. All laterites take time to develop and it is thought that those on New Caledonia began to form in the Miocene.

(b) *The Greenvale Nickel Laterite, north Queensland.* This deposit was discovered in 1966 as a result of the comparison of the geological environment with that of New Caledonia (Fletcher & Couper 1975). The section above the fresh serpentinite (which runs 0.28% NiO) is similar to that of the New Caledonian occurrences (Fig. 17.3). Nickel and cobalt are concentrated to ore grade in a laterite mantle covering about two thirds of the serpentinite. Erosion has removed the ore zone from the rest of the peridotite. Ore reserves run to 40 Mt averaging 1.57% Ni and 0.12% Co. The ore zone occurs mainly in the weathered serpentinite, often towards the top, and partially in the overlying limonitic laterite.

Supergene enrichment

Although it is more commonly applied to the enrichment of sulphide deposits, the term supergene enrichment has been extended by many workers to include similar

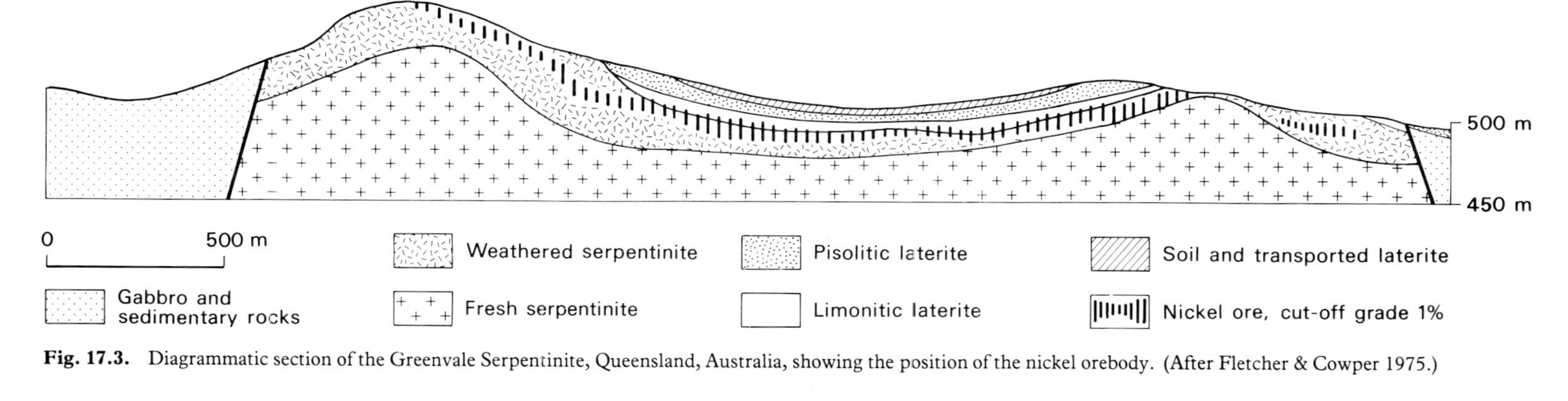

Fig. 17.3. Diagrammatic section of the Greenvale Serpentinite, Queensland, Australia, showing the position of the nickel orebody. (After Fletcher & Cowper 1975.)

processes affecting oxide or carbonate ores and rocks such as those of iron and manganese. In supergene sulphide enrichment the minerals of economic interest are carried down into hypogene (primary) ore where they are precipitated with a resultant increase in metal content, whereas in the case of iron and manganese ores it is chiefly the gangue material that is mobilized and carried away to leave behind a purer metal deposit.

SUPERGENE SULPHIDE ENRICHMENT

Surface waters percolating down the outcrops of sulphide orebodies oxidize many ore minerals and yield solvents that dissolve other minerals. Pyrite is almost ubiquitous in sulphide deposits and this breaks down to produce insoluble iron hydroxides (limonite) and sulphuric acid:

$$2FeS_2 + 15O + 8H_2O + CO_2 \rightarrow 2Fe(OH)_3 + 4H_2SO_4 + H_2CO_3 \quad \text{and}$$
$$2CuFeS_2 + 17O + 6H_2O + CO_2 \rightarrow 2Fe(OH)_3 + 2CuSO_4 + 2H_2SO_4 + H_2CO_3.$$

Copper, zinc and silver sulphides are soluble and thus the upper part of a sulphide orebody may be oxidized and generally leached of many of its valuable elements right down to the water table. This is called the zone of oxidation. The ferric hydroxide is left behind to form a residual deposit at the surface and this is known as a gossan or iron hat—such features are eagerly sought by prospectors. As the water percolates downwards through the zone of oxidation, it may, because it is still carbonated and still has oxidizing properties, precipitate secondary minerals such as malachite and azurite (Fig. 17.4).

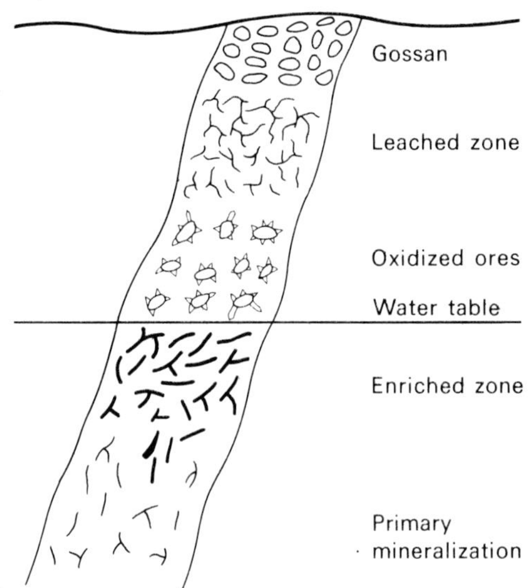

Fig. 17.4. Generalized section through a sulphide-bearing vein showing supergene enrichment. (Modified from Bateman 1950.)

Often, however, the bulk of the dissolved metals stays in solution until it reaches the water table below which conditions are usually reducing. This leads to various reactions which precipitate the dissolved metals and result in the replacement of primary by secondary sulphides. At the same time, the grade is increased and in this way spectacularly rich bonanzas can be formed. Typical reactions are as follows:

$$PbS + CuSo_4 \rightarrow CuS + PbSO_4 \quad \text{(Covellite + anglesite)},$$
$$5FeS_2 + 14CuSO_4 + 12H_2O \rightarrow 7Cu_2S + 5FeSO_4 + 12H_2SO_4 \quad \text{(Chalcocite)},$$
$$CuFeS_2 + CuSO_4 \rightarrow 2CuS + FeSO_4 \quad \text{(Covellite)}.$$

This zone of supergene enrichment usually overlies primary mineralization which may or may not be of ore grade. It is thus imperative to ascertain whether newly discovered near surface mineralization has undergone supergene enrichment, for, if this is the case, a drastic reduction in grade may be encountered when the supergene enrichment zone is bottomed. For this purpose a careful polished section study is often necessary.

Clearly, such processes require a considerable time for the evolution of significant secondary mineralization. They also require that the water table be fairly deep and that ground level is slowly lowered by erosion. This usually means that such phenomena are restricted to non-glaciated land areas.

Supergene enrichment has been important in the development of many porphyry copper deposits and a good example occurs in the Inspiration orebody of the Miami district, Arizona (Fig. 17.5). Primary ores of this district are developed along a granite-schist contact with most of the ore being developed in the schist. The unenriched ore averages about 1% Cu (Ransome 1919) and consists of pyrite, chalcopyrite and molybdenite. Supergene enrichment increased the grade up to as much as 5% in some places. The schist is more permeable than the granite and more supergene enrichment occurred within it. The enrichment shows a marked correlation with the water table (Fig. 17.5) where it starts abruptly. Downwards, it tapers off in intensity and dies out in primary mineralization. Chalcocite is the main secondary sulphide. It replaced both pyrite and chalcopyrite.

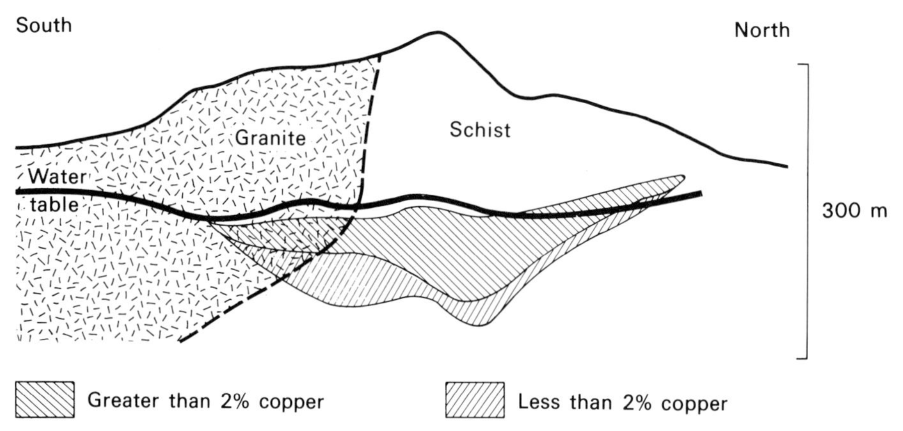

Fig. 17.5. Cross section through the Inspiration orebody at Miami, Arizona, showing the relationship between the high and low grade ores and the position of the supergene enriched zone (Cu over 2%) relative to the water table. (Modified from Ransome 1919.)

SUPERGENE ENRICHMENT OF LOW GRADE IRON FORMATION

Most of the world's iron ore is won from orebodies formed by the natural enrichment of banded iron formation (BIF). Through the removal of silica from the BIF the grade of iron may be increased by a factor of two to three times. Thus, for example, the Brookman Iron Formation of the Hamersley Basin in Western Australia averages 20-35% Fe but in the orebodies of Mount Tom Price it has been upgraded to dark blue hematite ore running 64-66% Fe.

The agent of this leaching is generally considered to be descending ground water, though a minority school in the past has argued the case for leaching by ascending hydrothermal water. In general, these orebodies show such a marked relationship to the present (or a past) land surface that there is little doubt that we are dealing with a process akin to lateritization. This relationship is clearly exemplified by the orebodies of Cerro Bolivar, Venezuela (Ruckmick 1963) which are shown in Fig. 17.6. These orebodies are developed in a tropical area having considerable relief

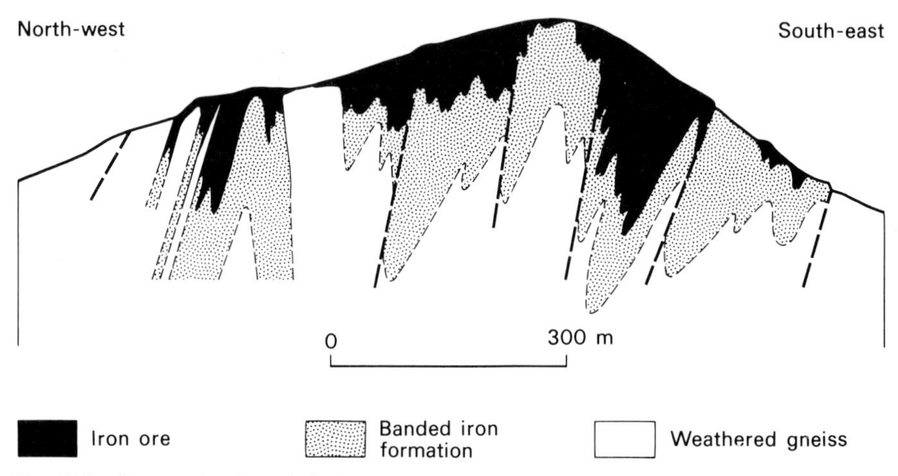

Fig. 17.6. Cross section through the iron orebodies of Cerro Bolivar, Venezuela. (After Ruckmick 1963.)

and thus the ground water passing through them can be sampled in springs emerging from their flanks. This water carries 10.5 ppm silica and 0.05 ppm iron. Its pH averages 6.1. Clearly, the rate of removal of silica is about two hundred times that of iron so that the iron tends to be left behind whilst the silica is removed. The iron is not entirely immobile and this probably accounts for the fact that many orebodies of this type consist of compact high grade material of low porosity. Obviously, the voids created by the removal of silica have been filled by iron minerals, normally hematite. This process seems to be taking place at Cerro Bolivar at the present time and Ruckmick has calculated that if the leaching process proceeded in the past at the same rate as it is today then the present orebodies would have required 24 Ma for their development.

That downward-moving waters were the agent of leaching is attested to by the frequency with which enriched zones of BIF occur in synclinal structures into which downward-moving ground water has been concentrated. The leaching is, of course, intensified if the BIF is underlain by impervious formations such as slate as in the Western Menominee District of Michigan and schist in the Middleback Ranges of South Australia. The BIF of the Middleback Ranges occurs in the Middleback Group. The rocks of this group have suffered folding along northerly axes followed by easterly cross-folding. This produced a number of domes and basins. The domes have largely been removed by erosion leaving isolated cross-folded synclinal areas scattered across an older basement. Downward-moving ground waters have produced orebodies by leaching the BIF in the keels of these synclines. The evidence for this mechanism of formation is particularly good at the Iron Monarch body. This orebody occupies a northward plunging syncline. The main northern down plunge termination of the orebody is a north-west trending dyke complex which appears to have acted as a dam to the downward-moving water preventing extensive leaching from proceeding further down the keel of the syncline (Fig. 17.7). For this reason, very little enrichment has occurred on the down plunge side.

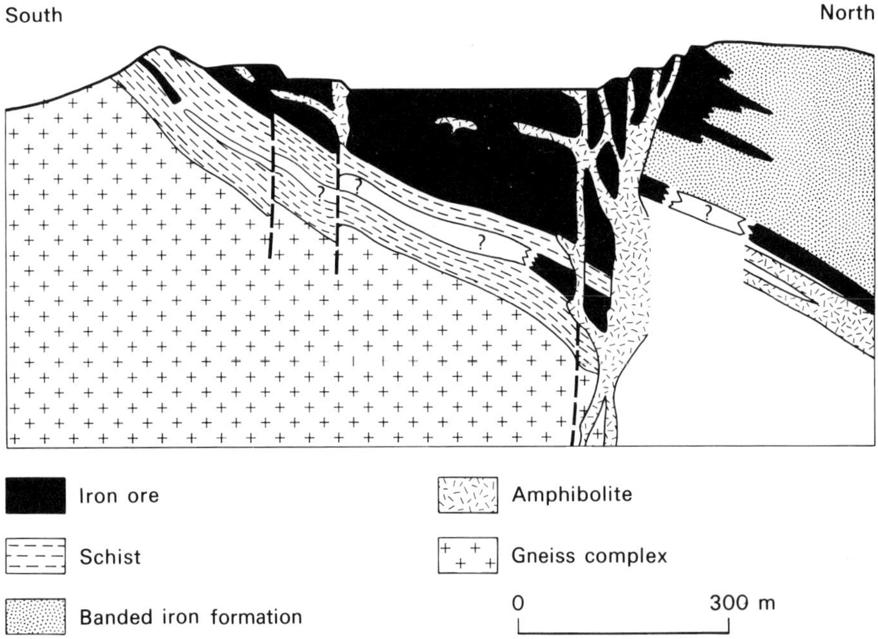

Fig. 17.7. Longitudinal section through the Iron Monarch orebody, South Australia. (Modified from Owen & Whitehead 1965.)

SUPERGENE ENRICHMENT OF MANGANESE DEPOSITS

Deep weathering processes akin to lateritization can also give rise to the formation of high grade manganese deposits. Although not comparable in size with the previously described sedimentary deposits (Chapter 16), they nevertheless form important accumulations of manganese. Large deposits of this type occur at Postmasburg in South Africa, and in Gabon, India, Brazil and Ghana.

Since manganese is more mobile than iron the problem arises as to why manganese is retained in the weathering profile. The reason is probably because in most cases the residual manganese deposits are formed on low grade manganiferous limestones and dolomites. These rocks are low in iron and silica, and manganese takes the place of iron in the laterite profile. The carbonates are easily dissolved (by comparison with silica) and the higher pH which probably prevailed in these environments would have tended to immobilize the manganese.

18

The Metamorphism of Ore Deposits

It is all too often forgotten that the majority of ore deposits occur in metamorphosed host rocks. Among these deposits the substantial proportion that are syndepositional or diagenetic in age must have been involved in the metamorphic episode(s) and consequently their textures and mineralogy may be considerably modified, often to the advantage of mineral separation processes. In addition, parts of the orebody may be mobilized and epigenetic features, such as cross-cutting veins, may be imposed on a syngenetic deposit. Ores in metamorphic rocks may be metamorph*ic* in the sense that their economic minerals have been largely concentrated by metamorphic processes, these include the pyrometasomatic deposits and some deposits of lateral segregation origin. On the other hand they may be metamorph*osed*, in the sense that they predate the metamorphism and have therefore been affected by a considerable change in pressure-temperature conditions which may have modified their texture, mineralogy, grade, shape and size—structural deformation often accompanies metamorphism. This chapter will be concerned entirely with metamorphosed ores.

It is important from the economic as well as the academic viewpoint to be able to recognize when an ore has been metamorphosed. Strongly metamorphosed ores may develop many similar features to high temperature epigenetic ores for which they may be mistaken. This can be very important to the exploration geologist, for in the case of an epigenetic ore, particularly one localized by an obvious structural control, the search for further orebodies may well be concentrated on a search for similar structures anywhere in the stratigraphic column, whereas repetitions of a metamorphosed syngenetic ore should be first sought for along the same or similar stratigraphic horizons. Notable examples of orebodies long thought to be high temperature epigenetic deposits, but now considered to be metamorphosed syndepositional ores, are those of Broken Hill, New South Wales and the Horne Orebody, Quebec. In both areas recent exploration having a stratigraphic basis has been successful in locating further mineralization.

Certain types of deposit, by virtue of their structural level of development or geological environment are rarely, if ever, seen in a metamorphosed state. Some examples are porphyry copper, molybdenum and tin deposits, ores of the carbonatite association and placer deposits. On the other hand, certain deposits such as the volcanic massive sulphide class have generally suffered some degree or other of metamorphism such that their original textures are commonly much modified.

Three types of metamorphism are normally recognized on the basis of field occurrence. These are usually referred to as contact, dynamic and regional metamorphism. Contact metamorphic rocks crop out at or near the contacts of igneous intrusions. In some cases the degree of metamorphic change can be seen to increase as the contact is approached. This suggests that the main agent of metamorphism in these rocks is the heat supplied by the intrusion. As a result, this type is sometimes referred to as thermal metamorphism. It may take the form of an entirely static heating of the host rocks without the development of any secondary structures such as schistosity or foliation, though these are developed in some metamorphic aureoles.

Dynamically metamorphosed rocks are typically developed in narrow zones such as major faults and thrusts where particularly strong deformation has occurred. Epigenetic ores, developed in dilatant zones along faults, often show signs of dynamic effects (brecciation, plastic flowage, etc.) due to fault movements during and after mineralization.

Regionally metamorphosed rocks occur over large tracts of the earth's surface. They are not necessarily associated with either igneous intrusions or thrust belts, but these features may be present. Research has shown that regionally metamorphosed rocks generally suffered metamorphism at about the time they were intensely deformed. Consequently they contain characteristic structures such as cleavage, schistosity, foliation or lineation which can be seen on both the macroscopic and microscopic scales, producing a distinctive fabric in rocks so affected.

How can we recognize when ores have been metamorphosed? One way is to compare their general behaviour with that of carbonate and silicate rocks which have undergone metamorphism. In contact and regionally metamorphosed areas the rocks generally show:

(1) the development of metamorphic textures;
(2) a change of grain size—usually an increase, and
(3) the progressive development of new minerals. In addition, as noted above, regionally metamorphosed rocks usually develop certain secondary structures. Let us examine each of these effects in turn.

Development of metamorphic textures

DEFORMATION

All three types of deformation, elastic, plastic and brittle, are important—elastic deformation largely because it raises the internal free energy of the grains and renders them more susceptible to recrystallization and grain growth. Plastic deformation occurs by primary (translation) gliding or by secondary (twin) gliding. In translation gliding, movement occurs along glide planes inside a grain without any rotation of the crystal lattice. Translation gliding can be readily induced in the laboratory in galena which has glide planes parallel to {100}. In twin gliding, rotation of part of the crystal lattice occurs so that it takes up a twinned position. Rotation is initially on the molecular scale, but with continued applied stress it spreads across the whole grain by rotation of one layer after another. The normal result is the development of polysynthetic twins. One example familiar to geologists is the secondary twinning of carbonates in deformed marbles. Most of the soft opaque minerals contain many potential glide planes, there are eight in galena and as a result the stresses which may affect these minerals during dynamic and regional metamorphism can lead to plastic flowage. Elevated temperatures facilitate such flowage, but flow can be induced at lower temperatures by increasing the applied pressure.

Brittle deformation takes the form of rupturing or shearing. Both generally follow lines of weakness in grains such as cleavages, twin planes, etc. Ruptures are often healed by new growth of the same or another mineral in the space created. On the other hand, softer minerals may flow into this space thus isolating the fragments of the ruptured mineral. This is particularly the case in polyphase grain aggregates made up of strong and weak minerals. At a given degree of deformation the stronger minerals such as pyrite and arsenopyrite may fail by rupturing whilst the softer minerals such as galena and the sulphosalts may flow. These processes can give rise to grain elongation and the development of schistose textures.

RECRYSTALLIZATION

The deformed state is one of high potential energy and, if the temperature is high enough, annealing will occur leading to a reduction of this high energy level. Of the various processes that take place during annealing, recrystallization is the most important. It consists of the replacement of strained grains by strain-free grains followed by grain growth, the new grains meeting in growth impingement boundaries which represent an unstable configuration. The presence of these grain boundaries leads to an increase in the free energy of the system over that which it would possess if no grain boundaries were present. This extra free energy is that of unsatisfied bonds at or near the surface of the grains; it is called the interfacial free energy.

The shapes of grains in a polycrystalline aggregate are governed by two main factors, firstly, the need to reduce the overall free energy level to a minimum, and secondly the requirement to fill space. The first requirement would be fulfilled by spherical grains but these would clearly not fill space. The microscopic examination of polished sections of artificially annealed metals and mineral sulphides shows that most grains meet three at a time at a point. Separation of these grains shows that they are bounded by a number of flat surfaces identical with those in a soap froth (Smith 1964). Inspection of such froths (which can be done by shaking up some soap solution in a plain tumbler) shows that in three dimensions the great majority of bubbles (and, by inference, grains too) meet in threes along lines. These lines are called triple junctions. When intersected by a surface (e.g. that of a polished section) these junctions appear as points (Fig. 18.1A). In a monomineralic aggregate the angles between the grain boundaries around such points are equal to or close to 120°, provided the section is normal to the triple junction. Such equilibrium grain configurations will normally be present in annealed (i.e. metamorphosed) ores. They are also present in some autoannealed ores such as chromite deposits in large lopoliths and at least one case of their occurrence in an epigenetic ore has been recorded (Burn 1971). In investigating such grain configurations in ores, it is usual to measure a number of these angles and to employ a frequency plot to determine whether they peak at 120° (Stanton 1972). Calculation of the standard deviation will produce a measure of the degree of perfection of the annealing. Very beautifully annealed mineral aggregates can be seen in the gold-copper ores of Mount Morgan (Lawrence 1972).

So far we have been concerned with monomineralic aggregates. In the case where a phase is is contact with two grains of a different phase the dihedral angle (θ) is no longer 120° (Fig. 18.1B) but some other constant angle which varies according to the composition of the phases. The values of a number of these

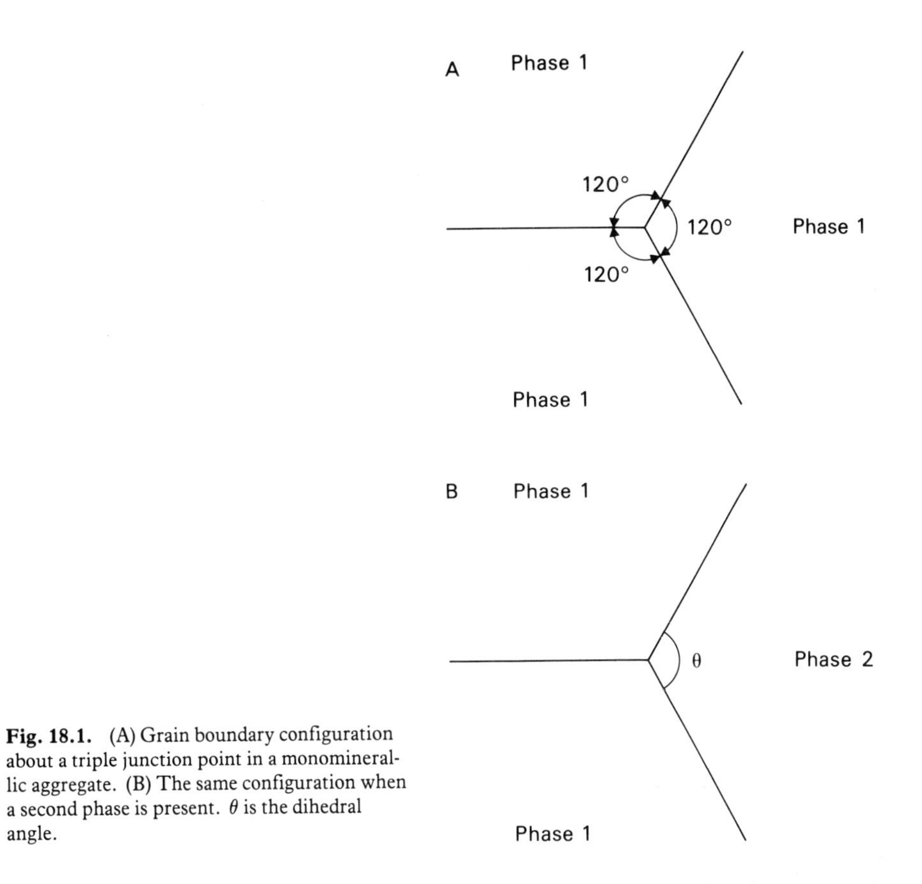

Fig. 18.1. (A) Grain boundary configuration about a triple junction point in a monomineralic aggregate. (B) The same configuration when a second phase is present. θ is the dihedral angle.

dihedral angles are known, mainly from the work of Stanton, and again measurements can be made to see if the grain aggregate has been annealed.

Increase in grain size

In an annealed aggregate, grain boundary energy is the dominant force leading to structural modification. An influx of heat will raise the energy level and if the temperature is high enough to permit diffusion, grain boundary adjustments will occur. Since the total grain boundary free energy is proportional to the grain boundary area, grain growth will occur in order to reduce the number of grains per unit volume. Grain growth does not go on indefinitely and it has been shown empirically by metallurgists that the stable average grain diameter is given by:

$$De^2 = C(T - T_0)$$

where De is the mean grain diameter, C and T_0 are constants for a given metal, and T is the absolute temperature. Thus, with increasing grades of metamorphism monomineralic aggregates in particular and indeed all grain aggregates should show an increase in grain size. Vokes (1968) has demonstrated that this is the case for metamorphosed volcanic massive sulphide deposits in Norway.

When recrystallization occurs under a directed pressure then the effects known as Riecke's Principle may become important. This is the phenomenon of pressure solution causing parts of grains to go into solution at points of contact, with the dissolved material being redeposited on those parts of grains which are not under such a high stress. This produces flat elongate grains which give the ore a schistose or gneissose appearance.

Textural adjustments and the development of new minerals

As Vokes (1968) has pointed out, because of the considerable ranges of pressure and temperature over which sulphide minerals are stable we do not find the progressive development of new minerals with increasing grade of metamorphism that are present in many classes of silicate rocks. Banded iron formations on the other hand do show the abundant development of new mineral phases (James 1955), as do many manganese deposits. Sulphide assemblages by comparison only show some minor mineralogical adjustments.

TEXTURAL ADJUSTMENTS

Two commonly noted phenomena are the reorganization of exsolution bodies and the destruction of colloform textures. Though solid solutions may be created during metamorphism by the diffusion of one or more elements into a mineral, the general tendency is for exsolution bodies to migrate to grain boundaries to form intergranular films and grains during the very slow cooling period which follows regional metamorphism.

A general decrease in the incidence of colloform textures with an increasing grade of metamorphism has been reported from pyritic deposits in the Urals, Japanese sulphide deposits and Canadian Precambrian volcanic massive sulphide deposits. Both these adjustments generally improve the ores from the point of view of the mineral processor. Exsolution bodies do not pass with their host grains into the wrong concentrate and the replacement of colloform textures by granular intergrowths leads to better mineral separation during grinding.

MINERALOGICAL ADJUSTMENTS

In contact metamorphism the main effects so far studied are those seen next to basic dykes cutting sulphide orebodies. These include the following reactions and changes.

Pyrite + chalcocite $\rightarrow$ chalcopyrite + bornite + S,
pyrite + chalcocite + enargite $\rightarrow$ chalcopyrite + bornite + tennantite + 4S,
the iron content of sphalerite increases,
copper diffuses into sphalerite,
marginal alteration of pyrite to produce pyrrhotite + magnetite.

The principal mineralogical change reported from regionally metamorphosed sulphide ores is an increase in the pyrrhotite/pyrite ratio with the appearance of magnetite at higher grades. The sulphur given off by the reactions involved in pyrrhotite and magnetite production may diffuse into the wall rocks promoting the extraction of metals from them and the formation of additional sulphide minerals.

Part III

Mineralization in Space and Time

19

Plate Tectonics and the Global Distribution of Ore Deposits

Mining geologists have for many decades attempted to relate various types of mineralization to large scale crustal structures. Prior to the development of plate tectonic theories of crustal evolution and deformation, the Hall-Dana theory of geosynclines and their deformation and uplift to form mountain chains dominated geological thought. Various schemes of pre-orogenic, syn-orogenic and post-orogenic stages of magmatism were proposed as integral parts of this orogenic cycle. It was noted by mining geologists that many ore deposits occur in geosynclinal regions and the hypothesis of distinct stages of magmatism linked to the evolution of geosynclines naturally led to the concept of accompanying stages of metallogenesis, since in the forties and fifties much emphasis was still placed on the magmatic-hydrothermal theory for the origin of the majority of metallic deposits.

When relating the genesis of orebodies to major tectonic features such as island arcs and continental mountain chains it is essential to know the age relations between the mineral deposits and their host rocks, that is whether the ores are syngenetic and thus part of the stratigraphic sequence, or whether they are epigenetic and therefore younger, perhaps much younger, than their host rocks. One of the reasons why a knowledge of age relations is important is that in the classical geosynclinal theory it was held that all the rocks of the geosyncline were formed within a subsiding linear trough. The plate tectonic theory tells us that some components of geosynclinal areas may have been transported for hundreds or even thousands of kilometres from their place of origin. If they contain syngenetic ore deposits then these must have evolved in a very different environment from that in which we now find them. The increasing trend towards syngenetic interpretations of many orebodies is therefore very important.

Geosynclines and plate tectonics

The geosynclinal concept has been modified by a number of writers to reconcile it with plate tectonic theory, e.g. Mitchell & Reading (1969), Dickinson (1971) and Hsü (1972). Some of the postulated processes of plate tectonics are shown in Fig. 19.1. On such a model of the earth's crust there are a number of linear areas of subsidence in which sedimentation may occur. These potential geosynclines are at or near plate junctures or within plate interiors where thinning of the continental crust allows of large-scale subsidence under sedimentary loading. The three principal situations at convergent junctures are shown in Fig. 19.1.

On the right is the Andean-type geosyncline on the margin of a continent with the oceanic plate dipping beneath it and a trench formed on the descending plate along the line of juncture. A more complex situation is present where the Benioff zone occurs in the ocean some distance from continental crust (left-hand side of diagram). Again a trench is present but now, after a gap of 75-275 km, there is a

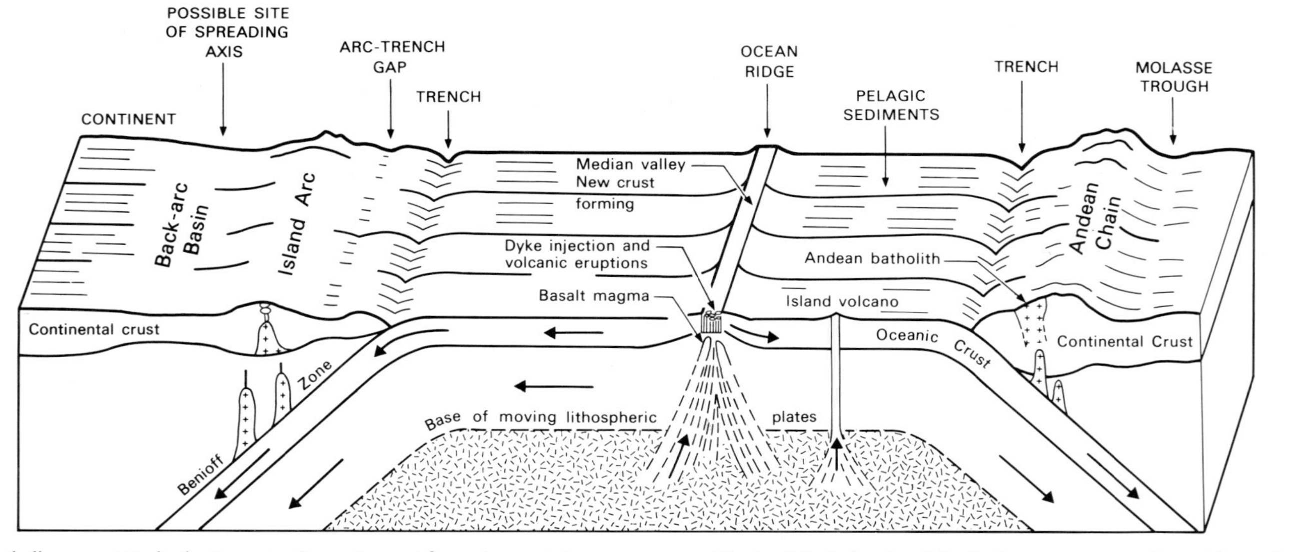

Fig. 19.1. Block diagram of the basic elements of oceanic crust formation and plate movements. On the right is the site of the Andean-type geosyncline with a molasse trough on the continental side, its deposits overlapping on to Precambrian crust and an arc-trench gap and trench on the oceanic side. On the left is the island arc-type geosyncline with its trench, troughs of deposition in the arc-trench gap, volcanic arc and (not shown) the back-arc wedges of clastic sedimentation. The back-arc basin represents the site of the Japan Sea-type of geosyncline which may or may not have its own spreading axis.

volcanic island chain. Clastic sedimentation may occur in the trench and in the arc-trench gap and an uplifted outer sedimentary arc may be present. This trench arc system is the Island Arc-type of geosyncline of Mitchell & Reading. In the back-arc area further sedimentation and igneous activity may occur and clastic wedges may be built out from the inner margin of the arc into this sea forming the Japan Sea-type of geosyncline. When the relative plate movement is largely lateral, as was the case for African and Eurasian plates during the Mesozoic and early Tertiary, we have what Hsü has termed the Mediterranean-type geosyncline.

Turning from plate margins we have the Atlantic-type geosyncline that forms on continental margins in the interior of plates. This produces the typical eugeosynclinal-miogeosynclinal couple as exemplified by recent sedimentation along the Atlantic coastline of north America. Another important area of sedimentation is the divergent junction zone during the early stages of continental break up. Modern environments of this sort are the East African Graben and the Red Sea; the term Red Sea-type geosyncline will be used for these. It is important from the ore environmental viewpoint to be aware of the various rock-types developed in these different settings and therefore the dominant rock-types of these various geosynclines are listed in Table 19.1.

Plate tectonics and mineralization

In discussing the mineralization which may be found in the six different types of geosyncline outlined above, the loci of rifting and growth of new crust will be considered first, as much of the material formed at these sites may be conveyed by various mechanisms, such as primary plate movement, reversal of plate movement and continental collision to geosynclinal areas of different types.

RED SEA-TYPE GEOSYNCLINES

These are initiated by the doming of continental areas which, due to stretching, develop three rift valleys that meet at a 120° triple junction (Burke & Dewey 1973). As is shown in Fig. 19.2 two of the rift valleys may combine to form a divergent plate boundary leading through the graben to the Red Sea stage of spreading whilst the third arm may only show partial development. This third arm may develop a considerable thickness of sediments, with some volcanics and igneous intrusions. These may be structurally deformed but the geological history of such zones is relatively simple. They do not often progress much beyond the graben stage and are called failed arms or aulacogens.

The initial stage of graben development is marked by deeply penetrating faults forming pathways to the mantle and giving rise to volcanism. This is usually of alkaline type sometimes with the development of carbonate lavas and intrusives (Fig. 19.3). Erosion of these may lead to the formation of soda deposits (e.g. Lakes Natron and Magadi in east Africa) and the intrusive carbonatites may carry a number of metals of economic interest (Chapter 10) as well as being a source of phosphorus and lime.

In the graben themselves sediments with or without volcanics may accumulate. In the east African rift valleys, red bed type deposits are common with conglomerate fans along the escarpments. Playa lakes have produced evaporites. The water of Lake Kivu is rich in zinc and precipitates sphalerite. The deposition of this and other metals is believed to be due to hydrothermal solutions which represent

Table 19.1. Common sedimentary and igneous rocks found in geosynclines of different types. (After Mitchell & Reading 1969 and Walton 1970, with modifications.)

Type of geosyncline	Shallow-water clastics	Carbonates	Tholeiites, pelagics and ophiolites	Tectonic mélange	Mature turbidites	Calc-alkaline volcanics and intrusions	Calc-alkaline turbidites	Molasse	Nature of basement
Andean									
mountain chain	rare	rare		rare		common (2)[3]		abundant (3)	continental
trench			abundant (1)[3]	abundant	rare to common (2)		rare to common (2)		oceanic
Island arc									
volcanic arc		locally abundant	rare (1)			abundant (2)	abundant (2)		oceanic
trench			abundant (1)	profuse			common (2)		oceanic
Japan Sea	abundant (3)	locally abundant	present if basement is oceanic		abundant (2)	tuffs (2)	rare to common (2)		oceanic or continental
Mediterranean[1]		abundant (1)	common	rare	abundant (2)	rare	rare	abundant (3)	continental + some oceanic
Atlantic[2]									
miogeosyncline	abundant (1)	abundant (2)							continental
eugeosyncline			common (1)		abundant (2) occasionally tholeiitic at base				oceanic
Red Sea	common (1) in early graben stages with evaporites		abundant (2) carbon- and metal-rich muds						oceanic (initially continental)

[1] Formed between laterally moving plates.
[2] Becomes Himalayan-type of geosyncline if involved in continental collision.
[3] General sequence of deposition.

N.B. For reasons of space this table omits the miogeosynclinal-type sequence of the arc-trench gap (cf., e.g. Dickinson 1971).

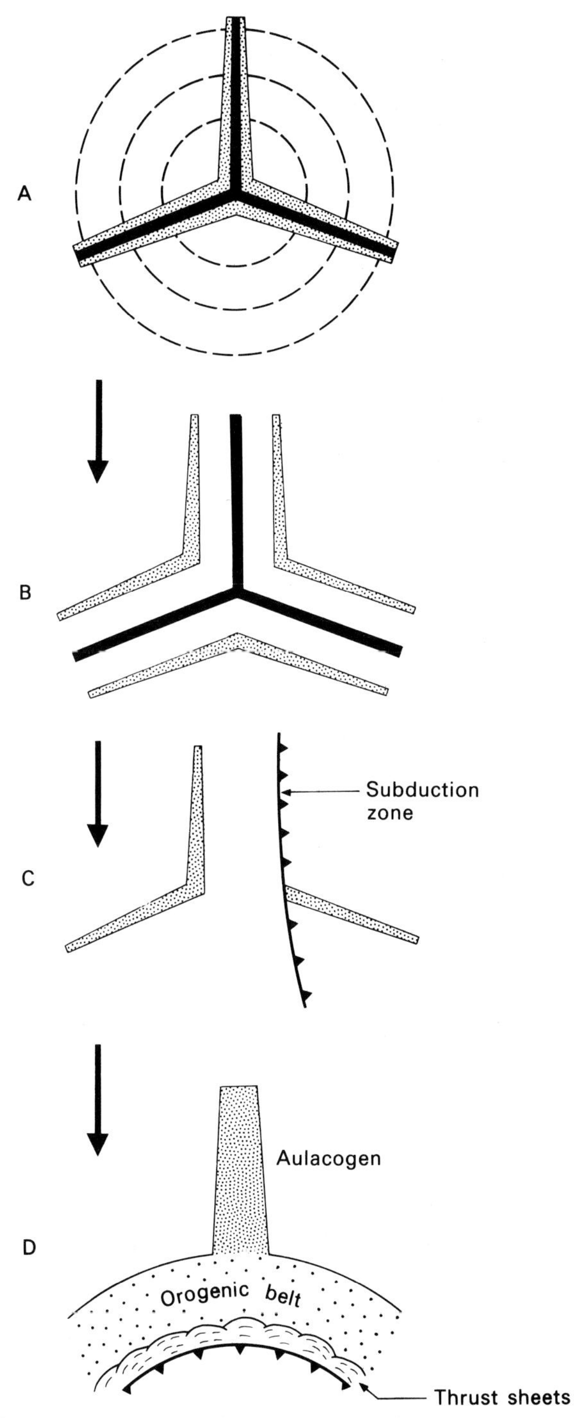

Fig. 19.2. Schematic diagrams showing the development of an aulacogen. (A) Development of three rift valleys with axial dykes in a regional dome. (B) The three rifts develop into accreting plate margins (e.g. early Cretaceous history of Atlantic Ocean-Benue Trough relationship). (C) One arm of system begins to close by marginal subduction causing deformation of sediments and volcanics. (D) Continental margin collides with subduction zone forming orogenic belt and the failed arm is preserved as an aulacogen. (After Burke & Dewey 1973.)

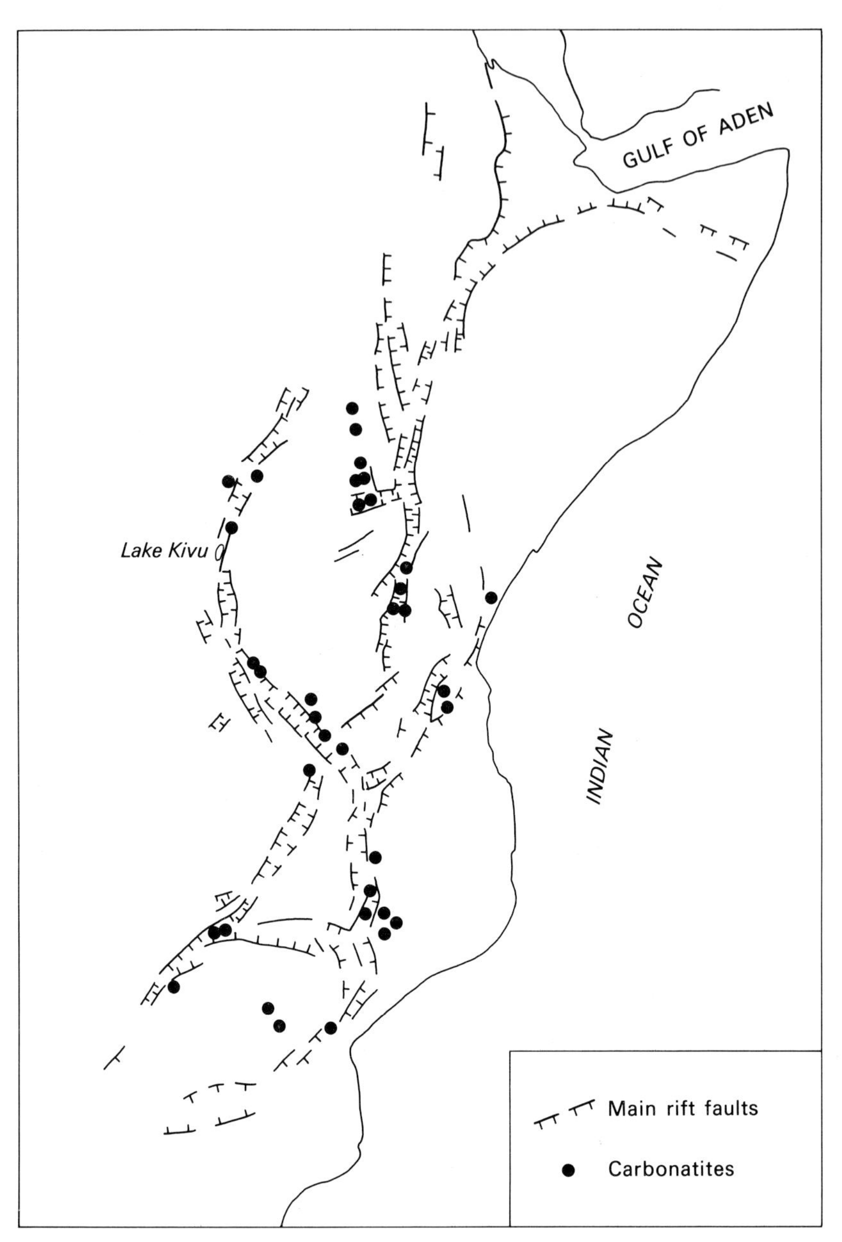

Fig. 19.3. Distribution of carbonatite intrusions relative to the East African Rift Valleys. (After Mitchell & Garson 1976.)

ground water that has been cycled through volcanics and sediments at times of high rainfall (Degens & Ross 1976).

Recently-formed aulacogens occur at both ends of the Red Sea. The Gulf of Suez is one of these. It contains a 4 km thick succession of Neogene salt, limestone and clastic sediments and in it lie the Ras Morgan Oilfields. Other young aulacogens also contain salt deposits and oilfields. Slightly older aulacogens are present on both sides of the Atlantic (Fig. 19.4). One occurs where the shoulder of Brazil fits into Africa. At this point spreading occurred on all three arms of a triple

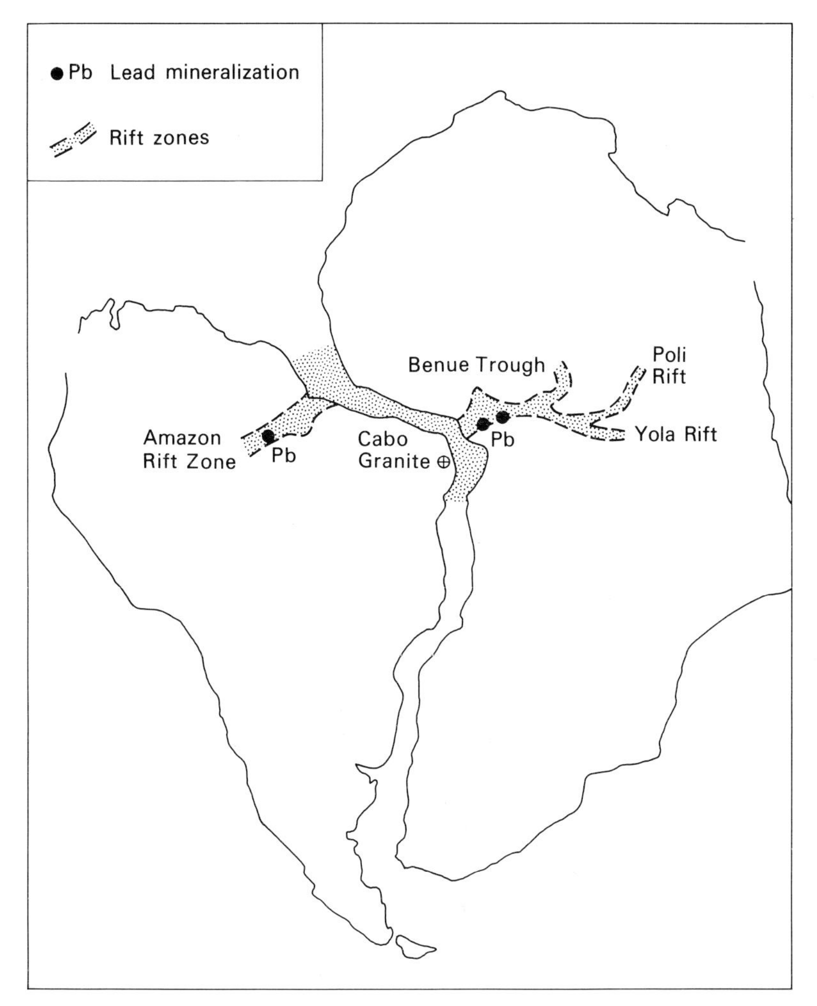

Fig. 19.4. Sketch map showing the locations of the Benue Trough, the Amazon Rift Zone and the lead mineralization within these aulacogens. (After Burke & Dewey 1973 and Mitchell & Garson 1976.)

junction 120-80 Ma ago. Then the Benue Trough closed with, Burke & Dewey suggested, a short-lived period of subduction, whilst the South Atlantic and Gulf of Guinea arms continued to spread. The Benue Trough is an aulacogen about 560 km long having a central zone of high Bouguer anomalies flanked on either side by elongate negative anomalies. This pattern is believed to be due to the presence of near-surface masses of crystalline basement and intermediate igneous rocks below the central zone. Lead-zinc-fluorite-baryte mineralization occurs in fractures in Lower Cretaceous limestone and similar lead mineralization is present in the Amazon Rift Zone. Tholeiitic igneous activity also took place and intrusions of diorite, gabbro and pyroxenite are present. The origin of the mineralization in both aulacogens is controversial. It may be associated with basic magmatism or with circulating brines which have passed through evaporite deposits.

The triple junction by the shoulder of Brazil is thought to have been generated by the Niger mantle plume. Plumes may cause melting of the continental crust forming granite intrusions, e.g. the Cabo Granite of Brazil. Now throughout the world the big tin-tungsten provinces are associated with subduction zones but there are some tin provinces which do not fit this pattern, e.g. those of Nigeria, Rondônia (Brazil) and the Sudan. Tin in these provinces is associated with sodic granites in ring complexes. Mitchell & Garson (1976) suggest that mantle plumes were responsible for the generation of these granites and their associated mineralization. Both the granitic magma and the associated metals were probably derived from the crust.

A number of Proterozoic aulacogens that carry more important mineralization than that in the above examples have been recognized. Burke & Dewey (1973) have identified a number of these in North America (Fig. 19.5). Those running perpendicular to the western margin of the continent carry great thicknesses of rocks of

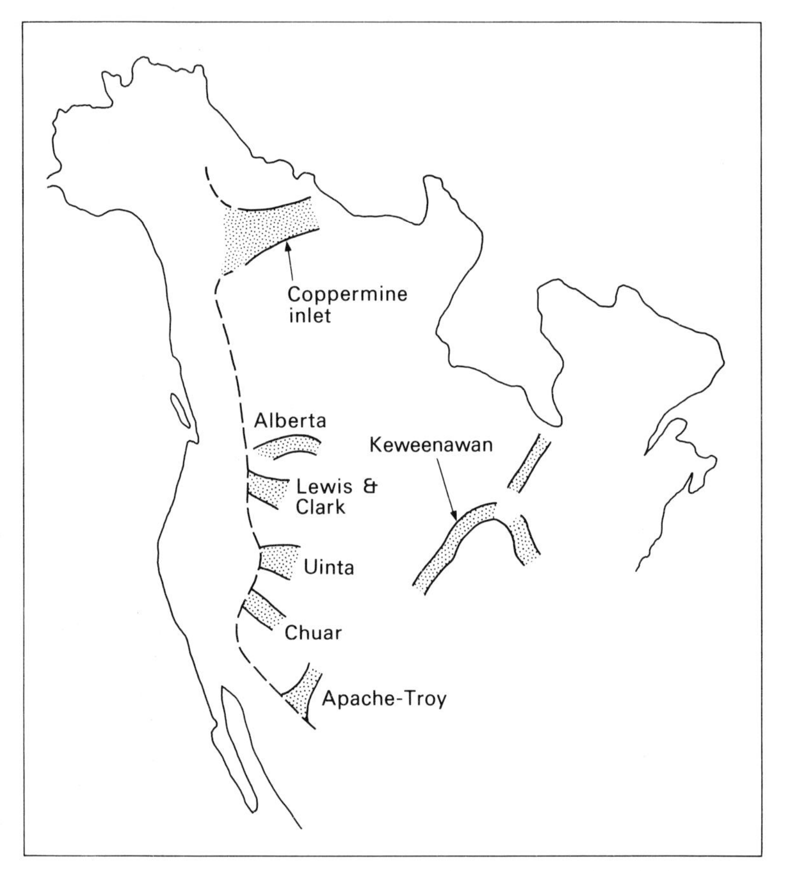

Fig. 19.5. Some of the Proterozoic aulacogens of North America. (After Burke & Dewey 1973.)

the Belt Series. The northernmost trough contains the Coppermine River Group consisting of more than 3 km of basalt flows with native copper mineralization overlain by greater than 4 km of sediments with evaporites near the top. Further south is the Alberta Rift which passes into British Columbia. It contains about 11 km of late Precambrian sediments in which important stratiform and epigenetic lead-zinc mineralization occur including the famous Sullivan Mine at Kimberly, B.C. Phosphorite deposits are also present. To the east, the Keweenawan Aulacogen, which is the same age as the Coppermine Aulacogen, also carries a considerable thickness of basalt and clastic sediments and again native copper mineralization is present. Other mining camps which are believed to lie within aulacogens include the copper and lead-zinc ores of Mount Isa, Queensland, and the Zambian Copperbelt (Raybould 1978). Red bed coppers are well developed in upper Palaeozoic aulacogens of the USSR and uranium mineralization is also present in some aulacogens.

Clearly, aulacogens are important carriers of mineral treasures but we must now look at the graben → Red Sea evolution. With further extension of the crust and the commencement of continental drift deep crustal flowage and tensional faulting will combine to thin the crust along the graben. At some stage during this process an opening to the sea may initiate marine conditions. Observations from the Rhine Graben, Mesozoic deposits along the Atlantic coastlines and the Miocene of the Red Sea region indicate that evaporite series of great thickness may form at this time. These evaporites contain halite as well as gypsum and therefore have a double economic importance.

The Red Sea is known to be floored by basalt and there is evidence that new oceanic crust is being formed along its median zone. On either side, pelagic sediments are forming, but the depths are not yet sufficiently great nor the other factors present which lead to phosphorite development as in areas of upwelling oceanic currents along some Atlantic and Pacific coasts where phosphorite deposits are forming at the present day. The recently formed carbonate miogeosynclinal area of Florida is well known for its phosphorites which yield by-product uranium. Whether such deposits can occur in Red Sea geosynclines depends on where one draws the boundary between these and the Atlantic-type situation into which they evolve.

The possible mode of development of oceanic crust along the median zone of the Red Sea is shown in Fig. 19.6. As this new crustal material moves away from the median zone, layer 1 of the oceanic crust, in the form of pelagic sediment, is added to it. As Sillitoe (1972a) has suggested, there is strong evidence that many Cyprus-type sulphide deposits are formed during this process of crustal birth. If slices of the oceanic crust are thrust into mélanges at convergent junctures or preserved in some other way then sections similar to that shown in Fig. 19.6 may be expected. This is the case where these cupriferous pyrite orebodies are found in Cyprus, Newfoundland, Turkey, Oman, etc. Francheteau *et al.* (1979) have recently described deposits of massive sulphides from the East Pacific Rise which occur on a spreading ridge away from transform faults. They show a number of similarities to the Cyprus deposits. Similar situations to the oceanic spreading ridges are found in back-arc basins and indeed many workers hold that the Troodos Massif of Cyprus was developed in such a milieu. In this massif non-economic nickel-copper mineralization and economic chromite deposits occur in the basic plutonic rocks.

Hydrothermal mineralization with the development of copper, zinc, silver and mercury has been reported from oceanic ridges in the Atlantic and Indian Oceans by Dmitriev *et al.* (1971). Surprisingly, tin mineralization was found in both ridges accompanied in one case by typical hydrothermal minerals normally associated with granitic environments such as tourmaline, topaz, fluorite and baryte. Cupriferous pyrite mineralization of stockwork-type has been reported by Bonatti *et al.* (1976) from the Mid-Atlantic Ridge. It occurs in metabasalts and they put forward evidence that it was generated by sea water solutions circulating through the oceanic

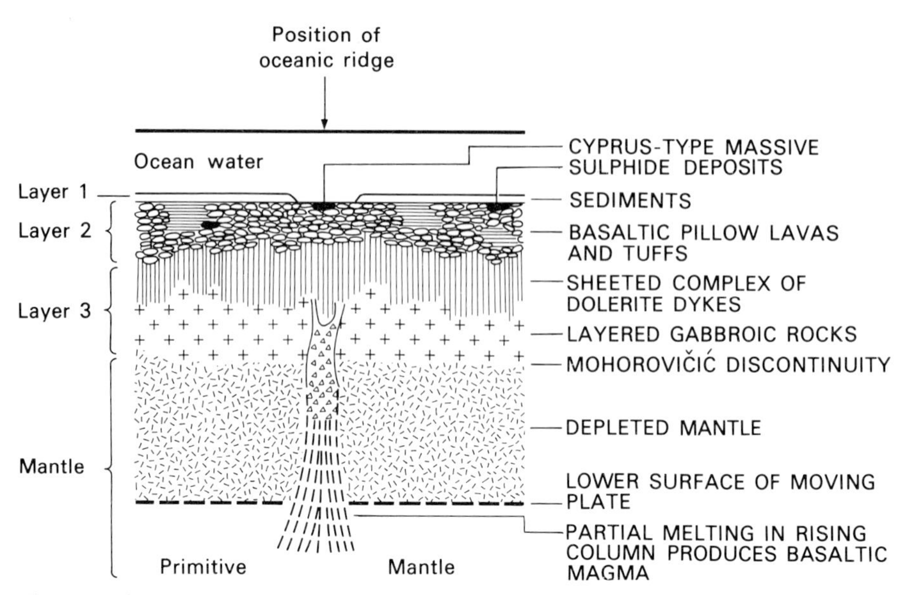

Fig. 19.6. Schematic representation of the development of oceanic crust along a spreading axis. The crustal layering and the possible locations of Cyprus-type massive sulphide deposits are shown. (After Cann 1970 and Sillitoe 1972a.)

crust in a manner similar to that described for the Cyprus deposits at the end of Chapter 4.

The median zone of the Red Sea is notable for the occurrence of hot brines and metal-rich muds in some of the deep basins. Those studied by Degens & Ross (1969, 1976) can be defined by the 2000 m contour. The largest of these deeps is the Atlantis II which is 14 km × 5 km and 170 m deep. At the base, multicoloured sediments usually 20 m thick, but up to 100 m, rest on basalt. Piston coring has demonstrated the presence of stratification and various sedimentary facies high in manganites, iron oxides and iron, copper and zinc sulphides. These sediments can run up to 20% Zn but the top 10 m in the Atlantis II deep run 1.3% Cu and 3.4% Zn. Feasibility studies are underway for their recovery. Above the sediments, the deeps contain brine solutions, these and the sediments run up to about 60°C and 30% sodium chloride. It has been suggested that the sediments and brines have originated either from the ascent of juvenile solutions, or from the recirculation of sea water which has leached salt from the evaporites flanking the Red Sea and passed through hot basalt from which it has leached metals. If this second mechanism is the sole or principal one, then metal-rich muds of this origin will only form when evaporites are sufficiently near to supply the required salt. This being the case, it is possible to outline areas of the present oceans which can be said, from the distribution of known evaporites, to have high or low probabilities of containing these deposits. A study of this nature has been made for the North Atlantic by Blissenbach & Fellerer (1973).

ISLAND ARC-TYPE GEOSYNCLINES

It is convenient to divide the mineral deposits in these geosynclines according to whether they were formed outside the arc-trench environment and transported to it by plate motion (allochthonous deposits), or whether they originated within this geosynclinal complex (autochthonous deposits) (Evans 1976b).

Allochthonous deposits.
Rocks and mineral deposits formed during the evolution of Red Sea-type geosynclines and large oceans may by various trains of circumstances arrive at the oceanic trench at the top of a subduction zone (Fig. 19.7). This material is largely subducted, when some of it may be recycled, whilst some is thrust into mélanges.

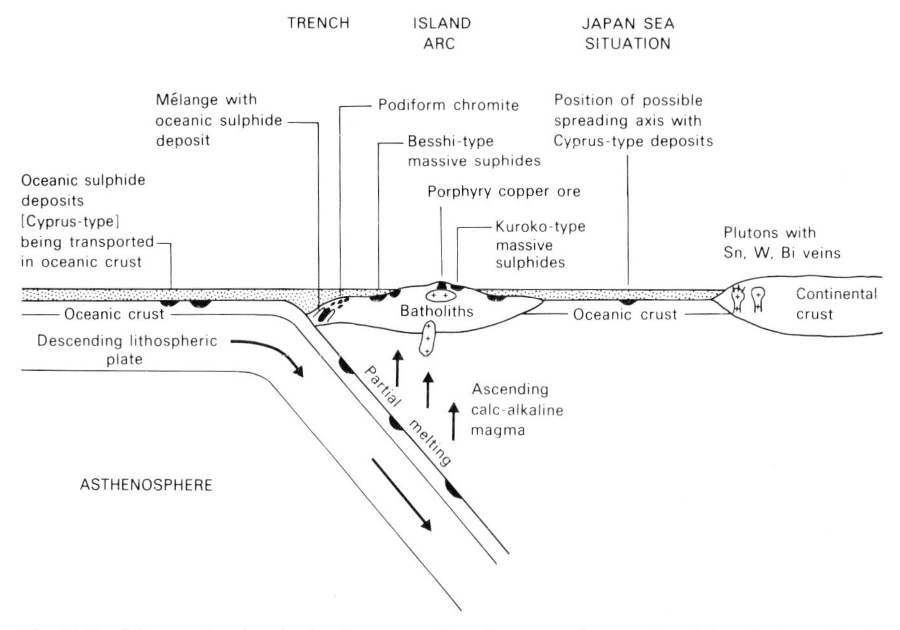

Fig. 19.7. Diagram showing the development and emplacement of some mineral deposits in an island arc and its adjacent regions. (After Sillitoe 1972a and 1972b with modifications.)

There is, however, only one area so far discovered with what appears to be Cyprus-type sulphide orebodies thrust into a mélange and that is north-western California where we find the Island Mountain deposit and some smaller occurrences in the Franciscan mélange. Cyprus-type massive sulphides are also likely to be present at the base of island arc-type geosynclinal sequences, whether these are the initial succession or a second or later one formed by the migration of the Benioff Zone, because in most cases the arc basement will have originated at an oceanic ridge.

Clearly, any other deposits formed in new oceanic crust may also eventually be mechanically incorporated into island-arc geosynclines. The most likely victims will be the chromite deposits of alpine-type peridotites and gabbros (Thayer 1964, 1967). Though some writers have proposed that these deposits were intruded partly as fluid magma and partly as crystal mush, Thayer (1969a, 1969b) has cited much evidence in favour of their being regarded as cumulates that originated in the upper mantle often at mid-ocean ridges. Recent evidence from the Papuan ultramafic belt is particularly compelling in this respect (Thayer 1971). Here, in a block apparently composed of obducted oceanic crust and upper mantle, rock units of pillowed basalts, gabbro and peridotite rest upon each other in thicknesses exceeding

2000 m for each unit. The plutonic rocks all show well-developed cumulate textures. Podiform chromitites are the most important magmatic deposits in the allochthonous peridotites (Thayer 1971). Platinum metal deposits are known in some of these, but they are more important as the source rocks for placer deposits of these metals. Economic deposits of podiform chromitite occur in present day arcs in Cuba, where they are found in dunite pods surrounded by peridotite, and on Luzon in the Philippines, again in dunite in a layered ultramafic complex. Numerous occurrences are present in fossil island arc successions (cf., e.g. Thayer 1964). According to Thayer (1971), the allochthonous 'high-temperature' lherzolitic peridotites and associated gabbroic 'granulites' of Green and the garnet peridotites of O'Hara contain no significant deposits of oxides or sulphides.

If we follow Thayer (1971) in excluding the Precambrian ultramafic sheets from the class of alpine-type plutonic rocks of the ophiolite suite, then it must be recorded that at present remarkably few magmatic sulphide segregations are known in these rocks. Uneconomic pyrrhotite-pentlandite accumulations are known in Cyprus but the only deposits so far being exploited are those of the Acoje Mine in the Philippines, where nickel and platinum sulphides occur. The accelerated pace of mineral exploration in present day arcs will eventually indicate whether such deposits are truly rare or whether many are still awaiting discovery. At the moment, the latter possibility appears to be the more plausible.

Autochthonous deposits.

In considering these deposits it is convenient to divide the geosynclinal development into three stages: initial or tholeiitic, main calc-alkaline stage and waning calc-alkaline stage.

(1) *Initial or tholeiitic stage.* The probable hydrous nature of such magmas and their derivation by partial melting of mantle material (Ringwood 1974) implies that cogenetic massive sulphide deposits may well be expected to form during this stage. However, no massive sulphide deposits have as yet been detected in the earliest tholeiitic sequences of island arcs, but such sequences must be considered as promising grounds for their discovery. Stanton (1972) has surmised that both the early basic lavas of the Solomon Islands with sulphide amygdules and disseminations belong to this stage.

(2) *Main calc-alkaline stage.* This is the main period of arc development. Considerable sedimentation and subaerial volcanicity may have occurred during the tholeiitic stage, but the main island arc building and plutonic igneous activity belong to this stage. Baker (1968) showed that whilst the early stage of island arc volcanism is dominated by basalt and basaltic-andesite, the more evolved arcs have andesite as the dominant volcanic rock. Ringwood (1974) has summarized the investigations which have shown that the early tholeiitic stage is succeeded by magmas having a calc-alkaline trend and which are probably ultimately derived from two sources—subducted oceanic crust and partial melting of the mantle wedge overlying the Benioff Zone. This will result in the diapiric uprise of wet pyroxenite from just above the Benioff Zone. The pyroxenite will undergo partial melting as it rises and the magmas so produced will fractionate as they rise to produce a wide range of magmas possessing calc-alkaline characteristics.

The exposure of much more rock to subaerial erosion will greatly increase the volume of sediment which reaches inter-island gaps, the arc-trench gap and the back-arc area. Reef limestones will then be common. Rapid erosion on the flanks of sub-aerial volcanoes, ignimbritic activity (related to the increase in silica content of the magmas) giving rise to submarine lahars, submarine slumping and so on, would all contribute to produce a vast volume of material for local sedimentation and for transportation by turbidity currents to the arc-trench gap and along submarine canyons into the trench itself. Within the arc-trench gaps local linear uplifts occur—the outer sedimentary arcs—together with troughs containing thick sedimentary sequences.

Volcanic massive sulphide, stockwork, pyrometasomatic and vein deposits are formed in this stage. Conformable sulphide orebodies of Besshi-type develop at this stage, accompanying the andesitic to dacitic volcanism. They occur in complex structural settings characterized by thick greywacke sequences as in the late Palaeozoic terrain of Honshu where the volcanism is basaltic to andesitic. The lack of ophiolitic components, the petrological association and the pyroclastic nature of the volcanics indicates a different environment from that of the Cyprus-type deposits. Sillitoe (1972a) suggested that the solutions which deposited these massive sulphide deposits might have come directly from subducted sulphide bodies (Fig. 19.7) with the metals in them being derived from the subducted bodies and layers 1 and 2 of the subducted ocean floor. Sillitoe (1972b) has further suggested that metal-bearing brines expelled from layers 1 and 2 would induce fusion in the wedge of mantle above the Benioff Zone, ascending with the resulting hydrous calc-alkaline magmas to form porphyry copper deposits when the volcanism is dominantly subaerial. Exploration in recent years has shown that many porphyry coppers are present in island arcs (Fig. 12.7) with, contrary to earlier ideas, both diorite and Lowell-Guilbert model types being present. There is no evidence that continental crust is required for their formation; therefore they can be sought for in any island arc with or without continental crust. The young age of the orebodies in some recent arcs indicates that they belong to the late stages of island arc evolution and perhaps overlap into the waning stage. Sillitoe's theory that the ultimate source of the metals in porphyry deposits is to be sought in subduction zones is attractive as it can account for the relatively uniform nature of porphyry copper deposits despite the great variety of their host rocks. On this very score, great difficulties attend one of the most popular alternative theories. The suggestion that brines, circulating through the plutons, obtain their metals from the immediate country rocks is difficult to sustain due to the complete lack of correlation of deposit geochemistry with country rock geochemistry. Sillitoe's theory is attractive on other grounds. It can account for the regional differences seen in these deposits (Chapter 12), relating them to differences in the contents of copper, molybdenum and gold in the subducted oceanic crust.

Pyrometasomatic skarn deposits are common at igneous-carbonate country rock contacts in the older and more complex arcs such as Indonesia, the Philippines and Japan. Many of the granites and granodiorites with which these deposits are associated also possess spatially related pegmatitic and vein mineralization and the resulting metallic association can be quite complex as in the Kitakami and Abukuma highlands of Japan. Pyrometasomatic and vein deposits of tin are well known from parts of the Indonesian arc. Less well known are the gold veins in the contact zones of diorites and granodiorites of the Solomon Islands. Other vein deposits are not common in the younger arcs, including those of mercury, which is surprising in view of their development in association with andesites in the Alpine orogenic belt, e.g. the eastern Carpathians. Cinnabar veins occur in the more complex arcs of the Philippines, Japan and New Zealand.

(3) *Waning calc-alkaline stage.* The massive sulphide deposits of Kuroko-type probably belong to the later stages of development of island arcs. They are associated with the more felsic stages of calc-alkaline magmatism and are marked by distinctive features that have been described in Chapter 13. Kuroko-type ores in ancient islands arc-type geosynclines include the Palaeozoic deposits of Captains Flat, New South Wales; Buchans, Newfoundland and Avoca, Ireland.

As Stanton (1972) has pointed out, manganese deposits of volcanic affiliation are very common in eugeosynclinal successions of all ages from Archaean to Recent and island arcs are no exception. Large numbers of deposits of subeconomic size occur in the Tertiary volcanic areas of Japan, Indonesia and the West Indies and in the Recent volcanic sequences of the larger south-western Pacific islands. These deposits have been formed *in situ* and are not tectonic slices of pelagic manganese deposits of the oceanic crust. They belong to this and the preceding stage of island arc evolution.

ANDEAN-TYPE GEOSYNCLINES

The Andean-type geosyncline in its type area lies along the western margin of South America where it abuts against the Nazca plate which is moving down a Benioff Zone beneath the present mountain chain. In this chain are developed belts of calc-alkaline volcanism and plutonism accompanied by tectonic deformation and crustal thickening. Thus, Andean and island arc geosynclines have much in common and it would result in tedious repetition if the evolutionary stages and the concomitant mineralization were dealt with in the same detail as had been accorded to island arcs.

The Andean chain itself is immensely rich in orebodies of various types and metals, but on the whole these show a plutonic-epigenetic affiliation rather than a volcanic-syngenetic one. This is no doubt to some extent a function of erosion. The profusion of metals may be related to proximity to the East Pacific Rise and the abundance of metals that are being added to the oceanic crust at this spreading axis. Sawkins (1972) has listed the main features of Andean Cordilleran deposits emphasizing their close relationship in time and space with calc-alkaline intrusives and their occurrence at high elevations (2000 m to 4000 m above sea level) at relatively shallow levels in the upper crust. He reminds us that this implies that erosion of the Andes down to present sea level would remove this entire suite of ore deposits.

Peterson (1970) has recorded a general zoning of metals across the Andes such that iron and copper are more important in the west and eastwards they give way to a lead-zinc belt and then to a tin belt. Wright & McCurry (1973) have interpreted this as showing that metals subducted in the oceanic crust have been remobilized at different depths along the Benioff Zone. Mitchell & Garson (1972) and Mitchell (1973) have shown that both the migration of Benioff Zones and their inclination may be important factors in determining whether certain elements are mobilized and carried up into the geosynclinal areas and also whether suitable igneous host rocks have been emplaced in which the metals will find a ready home and so form orebodies.

Mitchell & Garson (1972) have interpreted deposits of the tin-tungsten-bismuth-fluorite association present in belts near continental margins as being due to oceanward migration of Benioff Zones. For example, in west Malaysia large volumes of granite were formed during the Mesozoic above successive shallow-dipping Benioff Zones. The oceanwards migration of the Benioff Zone would then permit a rising stream of volatiles derived from descending crust at depths exceeding 200 km to ascend to the still hot but previously emplaced granites. Mitchell (1973) points out that the same sequence of plutonic and metallogenic events could result if the inclination of the Benioff Zone increased from shallow to steep.

JAPAN SEA-TYPE GEOSYNCLINES

In the back-arc regions behind island arcs a small ocean basin such as the Japan Sea or the Sea of Okhotsk may be present, which will have a number of possible complications (Table 19.1). For example, the basement may be oceanic or continental and, if oceanic, a spreading axis may be present. Its presence will presumably lead to the development of Cyprus-type sulphide deposits and chromitites. Black metalliferous mud deposits and iron-manganese deposits of volcanic affiliation are further possibilities. Other possible complications emerge when the sedimentary history is considered. On one side, the volcanic island arc will supply immature sediments and from the other (continental) side will come an entirely different mature suite of material. If the size of the basin of deposition allows, mixing of these materials of very different provenances may occur. On the continental side, conditions may permit the development of Minette-Clinton ironstones. On the arc side, conditions grade into the island arc-geosynclinal environment.

ATLANTIC-TYPE GEOSYNCLINES

Part of this geosyncline has of course, very early in its history, passed through the Graben and Red Sea stages discussed above. With this in mind, we may infer that in the basal portions of Atlantic-type geosynclines, along the line where the continental basement gives way to an oceanic one, extensive evaporite deposits laid down during the Red Sea stage may occur. If this is correct, metalliferous muds having distributions similar to those suggested by Blissenbach & Fellerer (1973) may be present. They consider that the embryonic North Atlantic of Jurassic-Cretaceous times probably possessed the ideal features for the formation of metalliferous muds: topographic depressions with concentrated thermal and exhalative activity in a narrow central zone within a small oceanic basin having closed northern and southern ends and evaporite successions flanking the central rift.

At the present time, the eastern seaboard region of North America is taken as a model for geosynclines with a plate interior setting (Dickinson 1971; Hsü 1972). The thick Mesozoic and Tertiary shelf carbonate sequence under the Floridan-Bahamas region with its thin oceanic pelagic equivalents floored by oceanic crust would, if involved in orogenic movements, form a typical miogeosynclinal-eugeosynclinal couple. This miogeosynclinal carbonate area is of course important along the Atlantic Coastal Plain from North Carolina to Florida for its Tertiary phosphorite deposits which also carry appreciable amounts of uranium. Atlantic-type geosynclines, by their very nature, are likely to remain undeformed for long periods.

MEDITERRANEAN-TYPE GEOSYNCLINES

As noted above, this geosynclinal type was defined by Hsü as originating in a plate margin setting where the plate motion was dominantly lateral. Subduction of oceanic crust is not, therefore, developed on any substantial scale and this had important results as far as the evolution of the Alps is concerned. It accounts for several problems of Alpine geology: the distinctive tectonic style, the subordinate development of tectonic mélanges, the longevity of the Alpine flysch trenches and the very small amount of calc-alkaline igneous activity of either plutonic or volcanic type. This has resulted in a complete absence of porphyry copper deposits and Besshi- and Kuroko-type massive sulphide deposits. Indeed, by comparison with the Andean belt or mature island arc terrains, the paucity of post-Hercynian mineral deposits in the Alps is most marked (Evans 1975). Despite the fairly common occurrence of ophiolites in parts of the Alps, this remark also applies to chromite deposits which are insignificant in size. The most important metalliferous mineralizations of the Alps are uranium-sulphide deposits in the Permian, that may have developed in a graben environment, and lead-zinc deposits in Triassic carbonate sequences of the eastern Alps. The latter appear to be syngenetic. The occurrence of baryte, celestine and anhydrite in the ores, the presence of evaporite beds in the host limestones and the general geological setting invite a comparison with the lead-zinc ores of the Red Sea region. Thus, as Evans (1975) has shown, by considering the possible nature of the evolving Alpine geosyncline from a plate tectonic point of view it is possible to account for the ore deposits which are present and these and other deposits likely to be developed in Mediterranean-type geosynclines, are the ones upon which the search for ore deposits should be concentrated in the post-Hercynian rocks of the Alpine region.

Space does not allow an extension of this chapter to deal with Precambrian deposits, but it is clear from the literature that plate tectonic interpretations are applicable, at least in part, to the problems of the geology of Precambrian shields and the genesis of many Precambrian orebodies in the light of this theory has already been considered by a number of workers. Some useful summaries are given in Windley (1977).

20

Ore Mineralization Through Geological Time

It is now well known to geologists that the earth, and its crust, in particular, have passed through an evolutionary sequence of changes throughout geological time (Windley 1977). These changes have been so considerable that we must expect them to have had some influence on the nature and extent of mineralization. Reference has already been made in this book (Chapter 6) to the association of most of the world's tin mineralization with Mesozoic and late Palaeozoic granites, to the virtual restriction of banded iron formation to the Precambrian and the bulk of it to the interval 2600-1800 Ma ago and to the importance of the Precambrian for nickel and orthomagmatic ilmenite deposits. In Chapter 9, it was noted that the lack of numerous Phanerozoic nickel sulphide deposits may be due to depletion of the mantle in sulphur during the Archaean, and in Chapter 13 attention was drawn to the fact that volcanic massive sulphide deposits show important geochemical changes with time. We will now examine such changes in the type and style of mineralization in a little more detail. These changes can be conveniently discussed in terms of the Archaean, Proterozoic and Phanerozoic intervals and the environments which prevailed during them.

The Archaean

This interval, 3800-2500 Ma ago, is notable both for the abundance of certain metals and the absence of others. Metals and metal associations developed in significant amounts include Au, Sb, Fe, Mn, Cr, Ni-Cu and Cu-Zn-Fe. Notable absentees are Pb, U, Th, Hg, Nb, Zr, REE and diamonds.

Two principal tectonic environments are found in the Archaean; the high grade regions and the greenstone belts. The former are not important for their mineral deposits which include Ni-Cu in amphibolites, e.g. Pikwe, Botswana, and chromite in layered anorthositic complexes e.g. Fiskenæsset (west Greenland). The greenstone belts on the other hand are very rich in mineral deposits whose diversity has been described by Watson (1976). The principal mineral deposits are related to the major rock groups of the greenstone belts and their adjoining granitic terrains as follows:

(1) ultramafic flows and intrusions: Cr, Ni-Cu;
(2) mafic to felsic volcanics: Au, Ag, Cu-Zn;
(3) sediments: Fe, Mn and baryte;
(4) granites and pegmatites: Li, Ta, Be, Sn, Mo, Bi.

CHROMITE

This is not common in greenstone belts but a very notable exception is present at Selukwe in Zimbabwe-Rhodesia. This is a very important occurrence of high grade chromite in serpentinites and talc-carbonate rocks intruded into schists which lie close to the Great Dyke. It resembles the podiform class of deposit, though it is in a very different tectonic environment from that in which this class is normally found.

NICKEL-COPPER

These deposits are mainly composed of massive and disseminated ores developed in or near the base of komatiitic and tholeiitic lava flows and sills as described in Chapter 9. Only four important fields occur, those of south-western Australia (Kalgoorlie belt), southern Canada (Abitibi belt), Zimbabwe-Rhodesia and the Baltic Shield of northern USSR. As these metals and their host rocks are probably mantle-derived this suggests the existence of metallogenic provinces controlled by inhomogeneities in the mantle, though the anomaly may not consist of an excess of nickel but rather a concentration of sulphur which led to the extraction of nickel from silicate minerals.

GOLD

Gold has been won in smaller or larger amounts from every greenstone belt of any size. Its occurrence is the principal reason for the early prospecting and mapping of these belts. The gold is principally in vein deposits cutting basic or intermediate igneous rocks—both intrusions and lava flows—but the more competent intrusives such as the Golden Mile Dolerite of Kalgoorlie are more important. Some gold deposits show an association with banded iron formation and these appear to have been deposited from subaqueous brines to form exhalites. The greatest concentration of gold mineralization occurs in the marginal zones of the greenstone belts near the bordering granite plutons and it decreases towards the centre of the belts. This suggests that it has been concentrated from the ultrabasic-basic volcanics by the action of thermal gradients set up by the intrusive plutons.

COPPER-ZINC

Volcanogenic massive sulphide deposits are very common in the Archaean especially in the Abitibi Orogen of southern Canada (Sangster & Scott 1976). These deposits are principally sources of copper and zinc but they are of Kuroko- rather than Besshi-type despite the fact that their lead content is normally very low. The virtual absence of lead mineralization from these greenstone belts may indicate that during the Archaean there had been insufficient time for much lead to be generated by the decay of uranium and thorium.

IRON

Banded iron formation is common throughout Archaean time but not in the quantities in which it appears in the Proterozoic. It is the Algoma type (Chapter 16) which is present.

The early to mid Proterozoic

The beginning of the Proterozoic about 2500 Ma ago was marked by a great change in tectonic conditions. The first stable lithospheric plates developed, although these seem to have been of small size. Their appearance permitted the formation of sedimentary basins, the deposition of platform sediments and the development of continental margin geosynclines.

GOLD-URANIUM CONGLOMERATES

The establishment of sedimentary basins allowed the formation of these deposits. The best known example is that of the Witwatersrand Basin with its widespread gold-uranium conglomerates (Chapter 16, Fig. 16.7), but other examples are known along the north shore of Lake Huron in Canada (Blind River area), at Serra de Jacobina in Brazil and at localities in Australia and Ghana. These deposits represent a unique metallogenic event which many feel has not been repeated because a reducing atmosphere was a *sine qua non* for the preservation of the detrital uranium minerals and pyrite.

SEDIMENTARY MANGANESE DEPOSITS

There appears to have been an appreciable concentration of manganese in carbonate sediments over the period 2300-2000 Ma ago. These concentrations are known in the Republic of South Africa, Brazil and India.

STRATIFORM LEAD-ZINC DEPOSITS IN CARBONATES

By about 1700 Ma ago, the CO_2 content in the hydrosphere had reached a level that permitted the deposition of thick dolomite sequences. In a number of localities these host syngenetic base metal sulphide orebodies such as those of McArthur River (Pb-Zn-Ag) and Mount Isa (Pb-Zn-Ag and separate Cu orebodies) in Australia and the Black Angel Mine in West Greenland. Various exhalative and biogenic origins have been suggested for these ores. They should not be confused with the carbonate-hosted lead-zinc deposits described in Chapter 15.

THE CHROMIUM-NICKEL-PLATINUM-COPPER ASSOCIATION

The presence of small crustal plates permitted the development of large-scale fracture systems and the intrusion at this time of giant dyke-like layered bodies, such as the Great Dyke of Zimbabwe-Rhodesia, and enormous layered stratiform igneous complexes like that of the Bushveld in South Africa. These are the repositories of enormous quantities of chromium and platinum with other important by-products (Chapter 8). Though similar intrusions occur in other parts of the world, the great concentration of chromium is in southern Africa and this has led some workers to postulate the presence of chromium-rich mantle beneath this region.

TITANIUM-IRON ASSOCIATION

About the middle of the Proterozoic many anorthosite plutons were emplaced in two linear belts which now lie in the northern and southern hemispheres when plotted on a pre-Permian continental drift reconstruction. A number of these carry ilmenite orebodies which are exploited in Norway and Canada. This was a unique magmatic event that has not been repeated. It suggests the gathering of reservoirs of magma in the top of the mantle which were able to penetrate upwards along deep fractures in the crust. The strength of the crust and the thermal conditions seem to have reached the point where magma could accumulate at the base of the crust in this extensive manner, rather like the accumulation of magma beneath present day rift valleys.

DIAMONDS

Diamantiferous kimberlites appear for the first time in the Proterozoic. This suggests that the geothermal gradients had decreased considerably permitting the development of thick lithospheric plates, because diamonds, requiring extreme pressure for their formation, cannot crystallize unless the lithosphere is at least 150 km thick.

BANDED IRON FORMATION (BIF)

The greatest development of BIF occurred during the interval 2600-1800 Ma ago (Goldich 1973). Although this rock-type is important in the Archaean it could not be developed on the large scale seen in the early Proterozoic because stable continental plates were not present. With the formation of these plates BIF could be laid down synchronously over very large areas in intra-plate basins and marginal miogeosynclines. The weathering of basic volcanics in the greenstone belts would have yielded ample iron and silica. If the atmosphere was essentially CO_2-rich the iron could have travelled largely in ionic solution. It is now suspected that iron-precipitating bacteria may have played an important part in depositing the iron and oxidizing it to the ferrous state as modern iron bacteria are able to oxidize ferrous iron at very low levels of oxygen concentration. Although BIF appears at later times in the Proterozoic, its development is very restricted compared with that in the early Proterozoic and this fall off in importance has been correlated by some workers with the evolution of an oxidizing atmosphere. In the Phanerozoic the place of BIF is taken by the Clinton and Minette ironstones.

Mid-late Proterozoic

HIGH GRADE LINEAR BELTS

It has been suggested (Piper 1974, 1975, 1976) that a supercontinent existed through much of Proterozoic time and Davies & Windley (1976) have plotted the trends of major high grade linear belts on this supercontinent showing that they lie on small circles having a common point of rotation. These linear belts affect middle to late Proterozoic as well as older rocks and they include shear belts, mobile belts and linear zones of transcurrent displacements of magnetic and gravity anomaly patterns. They contained some deep dislocations that penetrated right down to the mantle and which formed channelways for uprising magma. Nickel mineralization occurs in some of these belts, e.g. the Nelson River Gneissic Belt of Manitoba.

SEDIMENTARY COPPER

Watson (1973) has drawn attention to the anomalously high concentrations of copper in some late Proterozoic sediments in many parts of the world. These represent the oldest large sedimentary copper accumulations. Examples include the Katanga System of Zambia and Shaba (Chapter 13) and the Belt Series of the north-western USA. The formation of these copper concentrations was probably dependent on a supply of copper from the erosion of neighbouring basic volcanic rocks and the presence of an oxidizing atmosphere that allowed the chemical breakdown of copper sulphides during weathering and transportation.

SEDIMENTARY MANGANESE DEPOSITS

A second important period of manganese deposition occurred during the late Proterozoic and manganese-rich sediments were laid down on or along the margins of cratonic blocks. The most important deposits are in central India and Namibia. These deposits have been metamorphosed but the Indian examples, now spessartite-quartz rocks, appear to have originally been manganiferous argillaceous and arenaceous sediments.

TIN

Watson (1973) observed that tin mineralization does not appear in major quantities in the crust until the late Proterozoic where it is associated with high level alkaline and peralkaline anorogenic granite and pegmatites. This is particularly the case in Africa where these deposits lie in three north-south belts (Fig. 6.1). Another belt passes through the Rhodônia district of western Brazil.

The Phanerozoic

Towards the end of the Proterozoic a new tectonic pattern developed which gave rise to Phanerozoic fold belts formed by continental drift. Mineralization processes during this time interval tended to be concentrated, though by no means exclusively so, along such tectonic environments as rift valleys, aulacogens and associated domes, constructive and destructive plate margins and transform faults. This period of mineralization has been covered in the previous chapter.

> *'For as birds are born to fly freely through the air, so are fishes born to swim through the waters, while to other creatures Nature has given the earth that they might live in it, and particularly to man that he might cultivate it and draw out of its caverns metals and other mineral products'.*
>
> Georgius Agricola in
> *De Re Metallica,* 1556.

References

Anderson C. A. (1948) Structural Control of Copper Mineralization, Bagdad, Arizona. *Amer. Inst. Min. Metall. Engng Trans.*, **178**, 170-80.

Anderson G. M. (1975) Precipitation of Mississippi Valley-type Ores, *Econ. Geol.*, **70**, 937-42.

Anderson G. M. (1977) Thermodynamics and Sulfide Solubilities. In Greenwood H. J. (ed.), *Application of Thermodynamics to Petrology and Ore Deposits.* Mineralogical Association of Canada, Toronto.

Anhaeusser C. R. (1976) The Nature and Distribution of Archaean Gold Mineralization in Southern Africa. *Miner. Sci. Engng*, **8**, 46-84.

Annels A. E. (1979) Mufulira Greywackes and their Associated Sulphides. *Trans. Instn Min. Metall. (Sect. B: Appl. earth sci.)*, **88**, B15-B23.

Appel P. W. U. (1979) Stratabound Copper Sulphides in a Banded Iron-Formation and in Basaltic Tuffs in the Early Precambrian Isua Supracrustal Belt, West Greenland. *Econ. Geol.*, **74**, 45-52.

Badham J. P. N. (1978) Slumped Sulphide Deposits at Avoca, Ireland, and Their Significance. *Trans. Instn Min. Metall.* (*Sect. B: Appl. earth sci.*), **87**, B21-B26.

Baker P. E. (1968) Comparative Volcanology and Petrology of the Atlantic Island Arcs. *Bull. Volcanol.*, **32**, 189-206.

Barker D. S. (1969) North American Felspathoidal Rocks in Space and Time. *Bull. geol. Soc. Am.*, **80**, 2369-72.

Barnes H. L. (ed.) (1967a) *Geochemistry of Hydrothermal Ore Deposits.* Holt, Rinehart and Winston, New York.

Barnes H. L. (1967b) Sphalerite Solubility in Ore Solutions of the Illinois-Wisconsin District. *Econ. Geol. Monogr.*, **3**, 326-32.

Barnes H. L. (1975) Zoning of Ore Deposits: Types and Causes. *Trans. R. Soc. Edinburgh*, **69**, 295-311.

Barnes H. L. & Czamanske G. K. (1967) Solubilities and Transport of Ore Minerals. In Barnes H. L. (ed.), *Geochemistry of Hydrothermal Ore Deposits.* Holt, Rinehart and Winston, New York.

Barrett F. M., Binns R. A., Groves D. I., Marston R. J. & McQueen K. G. (1977) Structural History and Metamorphic Modification of Archaean Volcanic-type Nickel Deposits, Yilgarn Block, Western Australia. *Econ. Geol.*, **72**, 1195-223.

Barton P. B. & Skinner B. J. (1967) Sulphide Mineral Stabilities. In Barnes H. L. (ed.), *Geochemistry of Hydrothermal Ore Deposits*, 236-333. Holt, Rinehart and Winston, New York.

Barton P. B. & Toulmin P. (1963) Sphalerite Phase Equilibria in the System Fe-Zn-S between 580°C and 850°C. *Econ. Geol.*, **58**, 1191-2.

Bateman A. M. (1950) *Economic Mineral Deposits.* Wiley, New York.

Baumann L. (1965) Zur Erzführung und regionalen Verbreitung des 'Felsithorizontes' von Halbrücke. *Freib. Forsch.*, **C186**, 63-81.

Baumann L. (1970) Tin Deposits of the Erzgebirge. *Trans. Instn Min. Metall.* (*Sect. B: Appl. earth sci.*) **79**, B68-B75.

Baumann L. & Krs M. (1967) Paläomagnetische Altersbestimmungen an einigen Mineralparagenesen des Freiberger Lagerstättenbezirkes. *Geologie*, **16**, 765-80.

Bayley R. W. & James H. L. (1973) Precambrian Iron-Formations of the United States. *Econ. Geol.*, **68**, 934-59.

Bichan R. (1969) Origin of Chromite Seams in the Hartley Complex of the Great Dyke, Rhodesia. In Wilson H. D. B. (ed.), *Magmatic Ore Deposits, Econ. Geol.* Monograph **4**, 95-113.

Binda P. L. (1975) Detrital Bornite Grains in the Late Precambrian B Graywacke of Mufulira, Zambia. *Mineral. Deposita*, **10**, 101-7.

Blissenbach E. B. & Fellerer R. (1973) Continental Drift and the Origin of Certain Mineral Deposits. *Geol. Rundsch.*, **62**, 812-39.

Bonatti E., Guernstein-Honnorez B.-M. & Honnorez J. (1976) Copper-Iron Sulphide Mineralizations from the Equatorial Mid-Atlantic Ridge. *Econ. Geol.*, **71**, 1515-25.

Boyle R. W. (1959) The Geochemistry, Origin and Role of Carbon Dioxide, Sulfur, and Boron in the Yellowknife Gold Deposits Northwest Territories, Canada. *Econ. Geol.*, **54**, 1506-24.

Boyle R. W. (1970) Regularities in Wall-rock Alteration Phenomena Associated with Epigenetic Deposits. In Pouba Z. & Štemprok M. (eds), *Problems of Hydrothermal Ore Deposition,* 233-260. E. Schweizerbart'sche Verlagsbuchandlung, Stuttgart.
Brady L. L. & Jobson H. E. (1973) An Experimental Study of Heavy-mineral Segregation under Alluvial-flow Conditions. Prof. Pap. 562-K, *U.S. Geol. Surv.,* Washington.
Brocoum S. J. & Dalziel I. W. D. (1974) The Sudbury Basin, the Southern Province, the Grenville Front and the Penokean Orogeny. *Bull. geol. Soc. Am.,* **85**, 1571-80.
Brongersma-Sanders M. (1969) Permian Wind and the Occurrence of Fish and Metals in the Kupferschiefer and Marl Slate. *Proc. Inter-University Geol. Cong. 15th.,* Leicester, England, 61-72.
Brown A. C. (1978) Stratiform Copper Deposits—Evidence for their Post-sedimentary Origin. *Miner. Sci. Engng,* **10**, 172-81.
Brunt D. A. (1978) Uranium in Tertiary Stream Channels, Lake Frome Area, South Australia. *Proc. Australas. Inst. Min. Metall.,* No. **266**, 79-90.
Burke K. & Dewey J. F. (1973) Plume-generated Triple Junctions: Key Indicators in Applying Plate Tectonics to Old Rocks. *J. Geol.,* **81**, 406-33.
Burn R. G. (1971) Localized Deformation and Recrystallization of Sulphides in an Epigenetic Mineral Deposit. *Trans. Instn Min. Metall. (Sect. B: Appl. earth sci.),* **80**, B116-B119.
Burnham C. W. (1959) Metallogenic Provinces of the Southwestern United States and Northern Mexico. *Bull. State Bur. Mines Min. Resources, New Mexico,* **65**, 76 pp.
Burnie S. W., Schwarcz H. P. & Crocket J. H. (1972) A Sulfur Isotopic Study of the White Pine Mine, Michigan, *Econ. Geol.,* **67**, 895-914.
Callahan W. H. (1967) Some Spatial and Temporal Aspects of the Localization of Mississippi Valley-Appalachian Type Ore Deposits. In Brown J. S. (ed.), *Genesis of Stratiform Lead-zinc-barite-fluorite Deposits,* 14-19. Economic Geology Publishing Co., Lancaster, Pennsylvania.
Cann J. R. (1970) New Model for the Structure of the Ocean Crust. *Nature,* **266**, 928-30.
Carmichael I. S. E., Turner F. J. & Verhoogen J. (1974) *Igneous Petrology.* McGraw-Hill, New York.
Chětelat E. de (1947) La Geněse et l'ěvolution des gisements de nickel de la Nouvelle-Calédonie. *Bull. Soc. geol. Fr.,* ser. 5, **17**, 105-60.
Chivas R. & Wilkins W. T. (1977) Fluid Inclusion Studies in Relation to Hydrothermal Alteration and Mineralization at the Koloula Porphyry Copper Prospect, Guadalcanal. *Econ. Geol.,* **72**, 153-69.
Clark K. F. (1972) Stockwork Molybdenum Deposits in the Western Cordillera of North America. *Econ. Geol.,* **67**, 731-58.
COMRATE (Committee on Mineral Resources and the Environment) (1975) *Mineral Resources and the Environment.* National Academy of Sciences, Washington.
Coomer P. G. & Robinson B. W. (1976) Sulphur and Sulphate-oxygen Isotopes and the Origin of the Silvermines Deposits, Ireland. *Mineral. Deposita,* **11**, 155-69.
Craig J. R. & Scott S. D. (1974) Sulphide Phase Equilibria. In Ribbe P. H. (ed.), *Sulphide Mineralogy.* Min. Soc. Am., Short Course Notes, Vol. 1, Ch. 5.
Cuney M. (1978) Geologic Environment, Mineralogy and Fluid Inclusions of the Bois Noirs-Limouzat Uranium Vein, Forez, France. *Econ. Geol.,* **73**, 1567-610.
Danielson M. J. (1975) King Island Scheelite Deposits. In Knight C. L. (ed.), *Economic Geology of Australia and Papua New Guinea.* 592-98. Australas. Inst. Min. Metall., Parkville.
Davies F. B. & Windley B. F. (1976) Significance of Major Proterozoic High Grade Linear Belts in Continental Evolution. *Nature,* **263**, 383-5.
Deans T. (1950) The Kupferschiefer and the Associated Lead-zinc Mineralization in the Permian of Silesia, Germany and England. *Int. geol Congr.,* 18th., London, Rept., pt. 7, 340-52.
Degens E. T. & Ross D. A. (eds) (1969) *Hot Brines and Recent Heavy Metal Deposits in the Red Sea.* Springer, New York.
Degens E. T. & Ross D. A. (1976) Strata-bound Metalliferous Deposits Found in or near Active Rifts. In Wolf K. H. (ed.), *Handbook of Strata-Bound and Stratiform Ore Deposits,* Vol. 4, 165-202. Elsevier, Amsterdam.
De Vore G. W. (1955) The Role of Adsorption in the Fractionation and Distribution of Elements. *J. Geol.,* **63**, 159-90.
Dickinson W. R. (1971) Plate Tectonic Models of Geosynclines. *Earth Planet. Sci. Lett.,* **10**, 165-74.
Dietz R. S. (1964) Sudbury Structure as an Astrobleme. *J. Geol.,* **72**, 412-34.
Dixon C. J. (1979) *Atlas of Economic Mineral Deposits.* Chapman and Hall, London.
Dmitriev L., Barsukov V. & Udintsev G. (1971) Rift Zones of the Ocean and the Problem of Ore-Formation. *Proc. IMA-IAGOD Meetings '70,* Spec. Issue 3 (IAGOD Vol.), Soc. Mining Geol. Japan, 65-9.
Dunham K. C. (1959) Non-ferrous Mining Potentialities of the Northern Pennines. In *Future of Non-ferrous Mining in Great Britain and Ireland,* 115-147. Instn. Min. Metall., London.
Dunham K. C. (1964) Neptunist Concepts in Ore Genesis. *Econ. Geol.,* **59**, 1-21.

Dunsmore H. E. (1973) Diagenetic Processes of Lead-zinc Emplacement in Carbonates. *Trans. Instn Min. Metall. (Sect. B: Appl. earth sci.)*, **82**, B168-B173.
Dunsmore H. E. & Shearman D. J. (1977) Mississippi Valley-type Lead-zinc Orebodies: a Sedimentary and Diagenetic Origin. In, *Proceedings of the Forum on Oil and Ore in Sediments,* 189-205. Geology Dept., Imperial College, London.
Du Toit A. L. (1954) *The Geology of South Africa.* Oliver and Boyd, Edinburgh.
Dybdahl I. (1960) Ilmenite Deposits of the Egersund Anorthosite Complex. In Vokes F. (ed.), *Mines in South and Central Norway.* Guide to Excursion No. C10. *Int. geol. Congr.* 21st, Norden.
Edwards A. B. (1952) The Ore Minerals and Their Textures. *J. Proc. R. Soc. New South Wales,* **85**, 26-46.
Edwards A. B. (1960) *Textures of the Ore Minerals and their Significance,* Australas. Inst. Min. Metall., Melbourne.
Edwards A. B. & Lyon R. J. P. (1957) Mineralization at Aberfoyle Tin Mine, Rossarden, Tasmania. *Proc. Australas. Inst. Min. Metall.,* No. **181**, 93-145.
Ekiert F. (1958) Warunki geologiczne i objawy mineralizacji Cechsztynu w niecce Manfeldskiej. (In Polish with English and Russian summaries). *Warsaw Inst. Geologiczny Biul.,* **126**, 57-110.
El Shazly E. M., Webb J. S. & Williams D. (1957) Trace Elements in Sphalerite, Galena and Associated Minerals from the British Isles. *Trans. Instn Min. Metall.,* **66**, 241-71.
Evans A. M. (1975) Mineralization in Geosynclines—the Alpine Enigma. *Mineral. Deposita,* **10**, 254-60.
Evans A. M. (1976a) Genesis of Irish Base-metal Deposits. In Wolf K. H. (ed.), *Handbook of Strata-Bound and Stratiform Deposits,* Vol. 5, 231-55. Elsevier, Amsterdam.
Evans A. M. (1976b) Mineralization in Geosynclines. In Wolf K. H. (ed.), *Handbook of Strata-Bound and Stratiform Deposits,* Vol. 4, 1-29. Elsevier, Amsterdam.
Evans A. M. & Evans N. D. M. (1977) Some Preliminary Palaeomagnetic Studies of Mineralization in the Mendip Orefield. *Trans. Instn. Min. Metall. (Sect. B. Appl. earth sci.),* **86**, B149-B151.
Evans A. M. & Maroof S. I. (1976) Basement Controls on Mineralization in the British Isles. *Ming Mag. London,* **134**, 401-11.
Fleischer V. D., Garlick W. G. & Haldane R. (1976) Geology of the Zambian Copperbelt. In Wolf K. H. (ed.), *Handbook of Strata-Bound and Stratiform Ore Deposits,* Vol. 6, 223-352. Elsevier, Amsterdam.
Fletcher K. & Couper J. (1975) Greenvale Nickel Laterite, North Queensland. In Knight C. L. (ed.), *Economic Geology of Australia and Papua New Guinea, I. Metals,* Australas. Inst. Min. Metall., Parkville, 995-1001.
Forsythe D. L. (1971) Vertical Zoning of Gold-silver Tellurides in the Emperor Gold Mine, Fiji. *Proc. Australas. Inst. Min. Metall.,* No. **240**, 25-31.
Fournier R. O. (1967) The Porphyry Copper Deposit Exposed in the Liberty Open-pit Mine near Ely, Nevada. Part II. The Formation of Hydrothermal Alteration Zones. *Econ. Geol.,* **62**, 207-27.
Francheteau J. and 14 co-authors (1979) Massive Deep-sea Sulphide Ore Deposits Discovered on the East Pacific Rise. *Nature,* **277**, 523-8.
Frietsch R. (1978) On the Magmatic Origin of Iron Ores of the Kiruna Type. *Econ. Geol.,* **73**, 478-85.
Fyfe W. S. & Henley R. W. (1973) Some Thoughts on Chemical Transport Processes, with particular reference to Gold. *Miner. Sci. Engng,* **5**, 295-303.
Gandhi S. S. (1978) Geological Setting and Genetic Aspects of Uranium occurrences in the Kaipokok Bay-Big River Area, Labrador. *Econ. Geol.,* **73**, 1492-522.
Gee R. D. (1975) Regional Geology of the Archaean Nucleii of the Western Australian Shield. In Knight C. L. (ed.), *Economic Geology of Australia and Papua New Guinea - 1. Metals,* Mon, 5, 43-55. Australas. Inst. Min. Metall., Parkville.
Gilluly J. (1932) Geology and Ore Deposits of the Stockton and Fairfield Quadrangles, Utah. Prof. Pap. 173, *U.S. Geol. Surv.,* Washington.
Gilluly J., Waters A. C. & Woodford A. O. (1959) *Principles of Geology.* Freeman, San Francisco.
Glazkovsky A. A., Gorbunov G. I. & Sysoev F. A. (1977) Deposits of Nickel. In Smirnov V. I. (ed.), *Ore Deposits of the USSR Vol. II,* 3-79. Pitman, London.
Goldich S. S. (1973) Ages of Precambrian Banded Iron Formations. *Econ. Geol.,* **68**, 1126-34.
Goldsmith J. R. & Newton R. C. (1969) P-T-X Relations in the System $CaCO_3$-$MgCO_3$ at High Temperature and Pressures. *Am. J. Sci.,* **267A**, 160-90.
Goodwin A. M. (1973) Archaean Iron-formations and Tectonic Basins of the Canadian Shield. *Econ. Geol.,* **68**, 915-33.
Graf Jr. J. L. (1977) Rare Earth Elements as Hydrothermal Tracers during the Formation of Massive Sulphide Deosits in Volcanic Rocks. *Econ. Geol.,* **72**, 527-48.
Grant J. N., Halls C., Avila W. & Avila G. (1977) Igneous Geology and the Evolution of Hydrothermal Systems in some sub-volcanic Tin Deposits of Bolivia. In, *Volcanic Processes in Ore Genesis.* Spec. Pub. No. 7, Geol. Soc. London, 117-126.

Greig J. A., Baadsgaard H., Cumming G. L., Folinsbee R. E., Krouse H. R., Ohmoto H., Sasaki A. & Smejkal V. (1971) Lead and Sulphur Isotopes of the Irish Base Metal Mines in Carbonate Host Rocks. *Society Mining Geologists Japan Spec. Issue 2 Proc.* (IMA-IAGOD Mtgs, 1970, Joint Symp. Vol.), 84-92.

Gross G. A. (1965) Geology of Iron Deposits in Canada. I. General Geology and Evaluation of Iron Deposits. *Geol. Surv., Can. Econ. Geol. Rep.,* **22**.

Gross G. A. (1970) Nature and Occurrence of Iron Ore Deposits. In, *Survey of World Iron Ore Resources,* United Nations, New York, 13-31.

Groves D. I. & Solomon M. (1969) Fluid Inclusion Studies at Mount Bischoff, Tasmania. *Trans. Instn Min. Metall. (Sect. B: Appl. earth sci.),* **78**, B1-B11.

Groves D. I., Solomon M. & Rafter T. A. (1970) Sulphur Isotope Fractionation and Fluid Inclusion Studies at the Rex Hill Mine, Tasmania. *Econ. Geol.,* **65**, 459-69.

Grubb P. L. C. (1973) High-level and Low-level Bauxitization: a Criterion for Classification. *Miner. Sci. Engng,* **5**, 219-31.

Haas J. L. Jr (1971) The Effect of Salinity on the Maximum Thermal Gradient of a Hydrothermal System at Hydrostatic Pressure. *Econ. Geol.,* **66**, 940-6.

Hagner A. F. & Collins L. G. (1967) Magnetite Ore formed during Regional Metamorphism. *Econ. Geol.,* **62**, 1034-71.

Hall A. L. (1932) The Bushveld Igneous Complex of the Central Transvaal. *Geol. Surv. S. Africa,* Mem. **28**, 560 pp.

Hall W. E., Friedman I. & Nash J. T. (1974) Fluid Inclusion and Light Stable Isotope Study of the Climax Molybdenum Deposits, Colorado. *Econ. Geol.,* **69**, 884-901.

Harder E. C. & Greig E. W. (1960) Bauxite. In Gillson J. L. *et al.* (eds), *Industrial Minerals and Rocks,* Amer. Inst. Ming Eng., New York, 65-85.

Hawkins B. W. (1975) Mary Kathleen Uranium Deposit. In Knight C. L. (ed.), *Economic Geology of Australia and Papua New Guinea,* 398-402. Australas. Inst. Min. Metall., Parkville.

Hawley J. E. (1962) The Sudbury Ores: Their Mineralogy and Origin. *Can. Mineral,* **7**, i-xiv & 1-207.

Heaton T. H. E. & Sheppard S. M. F. (1977) Hydrogen and Oxygen Isotope Evidence for Sea Water-Hydrothermal Alteration and Ore Deposition, Troodos Complex, Cyprus. In, *Volcanic Processes in Ore Genesis,* Spec. Publ. No. 7, Geol. Soc. London.

Helgeson H. C. (1964) *Complexing and Hydrothermal Ore Deposition.* Pergamon Press, New York.

Heyl A. V. (1969) Some Aspects of Genesis of Zinc-lead-barite-fluorite deposits in the Mississippi Valley, U.S.A. *Trans. Instn Min. Metall. (Sect. B. Appl. earth sci.),* **78**, B148-B160.

Heyl A. V., Delevaux M. H., Zartman R. E. & Brock M. R. (1966) Isotopic Study of Galenas from the Upper Mississippi Valley, the Illinois-Kentucky and some Appalachian Valley Mineral Districts. *Econ. Geol.,* **61**, 933-61.

Heyl A. V., Landis G. P. & Zartman R. E. (1974) Isotopic Evidence for the Origin of Mississippi Valley-type Mineral Deposits: A Review. *Econ. Geol.,* **69**, 992-1006.

Hills E. S. (1953) Tectonic Setting of Australian Ore Deposits. In Edwards A. B. (ed.), *Geology of Australian Ore Deposits,* Australas. Inst. Min. Metall., Melbourne, 41-61.

Höll R. & Maucher A. (1976) The Strata-Bound Ore Deposits in the Eastern Alps. In Wolf K. H. (ed.), *Handbook of Strata-Bound and Stratiform Ore Deposits,* Vol. 5. 1-36. Elsevier, Amsterdam.

Hollister V. F. (1975) An Appraisal of the Nature of some Porphyry Copper Deposits. *Miner. Sci. Engng,* **7**, 225-33.

Hollister V. F., Potter R. R. & Barker A. L. (1974) Porphyry Type Deposits of the Appalachian Orogen. *Econ. Geol.,* **69**, 618-30.

Holmes A. & Holmes D. L. (1978) *Principles of Physical Geology.* Nelson, London.

Horikoshi E. & Sato T. (1970) Volcanic Activity and Ore Deposition in the Kosaka Mine. In Tatsumi T. (ed.), *Volcanism and Ore Genesis.* 181-95. University of Tokyo Press, Tokyo.

Hose H. R. (1960) The Genesis of Bauxites, the Ores of Aluminium. *Int. geol. Congr.* 21st. Pt. **16**, 237-47.

Hosking K. F. G. (1951) Primary Ore Deposition in Cornwall. *Trans. R. geol. Soc. Cornwall,* **18**, 309-56.

Howd F. H. & Barnes H. L. (1975) Ore Solution Chemistry IV. Replacement of Marble by Sulphides at 450°C. *Econ. Geol.,* **70**, 968-81.

Hsü K. J. (1972) The Concept of the Geosyncline, Yesterday and To-day. *Trans. Leicester Lit. Philos. Soc.,* **66**, 26-48.

Hughes F. E. & Munro D. L. (1965) Uranium Ore Deposits at Mary Kathleen. In McAndrew J. (ed.), *Geology of Australian Ore Deposits,* 256-63. 8th. Commonwealth Mining Metallurgical Congress, Melbourne.

Hutchinson R. W. (1973) Volcanogenic Sulfide Deposits and Their Metallogenic Significance. *Econ. Geol.,* **68**, 1223-46.

Ixer R. A. & Townley R. (1979) The Sulphide Mineralogy and Paragenesis of the South Pennine Orefield, England. *Mercian Geol.,* **7**, 51-64.

Jackson E. D. & Thayer T. P. (1972) Some Criteria for Distinguishing between Stratiform, Concentric and Alpine Peridotite-Gabbro Complexes. *Int. geol. Congr.,* 24th. session, section 2, 289-96.

Jackson S. A. & Beales F. W. (1967) An Aspect of Sedimentary Basin Evolution; the Concentration of Mississippian Valley-type Ores during Late Stages of Diagenesis. *Bull. Can. Pet. Geol.,* **15**, 383-433.

Jacobsen J. B. E. (1975) Copper Deposits in Time and Space. *Miner. Sci. Engng,* **7**, 337-71.

Jacobsen J. B. E. & McCarthy T. S. (1976) The Copper-bearing Breccia Pipes of the Messina District South Africa. *Mineral. Deposita,* **11**, 33-45.

Jahns R. H. (1955) The Study of Pegmatites. *Econ. Geol.* (**50th. Anniv. Vol.**), 1025-130.

James H. L. (1954) Sedimentary Facies of Iron-formation. *Econ. Geol.,* **49**, 235-93.

James H. L. (1955) Zones of Regional Metamorphism in the Precambrian of Northern Michigan. *Bull. geol. Soc. Amer.,* **66**, 1455-88.

James H. L. & Sims P. K. (1973) Precambrian Iron-formations of the World. *Econ. Geol.,* **68**, 913-4.

Jung W. & Knitzschke G. (1976) Kupferschiefer in the German Democratic Republic (GDR) with Special Reference to the Kupferschiefer Deposit in the Southeastern Harz Foreland. In, Wolf K. H. (ed.), *Handbook of Strata-Bound and Stratiform Ore Deposits,* Vol. 6, 353-406. Elsevier, Amsterdam.

Kesler S. E. (1968) Contact-localized Ore Formation at the Memé Mine, Haiti. *Econ. Geol.,* **63**, 541-52.

Kesler S. E. (1973) Copper, Molybdenum and Gold Abundances in Porphyry Copper Deposits. *Econ. Geol.,* **68**, 106-112.

Krs M. & Šťovíčková N. (1966) Palaeomagnetic Investigation of Hydrothermal Deposits in the Jáchymov (Joachimsthal) Region Western Bohemia. *Trans. Instn Min. Metall. (Sect. B, Appl. earth sci.),* **75**, B51-B57.

Kullerud G. (1953) The FeS-ZnS System: a Geological Thermometer. *Nor. geol. Tiddskr.,* **32**, 61-147.

Laberge G. L. (1973) Possible Biological Origin of Precambrian Iron-formations. *Econ. Geol.,* **68**, 1098-109.

Lang A. H. (1970) Prospecting in Canada. *Geol. Surv. Canada, Econ. Geol. Rep.,* **7**.

Lapham D. M. (1968) Triassic Magnetite and Diabase at Cornwall, Pennsylvania. In Ridge J. D. (ed.), *Ore Deposits of the United States 1933-1967,* Vol. 1, 72-94. Am. Inst. Min. Metall. Pet. Engrs, New York.

Lawrence L. J. (1972) The Thermal Metamorphism of a Pyritic Sulphide Ore. *Econ. Geol.,* **67**, 487-96.

Laznicka P. (1976) Porphyry Copper and Molybdenum Deposits of the USSR and their Plate Tectonic Settings. *Trans. Instn Min. Metall. (Sect. B: Appl. earth sci.),* **85**, B14-B32.

Le Bas M. J. (1977) *Carbonatite-nephelinite Volcanism.* Wiley, London.

Levin E. M., Robbins C. R. & McMurdie H. F. (1969) *Phase Diagrams for Ceramicists.* Am. Ceram. Soc., Columbus, Ohio.

Lindgren W. (1913) (second edition, 1933). *Mineral deposits.* McGraw-Hill, New York.

Lindgren W. (1924) Contact Metamorphism at Bingham, Utah. *Geol. Soc. Amer. Bull.,* **35**, 507-34.

Lowell J. D. (1974) Regional Characteristics of Porphyry Copper Deposits of the Southwest. *Econ. Geol.,* **69**, 601-17.

Lowell J. D. & Guilbert J. M. (1970) Lateral and Vertical Alteration Mineralization Zoning in Porphyry Ore Deposits. *Econ. Geol.,* **65**, 373-408.

Lusk J., Campbell F. A. & Krouse H. R. (1975) Application of Sphalerite Geobarometry and Sulphur Isotope Geothermometry to Ores of the Quemont Mine, Noranda, Quebec. *Econ. Geol.,* **70**, 1070-83.

Mainwaring P. R. & Naldrett A. J. (1977) Country Rock Assimilation and the Genesis of Cu-Ni Sulphides in the Water Hen Intrusion, Duluth Complex, Minnesota. *Econ. Geol.,* **72**, 1269-84.

Malyutin R. S. & Sitkovskiy I. N. (1968) Structural Features of the Gyumushlug Lead-zinc Deposit. *Geologiya Rudnykh Mestorozhdeniy,* **10**, 96-99. (In Russian.)

Martin J. E. & Allchurch P. D. (1975) Perseverance Nickel Deposit, Agnew. In Knight C. L. (ed.), *Economic Geology of Australia and Papua New Guinea - 1. Metals.* Mon. 5, 149-155. Australas. Inst. Min. Metall., Parkville.

Mason A. A. C. (1953) The Vulcan Tin Mine. In Edwards A. B. (ed.), *Geology of Australian Ore Deposits,* 718-721. Australas. Inst. Min. Metall., Melbourne.

Mertie J. B. (1969) *Economic Geology of the Platinum Metals.* Prof. Pap. 630, *U.S. Geol. Surv.,* Washington.

Meyer C. & Hemley J. J. (1967) Wall Rock Alteration. In Barnes H. L. (ed.); *Geochemistry of Hydrothermal Ore Deposits,* 166-235. Holt, Rinehart and Winston, New York.

Meyer C., Shea E. P., Goddard Jr. C. C. & Staff (1968) Ore Deposits at Butte, Montana. In Ridge J. R. (ed.), *Ore Deposits of the United States, 1933-1967,* Vol. II, 1373-416. Am. Inst. Min. Metall. Pet. Engns, New York.

Milovskiy G. A., Zlenko B. F. & Gubanov A. M. (1978) Conditions of Formation of Scheelite Ores in the Chorukh-Dayron Mineralized Area (As Revealed by a Study of Gas-liquid Inclusions). *Geochem. Int.,* **15**, 45-52.

Mitcham T. W. (1974) Origin of Breccia Pipes. *Econ. Geol.,* **69**, 412-13.

Mitchell A. H. G. (1973) Metallogenic Belts and Angle of Dip of Benioff Zones. *Nat. Phys. Sci.,* **245**, 49-52.

Mitchell A. H. G. & Garson M. S. (1972) Relationship of Porphyry Copper and Circum-Pacific Tin Deposits to Palaeo-Benioff Zones. *Trans. Instn Min. Metall.*, (*Sect. B: Appl. earth sci.*), **81**, B10-B25.

Mitchell A. H. G. & Garson M. S. (1976) Mineralization at Plate Boundaries. *Miner. Sci. Engng,* **8**, 129-69.

Mitchell A. H. G. & Reading H. G. (1969) Continental Margins, Geosynclines and Ocean Floor Spreading. *J. Geol.,* **77**, 629-46.

Moorbath S., O'Nions R. K. & Pankhurst R. J. (1973) Early Archaean Age for the Isua Iron Formation, West Greenland, *Nature,* **245**, 138-9.

Moore A. C. (1973) Carbonatites and Kimberlites in Australia: a Review of the Evidence. *Miner. Sci. Engng,* **5**, 81-91.

Naldrett A. J. (1973) Nickel Sulphide Deposits—Their Classification and Genesis, with Special Emphasis on Deposits of Volcanic Association. *Can. Inst. Min. Met. Trans.,* **76**, 183-201.

Naldrett A. J., Bray J. G., Gasparrini E. L., Podolsky T. & Rucklidge J. C. (1970) Cryptic Variation and the Petrology of the Sudbury Nickel Irruptive. *Econ. Geol.,* **65**, 122-55.

Naldrett A. J. & Cabri L. J. (1976) Ultramafic and Related Mafic Rocks: Their Classification and Genesis with Special Reference to the Concentration of Nickel Sulphides and Platinum Group Elements. *Econ. Geol.,* **71**, 1131-58.

Nash J. T. (1976) Fluid Inclusion Petrology—Data from Porphyry Copper Deposits and Applications to Exploration. *Prof. Pap. 907-D, U.S. Geol. Surv.,* Washington.

Nriagu J. O. (1971) Studies in the System PbS-NaCl-H_2S-H_2O: Stability of Lead (II) Thio Complexes at 90°C. *Chem. Geol.,* **8**, 299-310.

Ohmoto H. & Rye R. O. (1974) Hydrogen and Oxygen Isotopic Compositions of Fluid Inclusions in the Kuroko Deposits, Japan. *Econ. Geol.,* **69**, 947-53.

Olsen J. C., Shawe D. R., Pray L. C. & Sharp W. N. (1954) Rare-earth Mineral Deposits of the Mountain Pass District, San Bernardino County, Californis. *U.S. Geol Surv.,* Prof. Pap. 261.

Owen H. B. & Whitehead S. (1965) Iron Ore Deposits of Iron Knob and the Middleback Ranges. In McAndrew J. (ed.), *Geology of Australian Ore Deposits,* 301-8. Aust. Inst. Min. Metall., Melbourne.

Page R. W. & McDougall I. (1972) Ages of Mineralization in Gold and Porphyry Copper Deposits in the New Guinea Highlands. *Econ. Geol.,* **67**, 1034-48.

Palabora Mining Company Limited Mine Geological and Mineralogical Staff (1976) The Geology and the Economic Deposits of Copper, Iron and Vermiculite in the Palabora Igneous Complex: a Brief Review. *Econ. Geol.,* **71**, 177-92.

Park C. F. Jr. & MacDiarmid R. A. (1975) *Ore Deposits.* Freeman, San Francisco.

Peterson U. (1970) Metallogenic Provinces in South America. *Geol. Rundsch.,* **59**, 834-97.

Phillips W. J. (1973) Mechanical Effects of Retrograde Boiling and its Probable Importance in the Formation of Some Porphyry Ore Deposits. *Trans. Instn Min. Metall.* (*Sect. B: Appl. earth sci.*), **82**, B90-B98.

Piper J. D. A. (1974) Proterozoic Crustal Distribution, Mobile Belts and Apparent Polar Movement, *Nature,* **251**, 381-4.

Piper J. D. A. (1975) Proterozoic Supercontinent: Time Duration and the Grenville Problem. *Nature,* **256,** 519-20.

Piper J. D. A. (1976) Palaeomagnetic Evidence for a Proterozoic Supercontinent. *Phil. Trans. R. Soc. Lond.,* **A280**, 469-90.

Popelar J. (1972) Gravity Interpretation of the Sudbury Area. *Geol. Assoc. Canada Spec. Pap.* **10**, 103-15.

Preto V. A. (1978) Setting and Genesis of Uranium Mineralization at Rexspar. *Canad. Inst. Min. Bull.,* **71**, 82-8.

Pretorius D. A. (1975) The Depositional Environment of the Witwatersrand Goldfields: a Chronological Review of Speculations and Observations. *Miner. Sci. Engng,* **7**, 18-47.

Rackley R. I. (1976) Origin of Western States-Type Uranium Mineralization. In Wolf K. H. (ed.), *Handbook of Strata-Bound and Stratiform Deposits,* Vol. 7, 89-156. Elsevier, Amsterdam.

Radkevich E. A. (1972) The Metallogenic Zoning in the Pacific Ore Belt. *Int. geol. Congr.* 24th. Session, Sect. 4, 52-59.

Ramdohr P. (1969) *The Ore Minerals and Their Intergrowths.* Pergamon Press, Oxford.

Ransome F. L. (1919) The Copper Deposits of Ray and Miami, Arizona. *U.S. Geol. Surv.* Prof. Pap. **115**.

Raybould J. G. (1978) Tectonic Controls on Proterozoic Stratiform Copper Mineralization. *Trans. Instn Min. Metall.* (*Sect. B: Appl. earth sci.*), **87**, B79-B86.

Reynolds R. L. & Goldhaber M. B. (1978) Origin of a South Texas Roll-type Uranium Deposit: I. Alteration of Iron-titanium Oxide Minerals. *Econ. Geol.,* **73**, 1677-89.

Richards J. R. & Pidgeon R. T. (1963) Some Age Measurements on Micas from Broken Hill, Australia. *J. geol. Soc. Aust.,* **10**, 664-78.

Richter G. (1941) Geologische Gesetzmässigketten in der Metallführung des Kupferschiefers. *Arkiv für Lagerstättenforschung,* No. **73**.

Ringwood A. E. (1974) The Petrological Evolution of Island Arc Systems. *J. Geol. Soc. London,* **130**, 183-204.
Rittenhouse G. (1943) Transportation and Deposition of Heavy Minerals. *Geol. Soc. Am. Bull.,* **54**, 1725-80.
Robinson B. W. & Ohmoto H. (1973) Mineralogy, Fluid Inclusions and Stable Isotopes of the Echo Bay U-Ni-Ag-Cu Deposits, Northwest Territories, Canada. *Econ. Geol.,* **68**, 635-56.
Roedder E. (1972) Composition of Fluid Inclusions. Prof. Pap. 440-JJ. *U.S. Geol. Surv.*
Ronov A. B. (1964) Common Tendencies in the Chemical Evolution of the Earth's Crust, Ocean and Atmosphere. *Geochem. Int.,* **1**, 713-37.
Roscoe S. M. (1968) Huronian Rocks and Uraniferous Conglomerates of the Canadian Shield. *Geol. Surv. Can. Pap. 68-40.*
Ross J. R. & Hopkins G. M. F. (1975) Kambalda Nickel Sulphide Deposits. In Knight C. L. (ed.), *Economic Geology of Australia and Papua New Guinea - 1. Metals.* Mon, 5, 100-121. Australas. Inst. Min. Metall., Parkville.
Roy S. (1976) Ancient Manganese Deposits. In Wolf K. H. (ed.), *Handbook of Strata-Bound and Stratiform Deposits,* Vol. 7, 395-474. Elsevier, Amsterdam.
Rubey W. W. (1933) The Size Distribution of Heavy Minerals within a Waterlain Sandstone. *J. Sediment. Petrol.,* **3**, 3-29.
Ruckmick J. C. (1963) The Iron Ores of Cerro Bolivar, Venezuela. *Econ. Geol.,* **58**, 218-36.
Rye R. O. & Ohmoto H. (1974) Sulfur and Carbon Isotopes and Ore Genesis: A Review. *Econ. Geol.,* **69**, 826-42.
Sangster D. F. (1976) Carbonate-hosted Lead-zinc Deposits. In Wolf K. H. (ed.), *Handbook of Strata-Bound and Stratiform Deposits,* Vol. 6, 447-56. Elsevier, Amsterdam.
Sangster D. F. & Scott S. D. (1976) Precambrian, Strata-bound, Massive Cu-Zn-Pb Sulphide Ores in North America. In Wolf K. H. (ed.), *Handbook of Strata-Bound and Stratiform Ore Deposits,* Vol. 6, 129-222. Elsevier, Amsterdam.
Sato T. (1977) Kuroko Deposits: Their Geology, Geochemistry and Origin. In, *Volcanic Processes in Ore Genesis.* Spec. Pub. No. 7, Geol. Soc. London.
Saupe F. (1973) La Géologie du Gîsement de Mercure d'Almaden. *Sciences de la Terre, Mem.,* **29**.
Sawkins F. J. (1972) Sulphide Ore Deposits in Relation to Plate Tectonics. *J. Geol.,* **80**, 377-97.
Schneiderhöhn H. (1955) *Erzlagerstätten.* Gustav Fischer-Verlag, Stuttgart.
Schuiling R. D. (1967) Tin Belts on Continents around the Atlantic Ocean. *Econ. Geol.,* **62**, 540-50.
Scott S. D. (1973) Experimental Calibration of the Sphalerite Geobarometer. *Econ. Geol.,* **68**, 466-74.
Scott S. D. (1974) Experimental Methods in Sulphide Synthesis. In Ribbe P. H. (ed.), *Sulphide Mineralogy,* Min. Soc. Am. Short Course Notes, Vol. 1, Ch. 4.
Scott S. D. & Barnes H. L. (1971) Sphalerite Geothermometry and Geobarometry. *Econ. Geol.,* **66**, 653-69.
Selley R. C. (1976) *An Introduction to Sedimentology.* Academic Press, London.
Shaw D. M. (1954) Trace Elements in Pelitic Rocks. *Geol. Soc. Amer. Bull.,* **65**, 1151-82.
Shaw S. E. (1968) Rb-Sr Isotopic Studies of the Mine Sequence Rocks at Broken Hill. In Radmanovich M. & Woodcock J. T. (eds), *Broken Hill Mines - 1968, Australas. Inst. Min. Metall. Mongr.* Ser. **3**, 185-98.
Sheppard S. M. F. (1977) Identification of the Origin of Ore-forming Solutions by the Use of Stable Isotopes. In, *Volcanic Processes in Ore Genesis.* Spec. Publ. No. 7, Geol. Soc. London.
Shimazaki H. & MacLean W. H. (1976) An Experimental Study on the Partition of Zinc and Lead between the Silicate and Sulphide Liquids. *Mineral. Deposita,* **11**, 125-32.
Sillitoe R. H. (1972a) Formation of Certain Massive Sulphide Deposits at Sites of Sea-floor Spreading. *Trans. Instn Min. Metall. (Sect. B: Appl. earth sci.),* **81**, B141-B148.
Sillitoe R. H. (1972b) A Plate Tectonic Model for the Origin of Porphyry Copper Deposits. *Econ. Geol.,* **67**, 184-97.
Sillitoe R. H. (1973) The Tops and Bottoms of Porphyry Copper Deposits. *Econ. Geol.,* **68**, 799-815.
Sillitoe R. H., Halls C. & Grant J. N. (1975) Porphyry Tin Deposits in Bolivia. *Econ. Geol.,* **70**, 913-27.
Smith C. S. (1964) Some Elementary Principles of Polycrystalline Microstructure. *Metall. Rev.,* **9**, 1-48.
Solomon M. (1976) "Volcanic" Massive Sulphide Deposits and their Host Rocks - a Review and an Explanation. In Wolf K. H. (ed.), *Handbook of Strata-Bound and Stratiform Ore Deposits,* Vol. 6, 21-54. Elsevier, Amsterdam.
Souch B. E., Podolsky T. & Geological Staff (1969) The Sulphide Ores of Sudbury: Their Particular Relationship to a Distinctive Inclusion-bearing Facies of the Nickel Irruptive. In Wilson H. D. B. (ed.), *Magmatic Ore Deposits - A Symposium,* Econ. Geol. Mon. **4**, 252-61.
Spooner E. T. C. (1977) Hydrodynamic Model for the Origin of Ophiolitic Cupriferous Pyrite Ore Deposits of Cyprus. In, *Volcanic Process in Ore Genesis.* Spec. Publ. No. 7, Geol. Soc. London.

Spooner E. T. C., Bray C. J. & Chapman H. J. (1977) A Sea Water Source for the Hydrothermal Fluid which formed the Ophiolitic Cupriferous Pyrite Ore Deposits of the Troodos Massif, Cyprus. *J. geol. Soc. Lond.*, **134**, 395.

Stanton R. L. (1972) *Ore Petrology.* McGraw-Hill, New York.

Stanton R. L. (1978) Mineralization in Island Arcs with Particular Reference to the South-west Pacific Region. *Proc. Australas. Inst. Min. Metall.*, No. **268**, 9-19.

Steele I. M., Bishop F. C., Smith J. V. & Windley B. F. (1977) The Fiskenæsset Complex, West Greenland, Part III. *Grønlands geol. Unders. Bull.* **124**.

Stevenson J. S. (1961) Recognition of the Quartzite Breccia in the Whitewater Series, Sudbury Basin, Ontario. *Int. geol. Congr.*, 21st, Copenhagen, 1960, Rept. pt. **26**, 32-41.

Stevenson J. S. (1972) The Onaping Ash-flow Sheet, Sudbury, Ontario. *Geol Assoc. Canada Spec. Pap.* **10**, 41-8.

Sullivan C. J. (1948) Ore and Granitization. *Econ. Geol.*, **43**, 471-98.

Symons R. (1961) Operation at Bikita Minerals (Private) Ltd., Southern Rhodesia. *Trans. Instn Min. Metall.*, **71**, 129-72.

Taylor S. & Andrew C. J. (1978) Silvermines Orebodies, County Tipperary, Ireland. *Trans. Instn Min. Metall. (Sect. B: Appl. earth sci.)*, **87**, B111-B124.

Taylor S. R. (1955) The Origin of Some New Zealand Metamorphic Rocks as shown by their Major and Trace Element Compositions. *Geochim. et cosmoch. Acta*, **8**, 182-97.

Thayer T. P. (1964) Principal Features and Origin of Podiform Chromite Deposits, and some Observations on the Guleman-Soridag District, Turkey. *Econ. Geol.*, **59**, 1497-1524.

Thayer T. P. (1967) Chemical and Structural Relations of Ultramafic and Feldspathic Rocks in Alpine Intrusive Complexes. In Wyllie P. J. (ed.), *Ultramafic and Related Rocks.* 222-39. Wiley, New York.

Thayer T. P. (1969a) Gravity Differentiation and Magmatic Re-emplacement of Podiform Chromite Deposits. *Econ. Geol.*, Monogr., **4**, 132-46.

Thayer T. P. (1969b) Peridotite-gabbro Complexes as keys to Petrology of Mid-ocean Ridges. *Geol. Soc. Am. Bull.*, **80**, 1515-22.

Thayer T. P. (1971) Authigenic, Polygenic and Allogenic Ultramafic and Gabbroic Rocks as Hosts for Magmatic Ore Deposits. *Geol. Soc. Aust.*, Spec. Publ. 3, 239-51.

Thayer T. P. (1973) *Chromium.* In Probst D. A. & Pratt W. P. (eds), United States Mineral Resources, *Prof. Pap. 820,* 111-121. *U.S. Geol. Surv.*, Washington.

Theodore T. G. (1977) Selected Copper-bearing Skarns and Epizonal Granitic Intrusions in the Southwestern United States. *Geol. Soc. Malaysia,* Bull. **9**, 31-50.

Thurlow J. G. (1977) Occurrences, Origin and Significance of Mechanically Transported Sulphide Ores at Buchans, Newfoundland. In, *Volcanic Processes in Ore Genesis.* Spec. Publ. No. 7, Geol. Soc. London.

Titley S. R. (1978) Copper, Molybdenum and Gold Content of Some Porphyry Copper Systems of the Southwestern and Western Pacific. *Econ. Geol.*, **73**, 977-81.

Trendall A. F. (1968) Three Great Basins of Precambrian Iron Formation: a Systematic Comparison. *Bull. geol. Soc. Amer.*, **79**, 1527-44.

Trendall A. F. (1973) Iron Formations of the Hamersley Group of Western Australia: Type Examples of Varved Precambrian Evaporites. In, *Genesis of Precambrian Iron and Manganese Deposits,* 257-70. Proc. Kiev Symp. 1970, Unesco, Paris.

Trendall A. F. & Blockley J. G. (1970) The Iron Formation of the Precambrian Hamersley Group, Western Australia. *Bull. geol. Surv. W. Australia,* **119**.

Trudinger P. A. (1976) Microbiological Processes in Relation to Ore Genesis. In Wolf K. H. (ed.), *Handbook of Strata-Bound and Stratiform Deposits,* Vol. 2, 135-90. Elsevier, Amsterdam.

Turneaure F. S. (1960) A Comparative Study of Major Ore Deposits of Central Bolivia. Parts I and II. *Econ. Geol.*, **55**, 217-254 and 574-606.

Urabe T. & Sato T. (1978) Kuroko Deposits of the Kosaka Mine, Northeast Honshu, Japan - Products of Submarine Hot Springs on Miocene Sea Floor. *Econ. Geol.*, **73**, 161-79.

van Gruenewaldt G. (1977) The Mineral Resources of the Bushveld Complex. *Miner. Sci. Engng,* **9**, 83-95.

Varentsov I. M. (1964) *Sedimentary Manganese Ores.* Elsevier, Amsterdam.

Varentsov I. M. & Rakhmanov V. P. (1977) Deposits of Manganese. In Smirnov V. I. (ed.), *Ore Deposits of the USSR,* Vol. 1, 114-78. Pitman, London.

Vermaak C. F. (1976) The Merensky Reef—Thoughts on Its Environment and Genesis. *Econ. Geol.*, **71**, 1270-98.

Verwoerd W. J. (1964) South African Carbonatites and Their Probable Mode of Origin. *Ann. Univ. Stellenbosch.*, Ser. A, **41**, 115-233.

Viljoen M. J. & Viljoen R. P. (1969) Evidence of the Existence of a mobile extrusive peridotitic magma from the Komati Formation of the Onverwacht Group. *Geol Soc. South Africa* Spec. Pub. 2, 87-112.

Vokes F. M. (1968) Regional Metamorphism of the Palaeozoic Geosynclinal Sulphide Ore Deposits of Norway. *Trans. Instn Min. Metall.* (*Sect. B: Appl. earth sci.*), **77**, B53-B59.

Wallace S. R., MacKenzie W. B., Blair R. G. & Muncaster N. K. (1978) Geology of the Urad and Henderson Molybdenite Deposits, Clear Creek County, Colorado, with a Section on a Comparison of These Deposits with Those at Climax Colorado. *Econ. Geol.*, **73**, 325-68.

Walton E. K. (1970) *Geosynclinal Theory and Lower Palaeozoic Rocks in Scotland.* Second Tomkeieff Memorial Lecture, University of Newcastle upon Tyne, Newcastle upon Tyne.

Watson J. V. (1973) Influence of Crustal Evolution on Ore Deposition. *Trans. Instn Min. Metall. (Sect. B: Appl. earth sci.)*, **82**, B107-B114.

Watson J. V. (1976) Mineralization in Archaean Provinces. In Windley B. F. (ed.), 443-53. *The Early History of the Earth,* Wiley, London.

Wedepohl K. H. (1971) 'Kupferschiefer' as a Prototype of Syngenetic Sedimentary Ore Deposits. *Int. Assoc. on Genesis of Ore Deposits,* Tokyo-Kyoto, Japan, Proc. Spec. Issue **3**, 268-73.

White W. S. (1971) A Paleohydrologic Model for Mineralization of the White Pine Copper Deposit, Northern Michigan. *Econ. Geol.*, **66**, 1-13.

Windley B. F. (1977) *The Evolving Continents.* Wiley, London.

Winward K. (1975) Quaternary Coastal Sediments. In Marham N. L. & Basden H. (eds), *The Mineral Deposits of New South Wales,* 595-621. Geol. Surv., New South Wales, Dept. of Mines, Sydney.

Wright J. B. & McCurry P. (1973) Magmas, Mineralization and Seafloor Spreading. *Geol. Rundsch.*, **62**, 116-25.

Index

Page numbers in italic indicate references to figures

£16—
(£7.50 pbk)